T0189322

Conformal Geometry

Miao Jin · Xianfeng Gu · Ying He
Yalin Wang

Conformal Geometry

Computational Algorithms and Engineering Applications

Springer

Miao Jin
The Centre for Advanced Computer Studies
University of Louisiana
Lafayette, LA
USA

Xianfeng Gu
State University of New York
Stony Brook, NY
USA

Ying He
School of Computer Science
and Engineering
Nanyang Technological University
Singapore
Singapore

Yalin Wang
School of Computing, Informatics
and Decision Systems Engineering
Arizona State University
Tempe, AZ
USA

ISBN 978-3-030-09202-3 ISBN 978-3-319-75332-4 (eBook)
https://doi.org/10.1007/978-3-319-75332-4

Softcover re-print of the Hardcover 1st edition 2018

Printed on acid-free paper

This Springer imprint is published by the registered company Springer International Publishing AG part of Springer Nature
The registered company address is: Gewerbestrasse 11, 6330 Cham, Switzerland

For those who love geometry.

Preface

Conformal means angle preserving in mathematics. *Conformal geometry* studies *conformal structure* of general surfaces. Conformal structure is a natural geometric structure and a special atlas on surfaces such that angles among tangent vectors can be coherently defined on different local coordinate systems. Conformal structure governs many physics phenomena including heat diffusion and electric–magnetic fields.

Computational conformal geometry focuses on algorithmic study of conformal geometry and offers powerful tools to handle a broad range of geometric problems in engineering fields. It links modern geometry theories to real engineering applications.

The power of computational conformal geometry in engineering fields stems from the following fundamental reasons.

- Conformal geometry studies surface conformal structure. All surfaces in daily life have a natural conformal structure. Therefore, geometric algorithms based on conformal geometry benefit general surfaces.
- Conformal structure of a general surface is more flexible than Riemannian metric structure and more rigid than topological structure. It can handle large deformations, which Riemannian geometry cannot efficiently handle; it preserves a lot of geometric information during the deformation, whereas topological methods lose too much information.
- Conformal maps are easy to control. For example, the conformal maps between two simply-connected closed surfaces form a six-dimensional space; therefore by fixing three points, the mapping is uniquely determined. This fact makes conformal geometric method very valuable for surface matching and comparison.
- Conformal maps preserve local shapes; therefore, it is convenient for visualization purposes.
- All surfaces can be classified according to their conformal structures, and all the conformal equivalent classes form a finite-dimensional manifold. This manifold has rich geometric structures and can be analyzed and studied. In comparison,

the isometric classes of surfaces form an infinite-dimensional space, and it is really difficult to deal with.
- Computational conformal geometric algorithms are based on solving elliptic partial differential equations, which are easy to solve, and the solving process is stable; namely, the solution is insensitive to the noise of the input surfaces. Therefore, computational conformal geometry method is very practical for real engineering applications.
- In conformal geometry, all surfaces in daily life can be deformed to three canonical spaces, the sphere, the plane, or the disk (the hyperbolic space). In other words, any surface admits one of the three canonical geometries, spherical geometry, Euclidean geometry, or the hyperbolic geometry. Most digital geometric processing tasks in three-dimensional space can be converted to the task in these two-dimensional canonical spaces.

The book provides an overview of computational conformal geometry applied in engineering fields. We first briefly introduce the major concepts and theorems of conformal geometry in an intuitive way with a large number of illustrative images rendered by graphics tools.

In part I of the book, we detail the major computational algorithms in conformal geometry in an accessible way for computer scientists and engineers. We provide less abstract mathematical reasoning, but more intuitive explanations and implementation issues from the engineering point of view.

In part II of the book, we dedicate each chapter to a specific application field of computational conformal geometry including computer graphics, computer vision, geometric modeling, medical imaging, and wireless sensor networks. We discuss the fundamental problems, and how computational conformal geometric methods tackle them in a theoretically elegant and computationally efficient way in each field.

Computational conformal geometry is an emerging field. There are still a lot of challenging and open problems both in theory and in practice. Applying computational conformal geometric methods to broader applications and adapting them to real systems are still developing. The book will be of interest to senior undergraduates, graduates, and researchers in computer science, applied mathematics, and wide branches of engineering.

Lafayette, LA, USA — Miao Jin
Stony Brook, NY, USA — Xianfeng Gu
Singapore — Ying He
Tempe, AZ, USA — Yalin Wang

Contents

Chapter 1
Introduction

Abstract This chapter gives a brief introduction of the basic theoretic foundations and profound and beautiful structures of conformal geometry. They are provided in an intuitive way with illustrative images rendered by graphics tools.

Conformal means angle preserving in mathematics. *Conformal geometry* studies *conformal structure* of general surfaces. Conformal structure is a natural geometric structure and a special atlas on surfaces such that angles among tangent vectors can be coherently defined on different local coordinate systems. Conformal structure governs many physics phenomena, such as heat diffusion, electric-magnetic fields, etc.

Computational conformal geometry focuses on algorithmic study of conformal geometry and offers powerful tools to handle a broad range of geometric problems in engineering fields. It links modern geometry theories to real engineering applications.

We aim to give a thorough introduce of computational conformal geometry from a practical point of view. The book first briefly introduce the basic theoretic foundations and profound and beautiful structures of conformal geometry in an intuitive way with illustrative images rendered by graphics tools. Then the book presents computational algorithms of conformal geometry from the engineering point of view. The book provides a detailed discussion of computational conformal geometry applied in various engineering fields.

1.1 Riemann Mapping Theorem

A *conformal map* between two surfaces preserves angles. *Riemann mapping* theorem states that any simply connected surface with a single boundary, i.e., a topological disk, can be conformally mapped to a unit disk. Figure 1.1a shows a scanned 3D

M. Jin et al., *Conformal Geometry*, https://doi.org/10.1007/978-3-319-75332-4_1

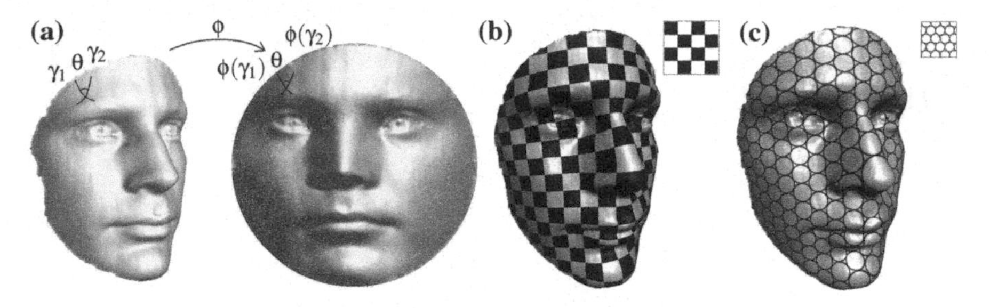

Fig. 1.1 Visualization of conformality using texture mapping in computer graphics: **a** A 3D face surface is conformally mapped to a unit disk. **b** A checker board texture is pulled back to the surface. **c** A circle packing texture is pulled back to the surface

human face, i.e., a topological disk surface denoted as S, mapped to a unit disk denoted as D by a conformal mapping $\phi : S \to D$. Suppose γ_1, γ_2 are two arbitrary curves on the face surface S, ϕ maps them to $\phi(\gamma_1)$, $\phi(\gamma_2)$. If the intersection angle between γ_1, γ_2 is θ, then the intersection angle between $\phi(\gamma_1)$ and $\phi(\gamma_2)$ is also θ. γ_1 and γ_2 can be chosen arbitrarily. Therefore, we say ϕ is conformal, meaning angle-preserving.

Conformality can be visualized using texture mapping technique in computer graphics. Figure 1.1b, c visualize the idea. A texture refers to an image on the plane. Based on the conformal map shown in Fig. 1.1a, we cover the planar disk by a checker board texture image and pull back the image onto the face surface. Since the mapping is conformal, all the right angles of the corners of the checker board are preserved on the human face as shown in Fig. 1.1b. If we replace the texture by a circle packing pattern, then planar circles are mapped to circles on the surface. The tangency relation among circles are preserved as shown in Fig. 1.1c.

1.2 Holomorphic Differential

A conformal map between two planar domains is a conventional analytic function, or holomorphic function. From this point of view, conformal mappings are the generalization of holomorphic functions, and Riemann surfaces are the generalization of complex plane. All surfaces in real life are real surfaces. The derivative of a analytic function is called a *holomorphic differential.* Holomorphic differentials can be defined on surfaces directly. Figure 1.2 visualizes two holomorphic differentials

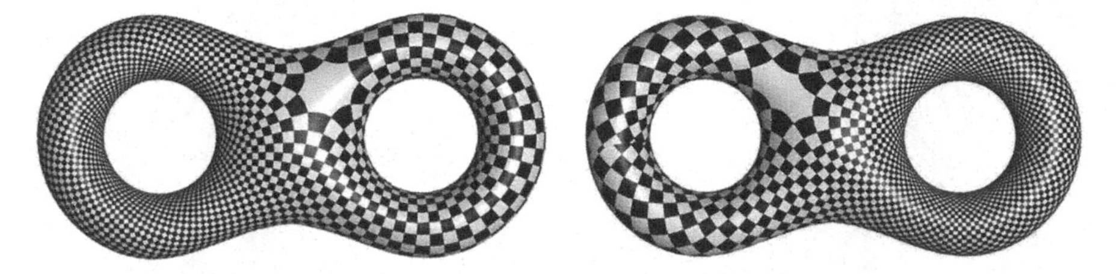

Fig. 1.2 Visualization of two holomorphic differentials defined on a genus two surface using texture mapping. Image from [132]

defined on a genus two surface using the same technique as the visualization of conformal mapping. By integrating the holomorphic differentials, we can locally map the surface to a plane. The mapping is conformal and visualized by checker board texture mapping.

1.3 Uniformization Theorem

A *covering space* of S is a space $\tilde{S}$ together with a continuous surjective map $h : \tilde{S} \to S$, such that for every $p \in S$ there exists an open neighborhood U of p such that $h^{-1}(U)$ (the inverse image of U under h) is a disjoint union of open sets in $\tilde{S}$ each of which is mapped homeomorphically onto U by h. The map h is called the *covering map*. A simply connected covering space is a *universal cover*.

Given a surface with a Riemannian metric, there exist an infinite number of metrics conformal to the original one. Uniformization theorem states that, among all of the conformal metrics, there exists a unique one that induces a constant Gaussian curvature. Moreover, the constant Gaussian curvature will be one of $\{+1, 0, -1\}$. We call such a metric the *uniformization metric* of the surface. According to Gauss-Bonnet theorem, the sign of the constant Gaussian curvature must match the sign of the Euler number of the surface: $+1$ for $\chi(S) > 0$, 0 for $\chi(S) = 0$, and -1 for $\chi(S) < 0$.

Figure 1.3 visualizes the uniformization theorem. We compute the uniformization metrics of three closed triangulated surfaces with genus numbers zero, one, and three, respectively as shown in the first row of Fig. 1.3. We then isometrically embed a portion of the universal covering space of each surface with its uniformization metric onto one of the three canonical surfaces: the *sphere* $\mathbb{S}^2$ for the genus zero surface with a positive Euler number, the *plane* $\mathbb{E}^2$ for the genus one surface with a zero Euler number, and the *hyperbolic space* $\mathbb{H}^2$ for the high genus surface with a negative Euler number as shown in the second row of Fig. 1.3.

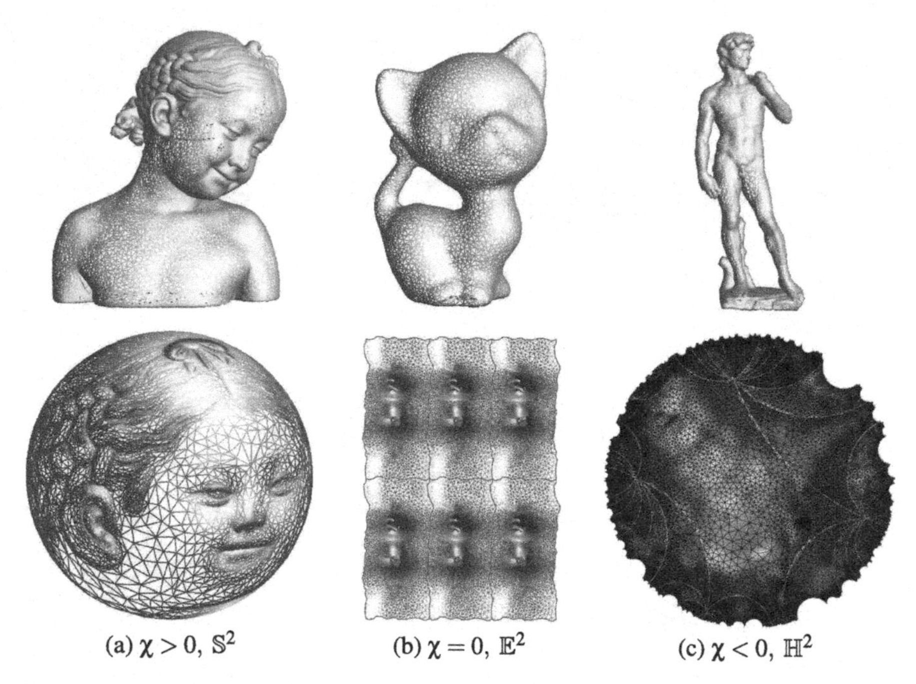

Fig. 1.3 **Uniformization Theorem**: each surface in $\mathbb{R}^3$ admits a uniformization metric, which is conformal to the original one and induces constant Gaussian curvature; the constant is one of $\{+1, 0, -1\}$ depending on the Euler characteristic number χ of the surface. Its universal covering space can be isometrically embedded onto one of the three canonical spaces: sphere, plane, or hyperbolic space with the uniformization metric. Image from [131]

1.4 Shape Space

Surfaces with the same topology are *conformally equivalent* or belong to the same conformal class if there exists a bijective conformal map between them. The conformal equivalence classes form a finite dimensional space that is called the Teichmüller space and the Modular space, i.e., the space of shapes, therefore, *shape space*.

A point in shape space represents a group of conformally equivalent surfaces, and a curve represents a deformation process. A surface has a set of unique coordinates in the space. The dimension of coordinates is determined by the topology of surface.

Specifically, a closed high genus surface with hyperbolic metric can be decomposed to $2g - 2$ pairs of pants, i.e., genus zero surfaces with three boundaries, by cutting the surface along $3g - 3$ geodesic loops. Two adjacent pairs of pants are glued together along the cutting geodesic loop with an angle, called twisting angle. The lengths of the cutting loops and the twisting angles give the coordinates of the surface in Teichmüller space, which are called *Fenchel–Nielsen coordinates*. The Fenchel–Nielsen coordinates uniquely determine the conformal structure of a closed

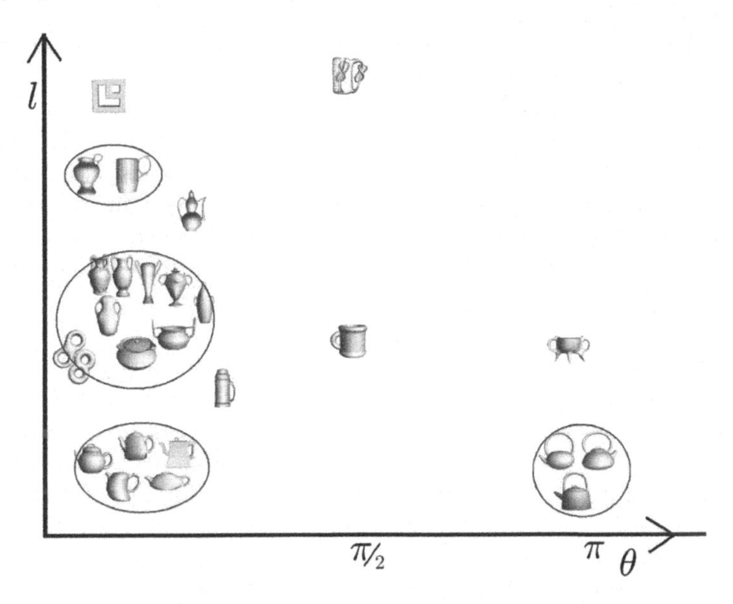

Fig. 1.4 Clustering of surfaces based on their Fenchel–Nielsen coordinates in Teichmüller space. The x-coordinate indicates the twisting angle, and the y-coordinate indicates the geodesic length. Surfaces are clustered based on both their twisting angle and geodesic lengths, with different groups marked with circles. Image from [130]

high genus surface. They can be treated as the fingerprint of the surfaces and applied for shape comparison and classification.

Figure 1.4 shows the clustering of a set of randomly chosen genus two surfaces based on their Fenchel–Nielsen coordinates in Teichmüller space. The x and y coordinates indicate the twisting angles and geodesic lengths, respectively. We first classify surfaces into three big groups based on the twisting angles, and then get more refined groups with marked circles when taking geodesic lengths into consideration.

1.5 Geometric Structures

A surface in general cannot be covered by one coordinate system. However, we can find a collection of open sets to cover the surface and map each one to a plane. We then use the planar coordinates as the local coordinate system of the corresponding open set. Such kind of local coordinate system is call an *atlas* of the surface. One point on the surface may be covered by multiple local coordinates. The transformation from one local coordinates system to another is called the *transition function* or *coordinates change*.

Suppose X is a topological space, G is the transformation group of X, a (X, G) structure is an atlas, such that the local coordinates are in X, and the transition functions are in G. For example, a spherical structure is an atlas, where all the local coordinates are on the sphere, all the transition functions are rotations.

According to Felix Klein's Erlangen program, different geometries study the invariants under different transformation groups. For example, suppose X is the Euclidean plane. If G is rigid motion, then the geometry is Euclidean geometry, and the invariants are lengths, angles, area etc. If G is affine transformation, then the geometry is affine geometry, and the major invariant is the ratio of three points on a line, parallelism. If G is real projective transformation, then the corresponding geometry is real projective geometry, and the major invariant is cross ratio of four points on a line.

If a surface admits a (X, G) structure, then the corresponding geometry can be defined on the surface directly. For example, in automobile industry and mechanics engineering fields, surfaces are represented as splines, i.e., piecewise rational polynomials. Conventional splines are defined on the Euclidean plane and constructed based on affine invariants. The fundamental problem in computer aided geometric design (CAD) field is to construct splines defined on general surfaces. If the surface admits an *affine structure*, then splines can be defined on the surface directly. Unfortunately, very few surfaces admits affine structure. For example, for closed surfaces, only those with genus one admits affine structure. This fact causes intrinsic difficulty for applications in CAD field.

Figure 1.5 visualizes the affine structures of two closed genus one surfaces. Figure 1.6 visualizes a closed genus zero David head model with its spherical and induced projective structures. Note that the projective structure is composed of six charts. Figure 1.7 visualizes the hyperbolic structure of a genus zero surface with three boundaries. Figure 1.8 visualizes the hyperbolic and real projective structures of a closed genus four surface.

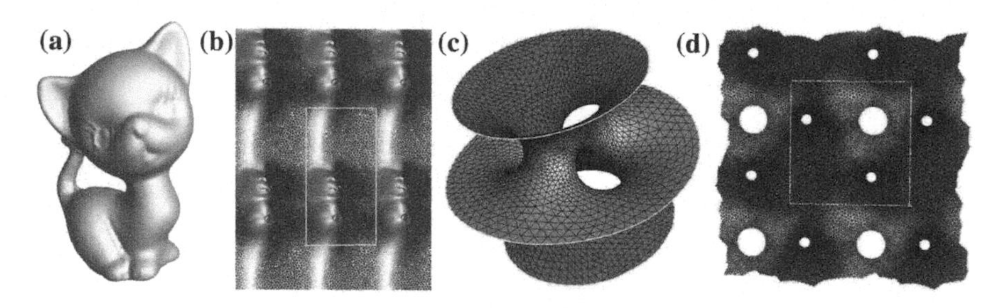

Fig. 1.5 Visualization of the affine structures of two genus one surfaces. Image from [135]

Fig. 1.6 Visualization of the spherical and induced projective structures of a genus zero surfaces. Note that the projective structure is composed of six charts. Image from [135]

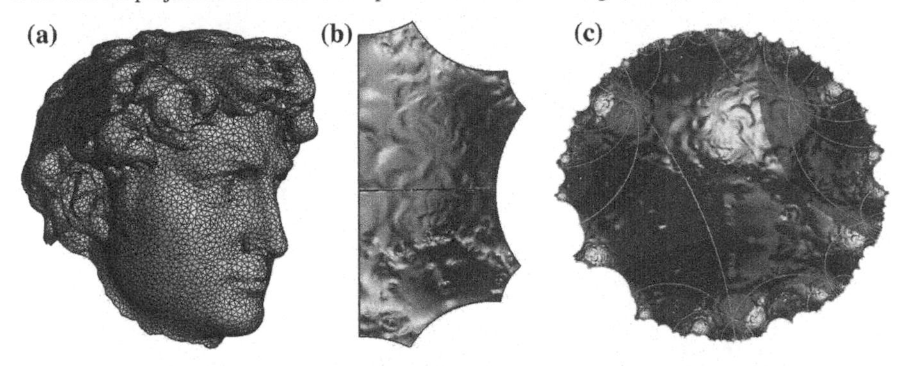

Fig. 1.7 Hyperbolic Structure of Genus Zero Surfaces with Three Boundaries: **a** Skull model. **b** One fundamental domain embedded in the Poincaré disk. **c** Its universal covering space embedded in the Poincaré disk. **d** David head model. **e** One fundamental domain embedded in the Poincaré disk. **f** Its universal covering space embedded in the Poincaré disk. Image from [135]

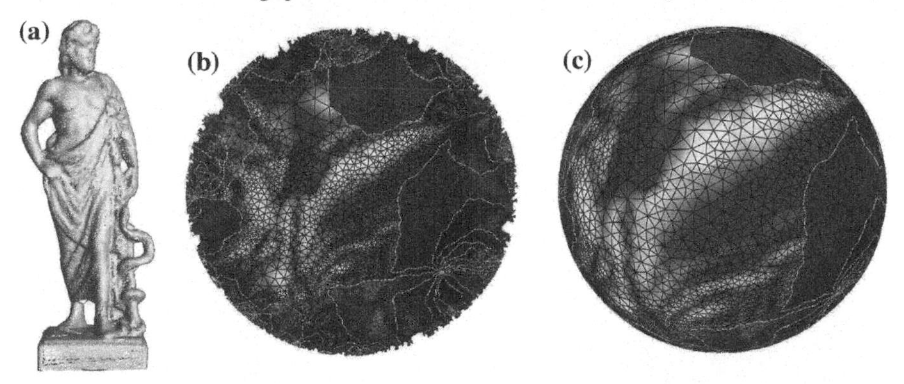

Fig. 1.8 Hyperbolic Structure and Real Projective Structure of Greek Sculpture Model. **a** Genus four Greek Sculpture model. **b** Hyperbolic structure induced from isometric embedding of Universal covering space in the Poincaré model. **c** Real projective structure induced from isometric embedding of Universal covering space in the Klein model. Image from [135]

Part I
Computational Algorithms

Chapter 2
Topological Algorithms

Abstract This chapter introduces halfedge data structure, a commonly used data structure in geometric software to efficiently represent piecewise linear triangular meshes, fundamental topological concepts and corresponding computational algorithms including cut graph, fundamental domain, homotopy, homology group basis, and canonical homotopy group generators.

In engineering fields, smooth surfaces are generally approximated by simplicial complexes (triangle meshes). The popularity of triangular meshes can be explained from both theory and practice. In theory, any surface with C^1 continuity can be triangulated. Simplicial homology and cohomology theories are also built on triangular meshes. In practice, all surfaces in real life can be digitized using $3D$ scanners, and the acquired point clouds can be easily triangulated to meshes. Modern graphics hardware supports triangular meshes directly. Most software tools support triangular meshes as the default data format.

A triangular mesh is a set of triangular faces coherently glued together and embedded in $\mathbb{R}^3$. Each vertex has a set of Euclidean coordinates in $\mathbb{R}^3$. Each face is a planar triangle. Each edge has a length induced from the Euclidean metric of $\mathbb{R}^3$. A mesh is flat everywhere except at vertices and edges. Topological information of a mesh is implied by its connectivity.

2.1 Halfedge Data Structure

Halfedge data structure is a commonly used data structure in geometric software to efficiently represent a triangular mesh.

A mesh denoted as $M = (V, E, F)$ consists of a list of *vertices* V, *non-oriented edges (edges)* E, and *oriented faces* F. For the convenience of programming, we use *half-edge* to refer the oriented edges and *edge* the non-oriented ones.

Suppose $V = \{v_0, v_1, \ldots, v_n\}$ is a list of vertices of Mesh M. A half-edge $h_{ij} = [v_i, v_j]$ connects two vertices v_i and v_j with v_i the *source vertex* and v_j the *target vertex*. The boundary of h_{ij} is

M. Jin et al., *Conformal Geometry*, https://doi.org/10.1007/978-3-319-75332-4_2

$$\partial[v_i, v_j] = v_j - v_i.$$

An oriented face $f_{ijk} = [v_i, v_j, v_k]$ consists three vertices, v_i, v_j, and v_k, ordered counter-clock-wisely with respect to the orientation. Further more, the boundary of each face has three consecutive half-edges,

$$\partial[v_i, v_j, v_k] = [v_i, v_j] + [v_j, v_k] + [v_k, v_i].$$

Given a closed mesh embedded in $\mathbb{R}^3$, the mesh divides the whole space to two volumes. One is finite, called the *inside volume*. The other is infinite, called the *outside volume*. The *normal* of an oriented face $f_{ijk} = [v_i, v_j, v_k]$ is the unit vector perpendicular to the face plane and expressed as

$$\mathbf{n} = \frac{(v_j - v_i) \times (v_k - v_i)}{|(v_j - v_i) \times (v_k - v_i)\|}. \tag{2.1}$$

Namely, the normal and $v_j - v_i$, $v_k - v_i$ satisfy the right hand rule. If a mesh is orientable, then all the normals are pointing outside. Most meshes we process are *oriented meshes*.

An open mesh has boundaries. An interior non-oriented edge $e_{ij} = [v_i, v_j]$ has two half-edges attached with opposite orientations, $h_{ij} = [v_i, v_j]$ and $h_{ji} = [v_j, v_i]$. A boundary edge has only one half-edge attached. When one walks along a boundary, i.e., a list of consecutive half-edges, the walking direction is consistent with the half-edge orientation. The surface is on the left hand side, and the hole is on the right hand side.

Definition 2.1 (*Dual Half-Edge*) Given an oriented edge $h_{ij} = [v_i, v_j]$, we call half-edge $h_{ji} = [v_j, v_i]$ the *dual half-edge* of h_{ij}.

Since halfedge data structure represents orientable surfaces only, we have the following requirement:
Half-edges attached to one interior edge must be dual to each other.

Definition 2.2 (*Boundary Edge*) Suppose an edge of mesh points to only one half-edge, then the edge is called a *boundary edge*. A half-edge without dual half-edge is called *boundary half-edge*.

2.2 Cut Graph and Fundamental Domain

Definition 2.3 (*Cut Graph*) A cut graph denoted as G in a mesh M consists a set of edges of M, such that M/G is a simply connected mesh.

A simply connected mesh is a topological disk.

To compute a cut graph of a given mesh M, we first construct a dual mesh of M, denoted as M', in the following way.

1. Each face f on M corresponds to a unique vertex of $v(f) \in M'$;
2. Each vertex v on M corresponds to a unique face of $f(v) \in M'$, suppose the faces adjacent to v are $f_1, f_2, \ldots, f_n$ sorted counter-clock-wisely, then

$$f(v) = [v(f_1), v(f_2), \ldots, v(f_n)],$$

3. Each edge $e \in M$ adjacent to faces f_i and f_j corresponds to an edge e'

$$e' = [v(f_i), v(f_j)].$$

Algorithm 1 computes a cut graph of a mesh with general topology.
input : A mesh M
output: A cut graph G of M

1. Construct a dual mesh of M, denoted as M';
2. Generate a minimal spanning tree T of the vertices of M';
3. $G = \{e|e' \notin T\}$.

Algorithm 1: Cut Graph

The basic idea of Algorithm 1 is to construct a topological disk D as big as possible. After removing D from M, i.e., $G = M/D$, there is only a one-dimensional graph G consisting of vertices and edges left. Note that a cut graph is not unique. It depends on both the chosen root vertex and the way to construct a minimal spanning tree T.

Algorithm 2 further prunes the branches of a cut graph such that there is no dangling vertex with valence one in the graph.
input : A cut graph G of M
output: A pruned cut graph G'
repeat
| remove the segment attached to v
until *there is no valence one node in* G;

Algorithm 2: Prune a Cut Graph

Figure 2.1 shows a computed cut graph of a genus two mesh using Algorithms 1 and 2.

A fundamental domain of a mesh M, denoted as D, is a topological disk consisting of all the faces of M. The computation of a fundamental domain of M is straightforward by simply cutting M along a cut graph G of M.

2.3 Homotopy and Homology Group Basis

Two loops γ and $\bar{\gamma}$ with the same base point on a surface M are *homotopic* to each other relative to the base point if there is a continuous map $h : [0, 1] \times [0, 1] \to M$ such

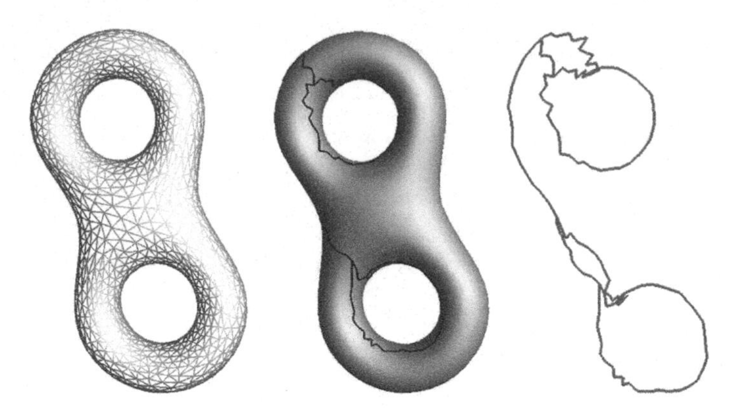

Fig. 2.1 A cut graph of a genus two surface

that $h(0, t) = \gamma(t), h(1, t) = \bar{\gamma}(t), and h(s, 0) = h(s, 1) = x$ for all $s, t \in [0, 1]$. A loop is contractible if it is homotopic to a constant point. In simple words, two loops are homotopic if they can deform to each other without leaving the surface. A loop is contractible if it can shrink to a point on the surface. The set of homotopy equivalence classes of loops with the base point denoted as x forms a group under concatenation, called the fundamental group and denoted $\pi_1(M, x)$. Fundamental groups of the same connected surface with different base points are isomorphic, so we denoted all as $\pi_1(M)$. A homotopy group basis, i.e., a set of homotpy group generators, is defined to be any set of $2g$ loops whose homotopy classes generate the fundamental group $\pi_1(M)$.

Theorem 2.4 *A set of homotpy group generators of G is also a set of homotopy group generators of M.*

Proof Let $D = M/G$, then D is a topological disk. Picke one point $p \in D$, Let $M_p = M/p$, then M is the union of D and M_p,

$$M = M_p \cup D,$$

The intersection of D and M_p is a topological annulus $D_p = D - p$,

$$D_p = D \cap M_p,$$

Suppose the generators of fundamental group of M_p, denoted as $\pi_1(M_p)$, is $\{\gamma_1, \gamma_2, \ldots, \gamma_n\}$, the annulus has one fundamental group generator, the fundamental group of the disk D is trivial, according to theorem $\pi_1(D \cup M_p) = \{\gamma_1, \gamma_2, \ldots, \gamma_n | r_1\}$, the equivalence relation in r_1 is caused by the generator of $\pi_1(D_p)$, suppose ∂D is a loop in M_p, and can be represented as

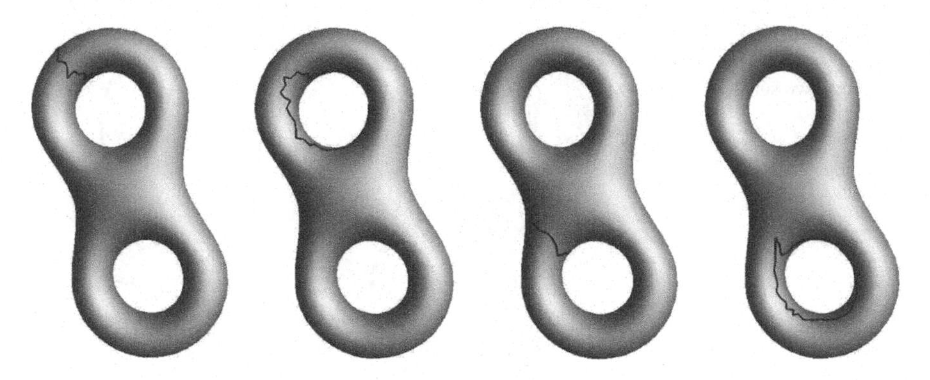

Fig. 2.2 A set of homology group generators of a genus two surface

$$\gamma_1 = s_1 s_2 \cdots s_m,$$

each s_k is one of γ_j or γ_j^{-1}. Therefore, the generatoros of M_p is equal to the generators of M.

Since G is a deformation retraction of M_p, the generators of $\pi_1(G)$ is also a set of generators of $\pi_1(M)$.

Algorithm 3 computes a set of homology group generators of a given mesh M denoted as $H_1(M, \mathbb{Z})$ by computing the independent loops of a cut graph G of M. Note that a homotopy basis is also a homology basis, but not vice versa unless the cycles in a homology basis have a common point. Figure 2.2 shows a set of homology group generators of a genus two surface computed using Algorithm 3.

input : A mesh M
output: A set of homology group generators of M
1. Compute a cut graph G of M;
2. Choose a root node r of G;
3. Use breadth first search method to construct a minimal spanning tree T of G. A unique path from the root node to each leaf v_i is computed, denoted as γ_i ;
3. Denote

$$G - T = \{e_1, e_2, \ldots, e_k\}.$$

Suppose e_k connecting v_i, v_j, then

$$l_k = \gamma_i, [v_i, v_j], \gamma_j^{-1}$$

form a loop. Denote $\{l_1, l_2, \ldots, l_k\}$ all the independent loops in G. They form a basis of the homology group $H_1(M, \mathbb{Z})$.

Algorithm 3: Homology Group Generators $H_1(M)$

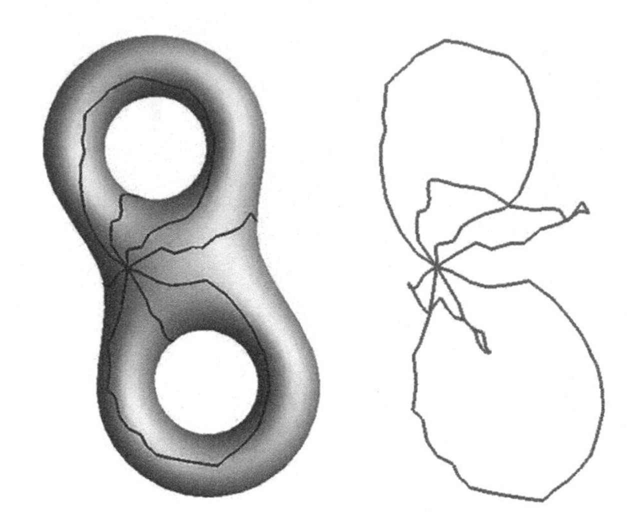

Fig. 2.3 A set of canonical homotopy group generators of a genus two surface

2.4 Canonical Homotopy Group Generators

Given a closed surface M of genus g, a set of loops

$$\{a_1, b_1, a_2, b_2, \dots, a_g, b_g\}$$

forms a set of canonical homotopy group generators, if

$$a_i \times b_i = 1, a_i \times a_j = 0, b_i \times b_j = 0, a_i \times b_j = 0$$

where $a_i \times b_j$ represents the (algebraic) intersection number. In practice, it is highly preferred to compute a set of canonical homotopy group generators. Many computations can be dramatically simplified by using a set of canonical homotopy group generators. Figure 2.3 shows a set of canonical homotopy group generators of a genus two surface.

If the tessellation of a mesh is too coarse, it is impossible to completely separate (a_i, b_i) from other (a_j, b_j)'s. In practice, we can always refine a mesh such that a set of separated generators exists. Algorithm 4 introduces a simple mesh subdivision method. The basic idea is to insert a new vertex on each edge and split each face to four smaller ones. We can continually subdivide a mesh to make it refiner and refiner.

input : A mesh M
output: A subdivided mesh $\bar{M}$

1.Suppose the vertex list of M is

$$V = \{v_1, v_2, \ldots, v_n\},$$

the edge list of M is

$$E = \{e_1, e_2, \ldots, e_m\},$$

the face list of M is F. Generate the vertex list of $\bar{M}$,

$$\bar{V} = V \cup E = \{v_1, v_2, \ldots, v_n, e_1, e_2, \ldots, e_m\}.$$

2. Suppose $f \in M$, $f = [v_1, v_2, v_3]$, edges $e_1 = [v_1, v_2]$, $e_2 = [v_2, v_3]$, and $e_3 = [v_3, v_1]$. Split f to four faces

$$\bar{f} = \{[v_1, e_1, e_3], [v_2, e_2, e_1], [v_3, e_3, e_2], [e_1, e_2, e_3]\}.$$

The face list of $\bar{M}$ is

$$\bar{F} = \cup_{f \in F} \bar{f}.$$

3. The subdivided mesh $\bar{M}$ is represetned as the vertex list and the face list,

$$\bar{M} = \{\bar{V}, \bar{F}\}.$$

Algorithm 4: Mesh Subdivision

With a refined mesh, we first compute a closed loop denoted as a_1. We slice the mesh along a_1 and get an open mesh denoted as M_1. The boundary of M_1 is

$$\partial M_1 = a_1 - a_1^{-1}.$$

We then randomly choose a vertex $v_1 \in a_1$ and find its dual vertex $v_1^{-1} \in a_1^{-1}$. We compute a shortest path from v_1 to v_1^{-1} with a requirement that it does not go through any boundary edge. The shortest path corresponds to a closed loop in M, denoted as b_1. We slice the mesh along $a_1 \cup b_1$ and get a new mesh denoted as M_2. The boundary of M_2 is

$$\partial M_2 = a_1 b_1 a_1^{-1} b_1^{-1}.$$

We compute the cut graph of M_2, denoted as G_2. We remove all boundary edges of G_2 and find a loop in $G_2/\partial M_2$, denoted as a_2. Then we slice M_2 along a_2 and get a new mesh denoted as M_3. The boundary of M_3 is

$$\partial M_3 = a_2 + a_2^{-1} + a_1 b_1 a_1^{-1} b_1^{-1}.$$

Similarly, we choose $v_2 \in a_2$ on M_3, $v_2^{-1} \in a_2^{-1}$ and compute the shortest path connecting v_2, v_2^{-1} without intersecting the boundary except at the starting and ending points. The shortest path corresponds to a loop in M, denoted as b_2.

We repeat this procedure to compute all the other pairs (a_j, b_j). In fact, each pair (a_j, b_j) corresponds to a handle, equivalent to locating one handle of a given mesh.

Algorithm 5 locates one handle and outputs a pair of loops that intersect at only one point. Algorithm 6 locates all handles and outputs a set of canonical homotopy group generators.

input : A mesh M
output: A pair of loops a, b that intersect at only one point

1. Compute a cut graph G of M;
2. Remove all boundary edges from G

$$G \leftarrow G - \partial M$$

If G is a tree, return the empty set;
3. Find the shortest loop in G, denoted as a;
4. Slice M along a, to get a new mesh

$$\bar{M} \leftarrow M - \{a\}, \partial \bar{M} = \partial M \cup a^+ \cup a^-,$$

5. choose a vertex $v^+ \in a^+$ and a vertex $v^- \in a^-$, compute a shortest path γ connecting v^+ and v^{-1}, such that the shortest path has no other intersections with the boundary of the mesh

$$\gamma \cap \partial \bar{M} = v^+ \cup v^-.$$

6. All the half edges on $\bar{M}$ has one-to-one correspondance with the half-edges on M, therefore γ corresponds to a closed loop on M, denoted as b. Then (a, b) is the output.

Algorithm 5: Locate one handle

input : A closed mesh M with high resolution and genus g
output: A set of canonical homotopy group generators $\{a_1, b_1, a_2, b_2, \ldots, a_g, b_g\}$

$M_1 \leftarrow M$;
repeat
  Locate a handle of M_k to obtain (a_k, b_k);
  $M_{k+1} \leftarrow M_k / \{a_k, b_k\}$
until *no more handles can be found*;
Output $\{a_1, b_1, a_2, b_2, \ldots, a_g, b_g\}$.

Algorithm 6: Canonical Homotopy Group Generators

Chapter 3
Harmonic Map

Abstract Conformal maps are harmonic, but harmonic maps are not necessarily conformal. This chapter introduces algorithms to compute conformal maps using harmonic maps on piecewise linear triangular meshes. Specifically, algorithms to compute conformal maps between two closed genus zero surfaces or a topological disk surface and a unit planar disk are detailed in this chapter.

3.1 Discrete Harmonic Energy

Denote M a triangular mesh with genus zero, i.e., either a topological disk or sphere surface.

Definition 3.1 (*Function Space*) All piecewise linear functions defined on M form a linear space, denoted by $C^{PL}(M)$.

Definition 3.2 (*Energy*) Suppose a set of string constants k_{ij} is assigned on each edge $[v_i, v_j]$ and $f \in C^{PL}$, the string energy is defined as:

$$E(f) = \sum_{[v_i,v_j]\in M} k_{ij}||f(v_i) - f(v_j)||^2 \tag{3.1}$$

By changing the string constants k_{ij} in the energy formula, we can define different string energies.

Definition 3.3 (*Tuette Energy*) If string constant $k_{ij} \equiv 1$, the string energy is known as the Tuette energy.

Definition 3.4 (*Harmonic Energy*) As shown in Fig. 3.1, if string constant $k_{ij} \equiv cot\alpha + cot\beta$ for interior edge $e = [v_i, v_j]$ with two adjacent faces $f = [v_i, v_j, v_k]$ and $f = [v_j, v_i, v_l]$, and $k_{ij} \equiv cot\alpha$ for boundary edge $e = [v_i, v_j]$ with one adjacent face $f = [v_i, v_j, v_k]$, the string energy is known as the harmonic energy.

A string energy is always a quadratic form. By carefully choosing the string coefficients, we make sure the quadratic form is *positive definite*. This will guarantee the convergence of the optimization process.

M. Jin et al., *Conformal Geometry*, https://doi.org/10.1007/978-3-319-75332-4_3

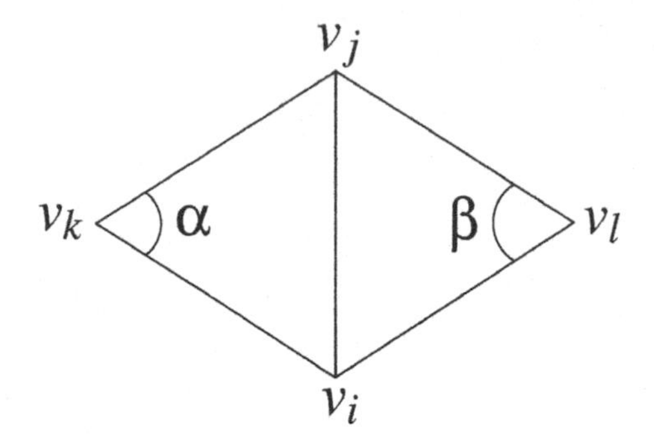

Fig. 3.1 String constant of harmonic energy

Definition 3.5 (*Laplace Operator*) Piecewise Laplacian is a linear operator

$$\Delta_{PL} : C^{PL} \to C^{PL}$$

on the space of piecewise linear functions on M, defined by the formula

$$\Delta_{PL}(f) = \sum_{[v_i, v_j] \in M} k_{ij}(f(v_j) - f(v_i)) \tag{3.2}$$

If f minimizes the string energy, then f satisfies the condition:

$$\Delta_{PL}(f) = 0.$$

Definition 3.6 (*Harmonic Map*) A map $\mathbf{f} : M_1 \to M_2$ is harmonic, if and only if $\Delta_{PL}(f)$ has only a normal component. Its tangential component is zero.

$$\Delta_{PL}(f) = (\Delta_{PL}(f))^{\perp} \tag{3.3}$$

From the theory of harmonic map, suppose $f : M \to \mathbb{R}$ is a function defined on the mesh, then the conventional harmonic energy is defined as

$$E(f) = \int_M |\nabla f|^2 d\sigma,$$

where $d\sigma$ is the area element of the surface, ∇f is the gradient of the function. Since a harmonic function is a harmonic map from M to $\mathbb{R}$, it has the property

$$\Delta_{PL}(f) = 0.$$

Theorem 3.7 *The conventional and discrete harmonic energies are consistent.*

3.2 Topological Disk Surfaces

Suppose M is a topological disk, a harmonic map from M to a convex planar domain is a diffeomorphism according to Rado's theorem.

Theorem 3.8 (Radó) *Assume $\Omega \subset \mathbb{E}^2$ is a convex domain with a smooth boundary $\partial\Omega$ and a metric surface $(S, \mathbf{g})$ is a simply connected domain with a single boundary. Given any homeomorphism $\phi : \partial S \to \partial\Omega$, then the harmonic map $u : S \to \Omega$, such that $u = \phi$ on ∂S, is a diffeomorphism.*

Furthermore, the harmonic map is unique.

Theorem 3.9 *Assume $\Omega \subset \mathbb{E}^2$ is a simply connected domain with a smooth boundary $\partial\Omega$ and a metric surface $(S, \mathbf{g})$ is simply connected with a single boundary. Given any homeomorphism $\phi : \partial S \to \partial\Omega$, if there exists a harmonic map $u : S \to \Omega$, such that $u = \phi$ on ∂S, then u is unique.*

The *Dirichlete boundary condition* can be set at the beginning of the algorithm. Harmonic map can then be directly computed by optimizing the harmonic energy.

Suppose

$$\mathbf{f} : M \to \mathbb{R}^2,$$

then the harmonic energy is

$$E(\mathbf{f}) = \sum_{[v_i, v_j] \in M} k_{ij} |f(v_i) - f(v_j)|^2,$$

with the boundary condition

$$\mathbf{f}|_{\partial M} = \mathbf{g}.$$

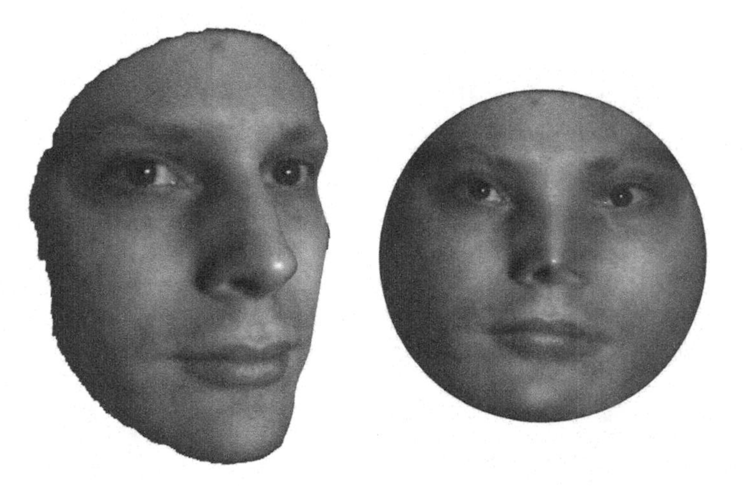

Fig. 3.2 Harmonic mapping of a 3D scanned face surface to a unit disk

The optimization of the harmonic energy is equivalent to solving the following linear system,

$$\Delta_{PL}\mathbf{f}(u) = \mathbf{0}, \forall u \notin \partial M$$

with all the boundary vertices fixed. It can be solved using Newton's method.

Algorithm 7 computes a harmonic map of a topological disk surface to a planar unit disk. Figure 3.2 shows a scanned triangular 3D human face mapped to a unit disk using Algorithm 7.

input : A simply connected mesh M with a single boundary
output: A harmonic map mapping M to a unit disk

Traverse the boundary of M and store the boundary vertices to a list

$$\partial M = \{v_0, v_1, \cdots, v_{n-1}\}.$$

Compute the length of the boundary:

$$s \leftarrow \sum_{i=0}^{n-1} l_{v_i,v_{i+1}},$$

where $v_n = v_0$ and $l_{v_i,v_{i+1}}$ is the edge length of $[v_i, v_{i+1}]$;
forall the $v_i \in \partial M$ **do**

$$s_i \leftarrow \sum_{j=0}^{i-1} l_{v_j,v_{j+1}}$$

$$\theta_i \leftarrow 2\pi \frac{s_i}{s}$$

$$f(v_i) \leftarrow (\cos\theta_i, \sin\theta_i)$$

end

Optimize the harmonic energy

$$E = \sum_{[v_i,v_j]\in M} k_{ij}(f(v_i) - f(v_j))^2,$$

using Newton's method with fixed boundary condition.

Algorithm 7: Harmonic Map of a Topological Disk Surface

3.3 Topological Sphere Surfaces

Theorem 3.10 *Harmonic maps between closed genus zero surfaces are conformal.*

The basic idea of the algorithm computing a harmonic map of a topological sphere surface to a unit sphere $\mathbb{S}^2$ is as follows: we first construct a degree-one map $\mathbf{h}$ between M and $\mathbb{S}^2$. Then we evolve $\mathbf{h}$ to minimize its harmonic energy until it becomes a harmonic map. The evolution of the map is a nonlinear heat diffusion process:

$$\frac{d\mathbf{f}(t)}{dt} = -\Delta\mathbf{f}(t). \tag{3.4}$$

Since $\mathbf{f}(M)$ is constrained to be on $\mathbb{S}^2$, we project $-\Delta\mathbf{f}$ onto the tangent space of $\mathbb{S}^2$. Specifically, suppose $\mathbf{f} : M_1 \to \mathbb{S}^2$, and denote the image of each vertex $v \in M$ as $\mathbf{f}(v)$. The normal of $\mathbf{f}(v)$ on $\mathbb{S}^2$ is $\mathbf{n}(\mathbf{f}(v))$. Define the normal component as:

Definition 3.11 The normal component

$$(\Delta\mathbf{f}(v))^{\perp} =< \Delta\mathbf{f}(v), \mathbf{n}(\mathbf{f}(v)) > \mathbf{n}(\mathbf{f}(v)), \tag{3.5}$$

where $<, >$ is the inner product in $\mathbb{R}^3$.

The nonlinear heat diffusion equation is

$$\frac{d\mathbf{f}(t)}{dt} = -(\Delta\mathbf{f}(v) - (\Delta\mathbf{f}(v))^{\perp}). \tag{3.6}$$

The solutions to a harmonic map (conformal map) from M to a unit sphere are not unique. Suppose $\mathbf{f}_1, \mathbf{f}_2 : M \to \mathbb{S}^2$ are two harmonic maps. Then $\phi = \mathbf{f}_2 \circ \mathbf{f}_1^{-1} : \mathbb{S}^2 \to \mathbb{S}^2$ is a Möbius transformation of the unit sphere, or equivalently, the extended complex plane $\bar{\mathbb{C}}$.

Definition 3.12 (*Möbius Transformation*) Mapping $\mathbf{f} : \bar{\mathbb{C}} \to \bar{\mathbb{C}}$ is a Möbius transformation if and only if

$$\mathbf{f} : z \to \frac{az + b}{cz + d}, \quad a, b, c, d \in \mathbb{C}, ad - bc = 1. \tag{3.7}$$

The Möbius tranformations of a unit sphere form a 6 dimensional Möbius transformation group. To ensure the convergence of the algorithm and the uniqueness of the solution, constraints need to be added. In practice we use the following *zero mass-center constraint.*

Definition 3.13 Mapping $\mathbf{f} : M \to \mathbb{S}^2$ satisfies the zero mass-center condition if and only if

$$\int_{\mathbb{S}^2} \mathbf{f} d\sigma_M = 0, \tag{3.8}$$

where σ_M is the area element on M.

All conformal maps from M to $\mathbb{S}^2$ satisfying the zero mass-center constraint are unique up to an Euclidean rotation group (3 dimensional). We use the Gauss map as the initial condition.

Definition 3.14 (*Gauss Map*) The Gauss map $g : M \to \mathbb{S}^2$ is defined as

$$g(v) = \mathbf{n}(v), v \in M, \tag{3.9}$$

$\mathbf{n}(v)$ is the normal at v.

Algorithm 8 computes a spherical conformal mapping of a closed genus zero surface to a unit sphere.

input : A closed gesnu zero mesh M, step length δt, energy difference threshold ε
output: $\mathbf{f} : M \to S^2$.

Compute the Gauss map $\mathbf{g} : M \to \mathbb{S}^2$ and set as the initial $\mathbf{f}$;
Compute harmonic energy E_0;
repeat
  forall the *vertex* $v \in M$ **do**
    Compute the absolute derivative

$$D\mathbf{f} = \Delta\mathbf{f} - \Delta\mathbf{f}^{\perp},$$

    Update $\mathbf{f}(v)$ by $\delta\mathbf{f}(v) \leftarrow -D\mathbf{f}(v)\delta t$;
  end
  Compute a Möbius transformation $\varphi : \mathbb{S}^2 \to \mathbb{S}^2$, such that the mass center of $\varphi \circ \mathbf{f}$ is also the sphere center;
  $E_0 \leftarrow E$;
  Compute harmonic energy E.
until $|E - E_0| < \varepsilon$;
Return $\mathbf{f}$

Algorithm 8: Conformal Spherical Mapping

The step of normalization using Möbius transformation in Algorithm 8 is non-linear and computationally expensive. In practice we use Algorithm 9 for normalization.

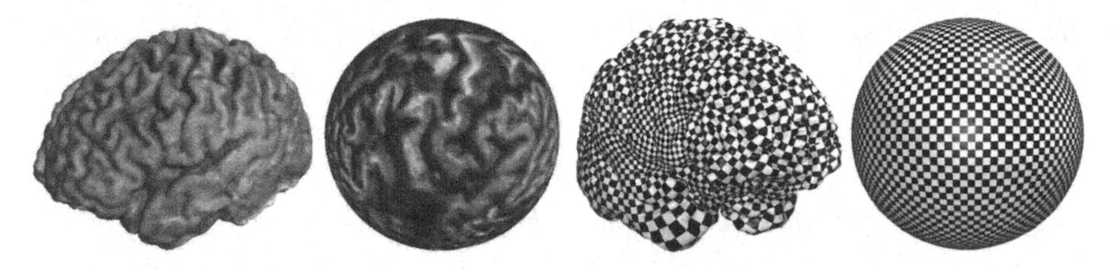

Fig. 3.3 Conformal spherical brain mapping. Image from (100)

input : A closed genus zero mesh M, a mapping of M to a unit sphere $\mathbf{f}: M \to \mathbb{S}^2$

output: Normalized mapping $\tilde{\mathbf{f}}$ such that the mass center is also the sphere center

Compute the mass center of $\mathbf{f}$;

$$\mathbf{c} \leftarrow \int_{\mathbb{S}^2} \mathbf{f} d\sigma_M.$$

forall the $v \in M$ **do**

$$\tilde{\mathbf{f}}(v) \leftarrow \mathbf{f}(v) - \mathbf{c}$$

end

forall the $v \in M$ **do**

$$\tilde{\mathbf{f}}(v) \leftarrow \frac{\tilde{\mathbf{f}}(v)}{|\tilde{\mathbf{f}}(v)|}.$$

end

Algorithm 9: Normalization

The approximation method in Algorithm 9 is good enough for practical purpose. With a carefully chosen step length, the energy decreases monotonically at each iteration.

Figure 3.3 demonstrates a real example of conformal spherical brain mapping using Algorithms 8 and 9. The brain surface is reconstructed from MRI images and represented as a triangular mesh.

Note that the initial map could be the Gauss map or any other degree-one map. However, in practice, the quality of initial map affects the efficiency and stability of the whole process.

3.4 Riemann Mapping

Conformal maps are harmonic, but harmonic maps are not necessarily conformal. Riemann mapping theorem claims the existence of conformal mapping between a topological disk surface and a unit planar disk.

Theorem 3.15 (Riemann Mapping Theorem) *Let Ω be a simply connected domain in $\mathbb{C}$ that is not all of $\mathbb{C}$. Then there exists a biholomorphic mapping*

$$\varphi : \Omega \to \mathbb{D}.$$

For each point $z_0 \in \Omega$, there is a unique such map φ such that

$$\varphi(z_0) = 0,\ \varphi'(z_0) \in \mathbb{R}^+.$$

The basic idea of the algorithm to compute Riemann mapping is as follows. Suppose M is a simply connected surface, ∂M is the boundary. We first double cover the surface, i.e., gluing with a copy of the surface along the boundaries to form a closed symmetric surface $\bar{M}$ with genus zero. Then we apply Algorithm 8 to find a conformal map of $\bar{M}$ to a unit sphere. We use Möbius transformations to adjust the conformal map such that ∂M is mapped to the equator of the unit sphere. We then use stereo-graphic projection to map the unit sphere onto the plane such that the lower hemisphere is conformally mapped to a unit disk.

A *stereo-graphic projection* is a conformal map, which maps the unit sphere $\mathbb{S}^2$ onto the whole complex plane $\mathbb{C}$.

$$\phi : \mathbb{S}^2 \to \mathbb{C},$$

Geometrically, we can imagine that we put a light source at the normal pole of the sphere, and lay down the screen as the tangent plane at the south pole, then the whole sphere surface is projected onto the screen.

$$\phi : (x, y, z) \to \left(\frac{2x}{2-z}, \frac{2y}{2-z}\right).$$

A Möbius transformation is a conformal map from the complex plane to itself,

$$\tau : \mathbb{C} \to \mathbb{C}.$$

By using Möbius transformation, we can map arbitrary three points $z_0, z_1, z_2 \in \mathbb{C}$ to $0, 1, \infty$,

$$\tau_1 = \frac{(z - z_0)(z_1 - z_2)}{(z - z_2)(z_1 - z_0)},$$

suppose w_0, w_1, w_2 is another set of three points, τ_2 maps them to $0, 1, \infty$, then

$$\tau_2^{-1} \circ \tau_1 : (z_0, z_1, z_2) \to (w_0, w_1, w_2).$$

Algorithm 10 computes the Riemann mapping of a topological disk surface to a unit disk.

input : A genus zero mesh M with a single boundary.
output: $\mathbf{h} : M \rightarrow \mathbb{H}^2$, $\mathbf{h}$ is conformal

Double covering M to $\bar{M}$;
Compute a harmonic map $\mathbf{f} : \bar{M} \rightarrow \mathbb{S}^2$;
Compute the stereo-graphic projection $\phi : \mathbb{S}^2 \rightarrow \mathbb{C}$;
Select three points v_0, v_1, v_2 on the boundary of M;
Use a Möbius transformation τ to map v_0, v_1, v_2 to $0, 1, i$;
$\mathbf{h} \leftarrow \tau \circ \phi \circ f$.

Algorithm 10: Riemann Mapping

Chapter 4
Harmonic and Holomorphic Forms

Abstract Differential forms are powerful tools to tackle many geometric problems on meshes including computing conformal maps. This chapter introduces the computational algorithms of cohomology and forms.

The concepts of chain and homotopy are intuitive and easy to visualize on triangular meshes. On the contrary, the concepts of cochains, cohomology, and forms are much more abstract because they are not geometric. However it is easier and simpler to tackle many geometric problems on meshes with differential forms. This chapter focuses on the computational algorithms of cohomology and forms.

4.1 Characteristic Forms

Suppose M is a mesh represented using half-edge data structure. A chain is represented as a list of half-edges

$$\gamma = \{h_1, h_2, \ldots, h_n\},$$

where $h_i = [v_{i_1}, v_{i_2}]$.

A simplicial 1-form is represented as a linear function defined on the half-edges

$$\omega : E \rightarrow \mathbb{R}^1, \omega([v_i, v_j]) = -\omega([v_j, v_i]).$$

The integration of a 1-form ω along a 1-chain γ is defined as

$$\int_\gamma \omega =< \omega, \gamma >= \sum_{[v_i,v_j]\in\gamma} \omega([v_i, v_j]).$$

M. Jin et al., *Conformal Geometry*, https://doi.org/10.1007/978-3-319-75332-4_4

Suppose M is a simply connected mesh and ω is a closed 1-form, then ω is exact, therefore, there exists a 0-form $f : M \to \mathbb{R}$, such that $\omega = df$. The 0-form f can be computed by integration. First, we select a root vertex v_0, and set $f(v_0)$ to be zero. Then we traverse all the vertices, then each vertex v_i has a unique path γ_i connecting to v_0, then we set

$$f(v_i) = \int_{\gamma_i} \omega.$$

The union of all such paths form a spanning tree T of M. The spanning tree depends on the implementation of traversal algorithm. Because ω is exact, f is independent of the choice of the spanning tree.

A closed 1-form has the property of $d\omega = 0$. Suppose $[v_0, v_1, v_2]$ is an arbitrary face, then

$$d\omega([v_0, v_1, v_2]) = \omega\partial[v_0, v_1, v_2] = \omega[v_0, v_1] + \omega[v_1, v_2] + \omega[v_2, v_0] = 0.$$

If we consider the mesh M is embedded in $\mathbb{R}^3$, then there exists a unique vector $\mathbf{w}$ on the plane spanned by vertices $\mathbf{v}_0, \mathbf{v}_1, \mathbf{v}_2$, such that

$$\begin{aligned} < \mathbf{w}, \mathbf{v}_1 - \mathbf{v}_0 > &= \omega([v_0, v_1]) \\ < \mathbf{w}, \mathbf{v}_2 - \mathbf{v}_1 > &= \omega([v_1, v_2]) \\ < \mathbf{w}, \mathbf{v}_0 - \mathbf{v}_2 > &= \omega([v_2, v_0]) \end{aligned}$$

The vector $\mathbf{w}$ has the following closed form

$$\mathbf{w} = -\omega[v_0, v_1]\mathbf{v}_2 \times \mathbf{n} - \omega[v_2, v_0]\mathbf{v}_1 \times \mathbf{n} - \omega[v_1, v_2]\mathbf{v}_0 \times \mathbf{n}$$

The formulae can be easily proved. Any closed 1-form ω corresponds to a vector valued 2-form

$$\mathbf{w} : F \to \mathbb{R}^3$$

The wedge product of two closed 1-forms is a 2-form, which can be constructed explicitly. Let ω_1 and ω_2 are the closed 1-forms, we first convert them to vector valued 2-forms, $\mathbf{w}_1$ and $\mathbf{w}_2$, then for an face $[v_i, v_j, v_k]$

$$\omega_1 \wedge \omega_2([v_i, v_j, v_k]) = \mathbf{w}_1 \times \mathbf{w}_2 \cdot \mathbf{n}A,$$

where A is the area of the face.

Suppose γ is a closed 1-chain, the characteristic 1-form ω of γ satisfies the following condition

$$\forall \tau, d\tau = 0, < \gamma, \tau >= \int_M \omega \wedge \tau = \sum_{[v_i, v_j, v_k] \in M} \omega \wedge \tau([v_i, v_j, v_k]).$$

The characteristic 1-form can be computed in the following way. We slice M along γ to form another mesh $\bar{M}$, such that $\bar{M}$ has two boundaries, the orientation of one boundary is consistent with that of γ, denoted as γ^+, the orientation of the other boundary is opposite to that of γ, denoted as γ^-,

$$\partial \bar{M} = \gamma^+ + \gamma^-.$$

We find a harmonic function $f : \bar{M} \to \mathbb{R}$, such that

$$\forall v \in \gamma^+, f(v) = 1, \forall v \in \gamma^-, f(v) = 0,$$

for all interior vertices

$$\forall v \notin \partial \bar{M}, \Delta f(v) = \sum_{[v,w] \in \bar{M}} k_{v,w}(f(v) - f(w)) = 0.$$

If all edge weights $k_{v,w}$ are positive, then such kind of harmonic function exists and is unique. Then the gradient of f, df is an exact one form on $\bar{M}$. Any half-edge h on M corresponds uniquely to a half-edge $\bar{h}$ on $\bar{M}$, we can define a closed 1-form on M

$$\omega(h) = df(\bar{h}).$$

Then ω is the characteristic 1-form of γ.

4.2 Cohomology Basis

Suppose M is a closed mesh with genus g. Algorithm 11 introduces a simple and efficient method to compute a set of 1-forms of M. They form the basis of the cohomology group $H^1(M, \mathbb{R})$.

input : A closed mesh of genus g
output: A set of basis of $H_1(M, \mathbb{R})$

$$\Omega = \{\omega_1, \omega_2, \ldots, \omega_{2g}\}$$

Compute a cut graph G of M;
Compute a spanning tree T of G;

$$G - T = \{e_1, e_2, \ldots, e_{2g}\}.$$

Slice M along G to get a topological disk $\bar{M}$;
The boundary of $\bar{M}$ is a list of vertices

$$\partial \bar{m} = \{v_0, v_1, \ldots, v_{n-1}\}$$

sorted in a cyclic order;
forall the $e_i \in G - T$ **do**
Find half-edges attached to e_i, h_i^+ and h_i^-;
suppose $\bar{h}_i^+ = [v_k, v_{k+1}]$ and $\bar{h}_i^- = [v_s, v_{s+1}]$;
Construct a function $f_i : \bar{M} \to \mathbb{R}$, such that

$$f(v_j) = \begin{cases} 1 & v_j \in \partial \bar{M}, k < j \leq s \\ 0 & otherwise \end{cases}$$

Define 1-form on M: $\omega_i(h) \leftarrow df_i(\bar{h})$;
end

Algorithm 11: Cohomology Basis $H^1(M, \mathbb{R})$

4.3 Harmonic 1-Form

In each cohomologous class, there are infinite closed forms. In practice, it is highly desirable to choose a unique representative for each cohomologous class. According to Hodge theory, harmonic forms are the best candidates.

Suppose ω is a closed 1-form, then locally ω is the gradient of some function $f : M \to \mathbb{R}$ (0-form), if f is harmonic, then ω is a harmonic 1-form. Namely, a harmonic 1-form satisfies the following equation

$$\Delta f(v) = \sum_{[v,w] \in M} k_{v,w} df([v, w]) = \sum_{[v,w] \in M} k_{v,w} \omega([v, w]) = 0. \quad (4.1)$$

Given a closed 1-form ω, we can add an exact 1-form df, such that $\omega + df$ is the unique harmonic 1-form cohomologous to ω. Then $f : M \to \mathbb{R}$ should satisfy the following condition,

Fig. 4.1 Visualize a basis of harmonic 1-form group on a genus two mesh

$$\sum_{[v,w]\in M} k_{v,w}(\omega([v,w]) + f(w) - f(v)) = 0. \tag{4.2}$$

The harmonic 1-form group is isomorphic to the cohomology group $H^1(M,\mathbb{R})$. Figure 4.1 visualizes a basis of harmonic 1-form group on a genus two mesh. Algorithm 12 computes a basis of harmonic 1-form group.

input : A closed mesh of genus g
output: Compute a set of basis of harmonic 1-form group
Compute a set of basis of cohomology group

$$\Omega = \{\omega_1, \omega_2, \ldots, \omega_{2g}\}.$$

forall the $\omega_i \in \Omega$ **do**
  Find $f_i : M \to \mathbb{R}$ by solving equation

$$\sum_{[v,w]\in M} k_{v,w}(\omega([v,w]) + f(w) - f(v)) = 0, \forall v \in M,$$

  $\omega_i \leftarrow \omega_i + df.$
end
Output Ω.

Algorithm 12: Harmonic 1-form Basis

4.4 Hodge Star Operator

Suppose S is a smooth surface, with an atlas $\{(U_\alpha, \phi_\alpha)\}$, then a differential 1-form has a local representation

$$\omega = f_\alpha dx_\alpha + g_\beta dy_\alpha,$$

where $f_\alpha(x_\alpha, y_\alpha)$ is a smooth surface. The hodge star operator

$$^*\omega = f_\alpha dy_\alpha - g_\beta dx_\alpha.$$

We say $^*\omega$ is conjugate to ω.

A 1-form is a linear functional on tangential vector fields. Suppose the surface has a Riemannian metric g, then we can associate each 1-form ω with a vector field $\underset{\approx}{\prec}$, such that for any vector field $\mathbf{v}$,

$$\omega(\mathbf{v}) =< \mathbf{w}, \mathbf{v} >_g .$$

where $<, >_g$ represents the inner product induced by the metric g. If ω is closed, then $\mathbf{w}$ is curl-free. If ω is harmonic, then $\mathbf{w}$ is both curl-free and divergence-free. Assume $^*\mathbf{w}$ is the vector field corresponding to $^*\omega$, then at each point,

$$^*\mathbf{w} = \mathbf{n} \times \mathbf{w}.$$

Namely, we rotate $\mathbf{w}$ at each point by 90 degree about the normal, then we get the vector field corresponding to the conjugate 1-form. In the above discussion, we use a Riemannian metric, but in fact, if S has a conformal structure, then we can define Hodge star also.

Suppose ω is a closed 1-form on a mesh M, we can convert it to the vector valued 2-form $\mathbf{w}$, then we rotate the vector on each face by 90 degree about the normal to get $^*\mathbf{w}$. But in this case, we can not convert $^*\mathbf{w}$ to 1-form directly. Suppose two faces $f_0 = [v_0, v_1, v_2]$ and $f_1 = [v_2, v_1, v_3]$, then

$$< {}^*\mathbf{w}(f_0), \mathbf{v}_2 - \mathbf{v}_1 > \neq < {}^*\mathbf{w}(f_1), \mathbf{v}_2 - \mathbf{v}_1 > .$$

This is the major error caused by the discrete approximation. Therefore, we take the average for $^*\omega$

$$^*\omega([v_1, v_2]) = \frac{1}{2}(< {}^*\mathbf{w}(f_0), \mathbf{v}_2 - \mathbf{v}_1 > + < {}^*\mathbf{w}(f_1), \mathbf{v}_2 - \mathbf{v}_1 >).$$

4.5 Holomorphic 1-Form

Suppose S is a Riemann surface, all the holomorphic 1-forms on S form a group. Each holomorphic 1-form ζ can be decomposed to real and imaginary parts. Both parts are real harmonic 1-forms,

$$\zeta = \omega + {}^*\omega\sqrt{-1}.$$

Therefore, we can approximate a holomorphic 1-form by computing harmonic 1-form ω and its conjugate $^*\omega$. Considering the approximation error of Hodge star operator, we adapt the following method.

Suppose M is a closed mesh with genus g, we first compute a set of basis of harmonic 1-form group,

$$\Omega = \{\omega_1, \omega_2, \ldots, \omega_{2g}\}.$$

Suppose ω is a harmonic 1-form, then ω can be represented as the linear combination of ω_k's:

$$\omega = \sum_{k=1}^{2g} \lambda_k \omega_k.$$

In a similar way, $^*\omega$ is also harmonic,

$$^*\omega = \sum_{k=1}^{2g} \mu_k \omega_k.$$

The linear coefficients $\{\mu_k\}$ can be computed as follows:

$$\int_M \omega_i \wedge {}^*\omega = \sum_k \mu_k \int_M \omega_i \wedge \omega_k, i = 1, 2, \ldots, 2g.$$

They form a linear system:

$$\begin{pmatrix} \int_M \omega_1 \wedge \omega_1 & \int_M \omega_1 \wedge \omega_2 & \cdots & \int_M \omega_1 \wedge \omega_{2g} \\ \int_M \omega_2 \wedge \omega_1 & \int_M \omega_2 \wedge \omega_2 & \cdots & \int_M \omega_2 \wedge \omega_{2g} \\ \vdots & \vdots & & \vdots \\ \int_M \omega_{2g} \wedge \omega_1 & \int_M \omega_{2g} \wedge \omega_2 & \cdots & \int_M \omega_{2g} \wedge \omega_{2g} \end{pmatrix} \begin{pmatrix} \mu_1 \\ \mu_2 \\ \vdots \\ \mu_2 \end{pmatrix} = \begin{pmatrix} \int_M \omega_1 \wedge {}^*\omega \\ \int_M \omega_2 \wedge {}^*\omega \\ \vdots \\ \int_M \omega_{2g} \wedge {}^*\omega \end{pmatrix}$$

It is easy to see that $\omega_k \wedge \omega_k = 0$, $\omega_i \wedge \omega_j = -\omega_j \wedge \omega_i$. The matrix on the left hand side can be easily computed, while the right hand side needs approximation.

For example, to compute $\omega_k \wedge {}^*\omega$, we first convert ω_k and ω to vector valued 2-forms $\mathbf{w}_k$ and $\mathbf{w}$, respectively. We then approximate $^*\mathbf{w}$ by $\mathbf{n} \times \mathbf{w}$. $\omega_k \wedge {}^*\omega$ can be approximated by

$$\omega_k \wedge {}^*\omega(f) = \mathbf{w}(f) \times {}^*\mathbf{w}(f) \cdot \mathbf{n}_f A_f,$$

where f is a face in M, $\mathbf{n}$ is the normal of f, and A is the area of f.

$$\int_M \omega_k \wedge {}^*\omega = \sum_{f \in M} \omega_k \wedge {}^*\omega(f) = \mathbf{w}(f) \times {}^*\mathbf{w}(f) \cdot \mathbf{n}_f A_f.$$

Therefore, we can solve the linear equation to obtain μ_k's. Algorithm 13 computes the conjugate harmonic 1-form.

input : A harmonic 1-form ω on a closed mesh M
output: The conjugate harmonic 1-form $^*\omega$
Compute a basis of harmonic 1-form group on M

$$\Omega = \{\omega_1, \omega_2, \ldots, \omega_{2g}\}.$$

forall the $\omega_k \in \Omega$ **do**
| Convert ω_k to a vector valued 2-form $\mathbf{w}_k$
end
Convert ω to a vector valued 2-form $\mathbf{w}$;
forall the *Face* $f \in M$ **do**

$$^*\mathbf{w}(f) \leftarrow \mathbf{n}_f \times \mathbf{w}(f)$$

end
forall the $\omega_k \in \Omega$ **do**
$s_k \leftarrow 0$;
forall the *face* $f \in M$ **do**

$$s_k \leftarrow s_k + \mathbf{w}_k(f) \times {}^*\mathbf{w}(f) \cdot \mathbf{n}_f$$

end
end
forall the $\omega_i, \omega_j \in \Omega, i < j$ **do**
Compute $\omega_i \wedge \omega_j$;
$a_{ij} \leftarrow \int_M \omega_i \wedge \omega_j$;
$a_{ji} \leftarrow -a_{ij}$
end
Solve linear equation

$$(a_{ij})(\mu_j) = (s_i),$$

Return

$$^*\omega \leftarrow \sum_{k=1}^{2g} \mu_k \omega_k.$$

Algorithm 13: Conjugate harmonic 1-form

The vector space of the holomorphic 1-form group of a genus g surface is of $2g$ real and g complex dimensions. Algorithms 14 and 15 compute the $2g$ real and g complex basis of holomorphic 1-form group of a genus g surface, respectively.

input : A closed genus g mesh M
output: A real holomorphic 1-form basis $\{\zeta_1, \zeta_2, \ldots, \zeta_{2g}\}$
Compute a harmonic 1-form basis of M

$$\Omega = \{\omega_1, \omega_2, \ldots, \omega_{2g}\};$$

forall the $\omega_k \in \Omega$ **do**
  Compute the conjugate of ω_k, ${}^*\omega_k$;
  $\zeta_k \leftarrow \omega_k + \sqrt{-1}{}^*\omega_k$;
end
$\{\zeta_1, \zeta_2, \ldots, \zeta_{2g}\}$ form a basis of the real vector space of holomorphic 1-form group.

Algorithm 14: Compute a real basis of holomorphic 1-form group

input : a closed mesh M of genus g
output: Holomorphic 1-form basis $\{\zeta_1, \zeta_2, \ldots, \zeta_g\}$
Compute a canonical homology group basis,

$$\Gamma = \{a_1, b_1, a_2, b_2, \ldots, a_g, b_g\}.$$

forall the $a_k \in \Gamma, b_k \in \Gamma$ **do**
  Compute the characteristic 1-form ω_k for a_k, η_k for b_k;
  Convert ω_k to harmonic 1-form;
  Convert η_k to harmonic 1-form;
end
forall the ω_k **do**
  Compute the conjugate of ω_k as ${}^*\omega_k$;

$$\zeta_k \leftarrow \omega_k + \sqrt{-1}{}^*\omega_k.$$

end
Then

$$\{\zeta_1, \zeta_2, \ldots, \zeta_g\}$$

form a basis of the holomorphic 1-form group.

Algorithm 15: Compute a complex basis of holomorphic 1-form group

Algorithm 16 computes a complex valued function $f : M \to \mathbb{C}$ based on one holomorphic 1-form and assigns the complex value as texture coordinates for each vertex.

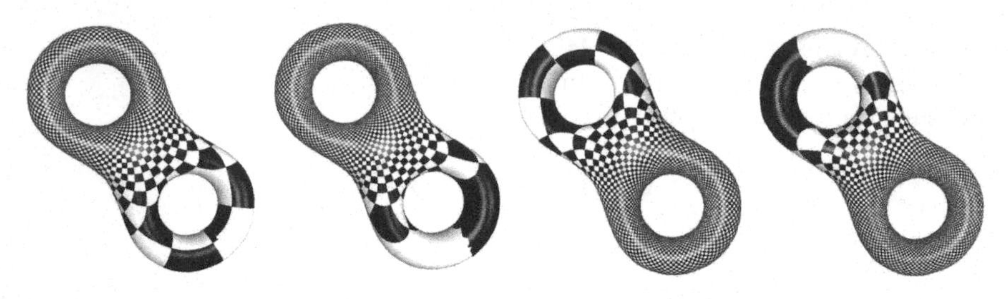

Fig. 4.2 Visualize a basis of holomorphic 1-form group on a genus two mesh

input : A closed genus g mesh M with a holomorphic 1-form
output: A complex valued function $f : M \to \mathbb{C}$
Compute a fundamental domain $\bar{M}$ of M;
Choose a root vertex $v_0 \in \bar{M}$;
Initialize a vertex queue $Q \leftarrow v_0$;
while *Q is not empty* **do**
  $v \leftarrow$ *pop* Q;
  forall the $[v, w] \in M$ **do**
    if *w has not accessed* **then**
      $f(w) \leftarrow f(v) + \omega([v, w])$ Insert w to Q.
    end
  end
end

Algorithm 16: Visualize holomorphic 1-form

Figure 4.2 visualizes a holomorphic 1-form group basis of a genus two triangular mesh with texture mapping using Algorithm 16.

Chapter 5
Discrete Ricci Flow

Abstract Surface Ricci flow is a powerful tool to design Riemannian metric of a surface such that the metric induces a user-defined Gaussian curvature function on the surface. The metric is conformal (i.e., angle-preserving) to the original one of surface. For engineering applications, smooth surfaces are approximated by discrete ones. This chapter introduces computational algorithms of Ricci flow on piecewise linear triangular meshes.

Ricci flow is a curvature flow method, introduced by Richard Hamilton in 1982 [109]. Ricci flow has been applied to the proof of Poincaré conjecture on three dimensional manifolds [209–211].

Surface Ricci flow is a powerful tool to design Riemannian metric of a surface such that the metric induces a user-defined Gaussian curvature function on the surface. The metric is conformal (i.e., angle-preserving) to the original one of surface. For engineering applications, the theory of Ricci flow on smooth surfaces needs to be generalized to discrete one. Computational algorithms of Ricci flow also need to be designed on piecewise linear triangular meshes.

Conformal mappings preserve angles and transform infinitesimal circles to infinitesimal circles with intersection angles among circles well-preserved. Figure 5.1 visualizes these properties. We texture map a checkerboard onto a bunny surface based on the inverse of a conformal mapping. All the corner angles are well-preserved. If we replace the checkerboard by a *circle packing pattern*, all the circles on texture are mapped to circles on the surface. The tangency relations among circles are preserved on the surface. A bridge connecting continuous and discrete conformal mapping is circle packing metric.

5.1 Discrete Background

In engineering fields, smooth surfaces are approximated by simplicial complexes (triangle meshes). Major concepts of differential geometry, such as metrics, con-

M. Jin et al., *Conformal Geometry*, https://doi.org/10.1007/978-3-319-75332-4_5

Fig. 5.1 Conformal mappings transform infinitesimal circles to infinitesimal circles, preserving the intersection angles among the circles. Image from [131]

formal deformation, and curvature in a continuous setting need be generalized to a discrete one.

Definition 5.1 (*Discrete Metric*) Suppose M is a triangular mesh, a *discrete metric* is a function defined on the non-oriented edges of M:

$$l : E \to \mathbb{R}^+,$$

such that for each triangle $[v_i, v_j, v_k]$, the triangle inequality holds

$$l([v_i, v_j]) + l([v_j, v_k]) > l([v_k, v_i]).$$

Let Γ be a function defined on the vertices, $\Gamma : V \to \mathbb{R}^+$, which assigns a radius γ_i to each vertex v_i. Similarly, let Φ be a function defined on the edges, $\Phi : E \to [0, \frac{\pi}{2}]$, which assigns an acute angle $\Phi([v_i, v_j])$ to each edge $[v_i, v_j]$ and is called a *weight* function on the edges. The pair of vertex radius and edge weight functions, (M, Γ, Φ), is called a *circle packing* of M.

Figure 5.2 visualizes a circle packing. Each vertex v_i has a circle with radius γ_i. For each edge $[v_i, v_j]$, the intersection angle ϕ_{ij} is defined by the two circles of v_i and v_j, which either intersect or are tangent.

If we assign a circle packing to a mesh, the circle packing naturally induces a circle packing metric.

Definition 5.2 (*Circle Packing Metric*) Suppose M is a triangular mesh, with a circle packing (M, Γ, Φ), then the *circle packing metric* is defined as

$$l([v_i, v_j]) = \sqrt{\gamma_i^2 + \gamma_j^2 + 2\cos\phi_{ij}\gamma_i\gamma_j}$$

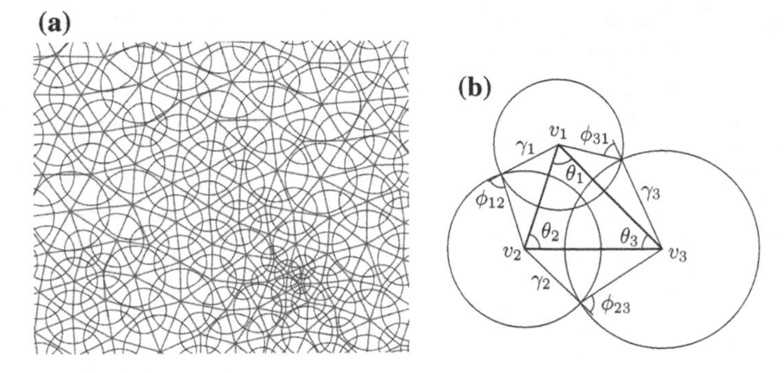

Fig. 5.2 Circle Packing Metric a Flat circle packing metric, **b** Circle packing metric on a triangle. Image from [131]

If all the intersection angles are acute, then the edge lengths induced by a circle packing satisfy triangle inequality.

If S is a smooth surface with a Riemannian metric $\mathbf{g} = (g_{ij})$, then a conformal metric is $e^{2u}\mathbf{g}$, where $u : S \to \mathbb{R}$ is a function on the surface. In discrete case, we can define conformal circle packing metric analogically.

Definition 5.3 (*Conformal Circle Packing Metric*) Suppose M is a triangular mesh with two different circle packing metrics (M, Γ_1, Φ_1) and (M, Γ_2, Φ_2), if

$$\Phi_1 \equiv \Phi_2,$$

then the two circle packing metrics are conformally equivalent.

Namely, two circle packings with the same intersection angles but different circle radii induce conformal circle packing metrics.

Proposition 5.4 *Let K be a combinatorial closed disk (simply connected,finite and with nonempty boundary). Then there exist an univalent circle packing $P \subset \mathbb{D}$ for K, where $\mathbb{D}$ is the planar unit disk, such that every boundary circle is tangent to the unit circle. Two such circle pakcings differ by a Möbius transformation.*

Such kind of circle packing is called the maximal circle packing of K.

Given a planar domain D, we can construct a sequence of circle packings $P_i = (T_i, \Gamma_i, \Phi_i)$, such that T_i is a regular planar tessellation by equilateral triangles with Φ_i constant and the radii shrinking to zero, $\lim_{k\to\infty} \gamma \to 0$. T_i exhausts D from interior eventually. For each circle packing $P_i = (T_i, \Gamma_i, \phi_i)$, we compute its maximal circle packing $\tilde{P}_i = (T_i, \tilde{\Gamma}_i, \phi_i)$. The discrete conformal map induced by the two circle packings is

$$\phi_i : P_i \to \tilde{P}_i,$$

Furthermore, suppose $p, q \in D$, we use a Möbius transformation to ensure $\phi_i(p)$ is the origin, $\phi_i(q)$ is on the real axis.

Thurston conjectured that discrete conformal map ϕ_i induced by circle packing converges to the real Riemann mapping $\phi_\infty : D \to \mathbb{D}$. Sullivan and Rodin proved Thurston's conjecture. Later, He and Andrew use circle packing proves the classicial Riemann mapping theorem. The whole literature of conventional complex analysis can be based on circle packing.

Definition 5.5 (*Discrete Gaussian Curvature*) The discrete Gaussian curvature K_i on a vertex $v_i \in M$ can be computed from the angle deficit,

$$K_i = \begin{cases} 2\pi - \sum_{f=[v_i,v_j,v_k]} \theta_i^{jk}, & v_i \notin \partial\Sigma \\ \pi - \sum_{f=[v_i,v_j,v_k]} \theta_i^{jk}, & v_i \in \partial\Sigma \end{cases} \tag{5.1}$$

where θ_i^{jk} represents the corner angle attached to vertex v_i in Face $f = [v_i, v_j, v_k]$, and $\partial\Sigma$ represents the boundary of Mesh M. The discrete Gaussian curvatures are determined by the discrete metrics.

Gauss-Bonnet theorem bridges differential geometry and topology. It also holds for discrete mesh as formulated in the discrete Gauss-Bonnet theorem.

Theorem 5.6 (Discrete Gauss-Bonnet Theorem) *Suppose M is a closed triangular mesh with a discrete metric, then*

$$\sum_{v \in M} K(v) = 2\pi\chi(M),$$

where $\chi(M)$ is the Euler number of the mesh, $\chi(M) = V + F - E$.

Proof The summation of three angles of each face equals to π. The total corner angles of a mesh equals to

$$\sum \alpha = F\pi.$$

The total discrete Gauss curvature

$$\sum_{v \in M} K(v) = \sum_{v \in M} 2\pi - \sum \alpha = 2V\pi - F\pi$$

Each face has three edges, edge edge is shared by two faces,

$$3F = 2E,$$

Therefore

$$\sum_{v \in M} K(v) = 2\pi(V + F - E).$$

Then discrete Gauss-Bonnet theorem also holds for meshes with boundaries. Let $\bar{M}$ is the double covering of M, i.e., gluing M with its copy along the corresponding boundaries, then the following relation holds:

$$\sum_{v \in M} K(v) = \frac{1}{2} \sum_{\bar{v} \in \bar{M}} K(\bar{v}).$$

Then

$$\chi(M) = \frac{1}{2} \chi(\bar{M}).$$

This result shows that the total Gauss curvature is independent of the choice of the metric.

5.2 Discrete Surface Ricci Flow

Suppose S is a smooth surface with Riemannian metric $\mathbf{g} = (g_{ij})$. Ricci flow deforms the metric $\mathbf{g}(t)$ according to its induced Gaussian curvature $K(t)$, where t is a time parameter

$$\frac{dg_{ij}(t)}{dt} = -2K(t)g_{ij}(t), \tag{5.2}$$

with a constraint that the total surface area is preserved. If we represent the Riemannian metric in the following form:

$$\mathbf{g}(t) = e^{2u(t)}\mathbf{g}(0),$$

then the surface Ricci flow can be written as:

$$\frac{du(t)}{dt} = -K(t). \tag{5.3}$$

From a physical sense, a curvature evolution induced by Ricci flow is like a heat diffusion on the same surface, as follows,

$$\frac{dK(t)}{dt} = -\Delta_{g(t)} K(t),$$

where $\Delta_{g(t)}$ is the Laplace-Beltrami operator induced by the metric $\mathbf{g}(t)$.

The Ricci flow can be easily modified to compute a metric with a prescribed curvature $\bar{K}$. The flow becomes

$$\frac{du(t)}{dt} = 2(\bar{K} - K)g_{ij}(t), \tag{5.4}$$

with the area preserved constraint.

The discrete surface Ricci flow is an exact analogy of the smooth surface Ricci flow as described in Eq. (5.4). Given a mesh M with a circle packing metric (M, Γ, Φ), denote K_i the current Gaussian curvature and $\bar{K}_i$ the prescribed Gaussian curvature of v_i. The discrete Ricci flow is defined as:

$$\frac{du_i(t)}{dt} = (\bar{K}_i - K_i(t)). \tag{5.5}$$

Definition 5.7 (*Convergence*) A solution to Eq. (5.5) *exists* and is *convergent*, if

1. $\lim_{t\to\infty} K_i(t) = \bar{K}_i, \forall i$,
2. $\lim_{t\to\infty} \gamma_i(t) = \bar{\gamma}_i \in \mathbb{R}^+, \forall i$.

A solution *converges exponentially* if there are constants c_1, c_2, so that for all time $t \geq 0$,

$$|K_i(t) - \bar{K}_i| \leq c_1 e^{-c_2 t},$$

and

$$|\gamma_i(t) - \bar{\gamma}_i| \leq c_1 e^{-c_2 t}.$$

Chow and Luo proved that the discrete Ricci flow is exponentially convergent.

Theorem 5.8 *Suppose (M, Φ) is a closed weighted mesh. Given any initial circle packing metric based on the weighted mesh, the solution to the discrete Ricci flow in the Euclidean geometry with the given initial value exists for all time and converges exponentially fast.*

For a given weighted mesh (M, Φ), we say a Gaussian curvature $K : V \to \mathbb{R}$ is *admissible*, if there exists a circle packing metric (M, Φ, Γ), which induces Gaussian curvature equals to K.

Theorem 5.9 *Suppose (M, Φ) is a weighted mesh, $\Phi : E \to [0, \frac{\pi}{2})$, I is a subset of vertices V. Then*

$$\sum_{i\in I} K_i > - \sum_{(e,v)\in Lk(I)} (\pi - \Phi(e)) + 2\pi\chi(F_I).$$

where F_I is the subcomplex consisting of cells whose vertices are in I and $Lk(I) = \{(e, v)| e\ is\ an\ edge\ so\ that\ e \cap I = \emptyset\ and\ the\ vertex\ v \in I\ so\ that\ e, v\ form\ a\ triangle\}$.

Algorithm 17 computes discrete surface Ricci flow using negative gradient method. Note that we compute an initial circle packing metric that approximates the originally induced Euclidean one. At each step, we normalize to ensure

$$\sum_{v_i\in M} \log \gamma_i = 0,$$

which equals to the area preserved constraint in smooth surface Ricci flow.

input : A mesh M embedded in $\mathbb{R}^3$, target curvature $\bar{K}$, curvature error threshold ε
output : A circle packing metric (M, Γ, Φ) that induces $\bar{K}$.

Compute an initial circle packing metric (M, Γ_0, ϕ);
while $max|K_i - \bar{K}_i| > \varepsilon$ **do**
  forall the *edge* $e = [v_i, v_j] \in M$ **do**
    Compute the edge length

$$l_{ij} \leftarrow \sqrt{\gamma_i^2 + \gamma_j^2 + 2\gamma_i\gamma_j \cos\phi_{ij}}$$

  end
  forall the *Corner in mesh* $c_i \in [v_i, v_j, v_k]$ **do**
    Compute the corner angle c_i

$$c_i \leftarrow \cos^{-1} \frac{l_{ij}^2 + l_{ki}^2 - l_{jk}^2}{2l_{ij}l_{ik}}.$$

  end
  forall the *Vertex* $v_i \in M$ **do**
    Compute the Gaussian curvature of v_i;

$$K_i \leftarrow 2\pi.$$

    forall the *Corners c attached to* v_i **do**

$$K_i \leftarrow K_i - c.$$

    end
  end
  forall the *Vertex* $v_i \in M$ **do**

$$\gamma_i \leftarrow \gamma_i + \gamma_i \times (\bar{K}_i - K_i)$$

  end

$$s \leftarrow 0,$$

  forall the *Vertex* $v_i \in M$ **do**

$$s \leftarrow s + \log \gamma_i,$$

  end
  forall the *Vertex* $v_i \in M$ **do**

$$\gamma_i \leftarrow \gamma_i \times \exp -\frac{s}{|V|}.$$

  end
end

Algorithm 17: Negative gradient descent method of discrete Surface Ricci Flow

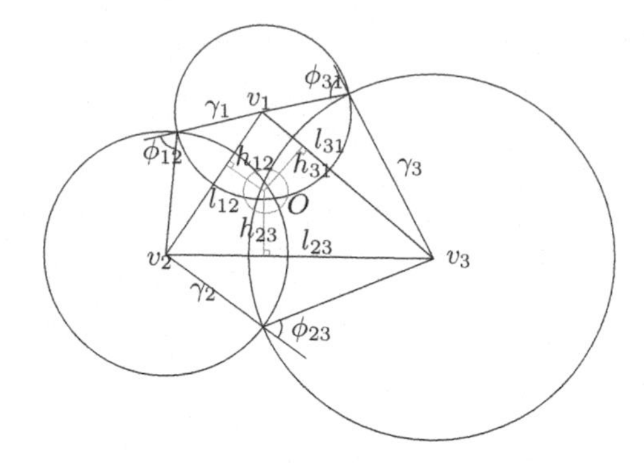

Fig. 5.3 Circle packing metric on a triangle. Image from [131]

An initial circle packing metric can be easily constructed from a triangulation with a good quality. If the quality of triangulation of a mesh is bad with too many skinny triangles or obtuse angles, re-meshing is necessary in practice.

Discrete Surface Ricci flow in Eq. (5.5) is in fact the negative gradient flow of a convex energy. The desired circle packing metric is the optimum point of the energy.

Figure 5.3 illustrates the circle packing metric of a triangle $[v_1, v_2, v_3]$. Each vertex associates with a circle $c(v_i, \gamma_i)$. Each pair of circles share a common chord. The three common chords intersect at a point denoted as O, called the *center* of triangle. The center can be obtained in a different way. There exists a unique circle orthogonal to the three circles, the center of the circle is also at O. The distance from O to the edge e_{ij} is h_{ij}.

Let $u_i = \log \gamma_i$, it can be verified by direct calculation that

$$\frac{\partial \theta_j}{\partial u_i} = \frac{\partial \theta_i}{\partial u_j} = h_{ij}.$$

Since $\theta_1 + \theta_2 + \theta_3 = \pi$, we have

$$\frac{\partial \theta_i}{\partial u_i} = -\frac{\partial \theta_j}{\partial u_i} - \frac{\partial \theta_k}{\partial u_i} = -\frac{\partial \theta_i}{\partial u_j} - \frac{\partial \theta_i}{\partial u_k}. \tag{5.6}$$

We can define edge weight w_{ij} now. Suppose two faces $[v_i, v_j, v_k]$ and $[v_j, v_i, v_l]$ sharing an edge $[v_i, v_j]$, then O_k is the center in $[v_i, v_j, v_k]$, O_l is the center in $[v_j, v_i, v_l]$, h_{ij}^k is the distance from O_k to Edge $[v_i, v_j]$, h_{ji}^l is the distance from O_l to Edge $[v_j, v_i]$, then

$$w_{ij} = h_{ij}^k + h_{ji}^l.$$

If e_{ij} is a boundary edge and attaches to only one face f_{ijk}, then

$$w_{ij} = h_{ij}^k.$$

From Eq. (5.6), it is easy to show that

$$\frac{\partial K_i}{\partial u_j} = \frac{\partial K_j}{\partial u_i}. \tag{5.7}$$

Now let (M, Φ) be a weighted mesh, we consider all possible circle packing metrics, namely, all the possible radii functions $\Gamma : V \to \mathbb{R}^+$. Suppose the vertex set of M is $(v_1, v_2, \ldots, v_n)$, then the corresponding radii is

$$\gamma = (\gamma_1, \gamma_2, \ldots, \gamma_n).$$

For the convenience of discussion, we take the logrithm of the radii,

$$\mathbf{u} = (u_1, u_2, \ldots, u_n).$$

All possible **u**s form the n dimensional space homeomorphic to $\mathbb{R}^n$, we call it the *u-space*. We add one normalization constraint,

$$\sum_{k=1}^{n} u_k = 0,$$

then all the normalized **u**s form an $n-1$ dimensional space homeomorphic to $\mathbb{R}^{n-1}$. We call it *the normalized u-space*.

We define 1-form ω in the u-space,

$$\omega = \sum_{i=1}^{n} K_i du_i,$$

then

$$d\omega = \sum_{[v_i, v_j] \in M} \left(\frac{\partial K_i}{\partial u_j} - \frac{\partial K_j}{\partial u_i} \right) du_j \wedge du_i.$$

$d\omega$ is zero according to Eq. (5.7). Therefore, ω is a closed 1-form. Since u-space is simply connected, ω is also exact. We choose a special metric

$$\mathbf{u}_0 = (0, 0, \ldots, 0)$$

and define the following energy function

$$E(\mathbf{u}) = \int_{\mathbf{u}_0}^{\mathbf{u}} \sum K_i du_i. \tag{5.8}$$

The energy is independent of the choice of an integration path. The gradient of energy is

$$\nabla E = \left(\frac{\partial E}{\partial u_1}, \frac{\partial E}{\partial u_1}, \frac{\partial E}{\partial u_2}, \ldots, \frac{\partial E}{\partial u_n}\right) = (K_1, K_2, \ldots, K_n).$$

Therefore, discrete surface Ricci flow in Eq. (5.5) is exactly the negative gradient flow of the energy.

We can further compute the Hessian matrix of E with respect to $\mathbf{u}$. If $[v_i, v_j]$ is an edge on the mesh M, then

$$\frac{\partial^2 E(\mathbf{u})}{\partial u_i \partial u_j} = \frac{\partial K_i}{\partial u_j} = -w_{ij}.$$

Otherwise $\frac{\partial^2 E(\mathbf{u})}{\partial u_i \partial u_j} = 0$, furthermore

$$\frac{\partial^2 E(\mathbf{u})}{\partial u_i^2} = \frac{\partial K_i}{\partial u_i} = \sum_{[v_i, v_j] \in M} w_{ij}.$$

In the normalized u-space, the Hessian matrix is diagonal dominant, so it is positive definite. Hence, the energy $E(\mathbf{u})$ has a unique global minimum.

We denote the Hessian matrix as $\Delta(\mathbf{u})$, then from above calculation, we get

$$d\mathbf{k} = \Delta(\mathbf{u}) d\mathbf{u}.$$

For discrete Ricci flow $dd\mathbf{u} = -\mathbf{k}dt$, we get

$$\frac{d\mathbf{u}}{dt} = -\Delta(\mathbf{u}) d\mathbf{u}.$$

Therefore, the discrete Gaussian curvature evolves like a heat diffusion process.

In order to compute a metric satisfying the prescribed curvature $\bar{K}$, the energy can be reformulated as

$$E(\mathbf{u}) = \int_{\mathbf{u}_0}^{\mathbf{u}} \sum_{i=1}^{n} (K_i - \bar{K}_i) du_i.$$

The energy can be minimized efficiently using Newton's method. The desired metric is the global minimizer of the energy.

Algorithm 18 computes discrete surface Ricci flow using Newton's method.

input : A mesh M embedded in $\mathbb{R}^3$, target curvature $\bar{K}$, curvature error threshold ε

output: A circle packing metric (M, Γ, Φ) that induces $\bar{K}$.

Compute an initial circle packing metric (M, Γ_0, ϕ);
Compute the initial curvature K.

$$\mathbf{u} \leftarrow 0.$$

while $max|K_i - \bar{K}_i| > \varepsilon$ **do**
  forall the *edge* $e_{ij} \in M$ **do**
    Compute the edge weight $w_{ij}(\mathbf{u})$ to form the Hessien matrix.
  end

$$\begin{aligned} d\mathbf{u} &\leftarrow H^{-1}(\bar{K} - K) \\ \mathbf{u} &\leftarrow \mathbf{u} + d\mathbf{u} \\ \mathbf{k} &\leftarrow K(\mathbf{u}). \end{aligned}$$

end
$\bar{\mathbf{u}} \leftarrow \mathbf{u}$

Algorithm 18: Newton's method of discrete surface Ricci flow

5.3 Isometric Planar Embedding

Once a flat metric is computed, we can isometrically embed a mesh into plane $\mathbb{R}^2$. In theory, this step is trivial. In practice, it is challenging to compute planar embedding accurately for a very large-size mesh because of the accumulation of numerical error.

For a large-size mesh, we formulate the planar embedding as a variational problem, and solve it using optimization method. This approach is more robust and accurate than a direct embedding.

Suppose M is a topological disk (or a finite portion of the universal covering space of a mesh), with a flat metric $l : E \to \mathbb{R}^+$. We denote the map as

$$\phi : M \to \mathbb{R}^2, \phi(v_i) = (u_i, v_i),$$

such that

$$|\phi(v_i) - \phi(v_j)| = l([v_i, v_j]), \forall [v_i, v_j] \in M.$$

Randomly pick a face of M, denoted as $f = [v_i, v_j, v_k]$, we can embed it in a canonical way in the xy-plane, such that

$$(x_i, y_i) = (0, 0), (x_j, y_j) = (l([v_i, v_j]), 0), (x_k, y_k) = (l([v_k, v_i]) \cos\theta_i, l([v_k, v_i]) \sin\theta_i).$$

where

$$\theta_i = cos^{-1} \frac{l([v_i, v_j])^2 + l([v_k, v_i])^2 - l([v_j, v_k])^2}{2l([v_i, v_j])l([v_k, v_i])}.$$

Let

$$\mathbf{s}_i = [(x_k, y_k) - (x_j, y_j)] \times \mathbf{n}, \mathbf{s}_j = [(x_i, y_i) - (x_k, y_k)] \times \mathbf{n}, \mathbf{s}_k = [(x_j, y_j) - (x_i, y_i)] \times \mathbf{n},$$

where $\mathbf{n} = (0, 0, 1)$ is the face normal.

Then $\phi : M \rightarrow \mathbb{R}^2, \forall (u_i, v_i) \in M$, ϕ has two components (ϕ_1, ϕ_2), $\phi_k : M \rightarrow \mathbb{R}$, ϕ_ks are piecewise linear functions on the mesh. Then

$$\nabla\phi_1(f) = u_i\mathbf{s}_i + u_j\mathbf{s}_j + u_k\mathbf{s}_k, \nabla\phi_2(f) = v_i\mathbf{s}_i + v_j\mathbf{s}_j + v_k\mathbf{s}_k,$$

If ϕ is an isometric map, then

$$\nabla\phi_2(f) = \mathbf{n} \times \nabla\phi_1(f).$$

In practice, we fix the image of one face and minimize the following energy

$$E(\phi) = \sum_{f \in M} |\nabla\phi_2(f) - \mathbf{n} \times \nabla\phi_1(f)|^2,$$

to obtain ϕ

$$\phi = \min_{\phi} E(\phi).$$

The variables of the energy $E(\phi)$ are the planar images of all vertices $\{(u_i, v_i)\}$. Since $E(\phi)$ is a quadratic form, optimizing $E(\phi)$ is a linear problem.

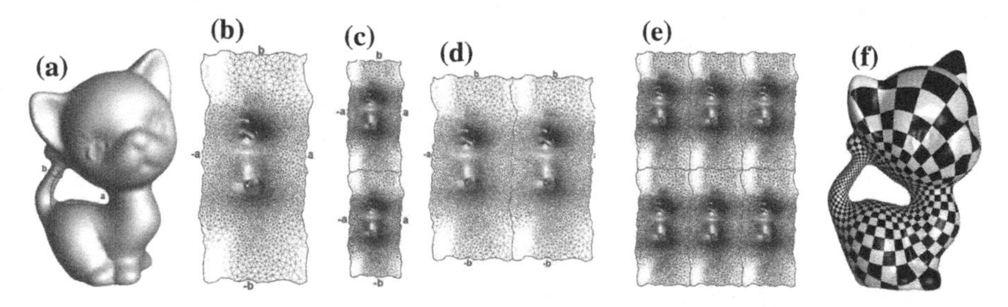

Fig. 5.4 Isometric Planar Embedding a A genus one kitten model marked with a set of canonical fundamental group generators a and b. **b** A fundamental domain is conformally flattened onto the plane, marked with four sides $aba^{-1}b^{-1}$. **c** One translation moves the side b to b^{-1}. **d** The other translation moves the side a to a^{-1}. **e** The layout of the universal covering space of the kitten mesh on the plane, which tiles the plane. **f** The conformal parameterization is used for the texture mapping purpose. A checkerboard texture is placed over the parameterization in **b**. The conformality can be verified from the fact that all the corner angles of the checkers are preserved. Image from [131]

Figure 5.4 demonstrates the isometric embedding of a genus one kitten triangular mesh model. We set the target curvature to zero everywhere and compute the conformal flat metric using Newton's method of discrete surface Ricci flow. We then isometrically embed both the fundamental domain of the mesh and a portion of its universal covering space onto the plane. Note that a pair of copies of the fundamental domains differ a rigid translation in plane. α and β are the generators of the rigid translation group as shown in Fig. 5.4c, d.

5.4 Hyperbolic Ricci Flow

In previous discussions, we assume a surface is embedded in three dimensional Euclidean space $\mathbb{R}^3$. Therefore the computation of length and angle is based on *the Euclidean background geometry*, i.e., computed using Euclidean geometry.

We generalize the method to *the hyperbolic background geometry*. Given a mesh M, we treat each face as a hyperbolic triangle and each edge as a geodesic (a hyperbolic line). Both corner angles and edge lengths are computed using the hyperbolic cosine law.

Here are the cosine laws for triangles with different background geometries:

$$\begin{array}{lll} l_k^2 & = l_i^2 + l_j^2 - 2 l_i l_j \cos\theta_k & \mathbb{E}^2; \\ \cosh l_k^2 & = \cosh l_i \cosh l_j + \sinh l_i \sinh l_j \cos\theta_k & \mathbb{H}^2; \\ \cos l_k^2 & = \cos l_i \cos l_j - \sin l_i \sin l_j \cos\theta_k & \mathbb{S}^2. \end{array}$$

For a mesh in hyperbolic background geometry with circle packing metric, we assume an edge $[v_i, v_j]$ connects two vertices with radii γ_i and γ_j, respectively and the intersection angle between the two circles are ϕ_{ij}. The edge length $l([v_i, v_j])$ can be computed as

$$\cosh l([v_i, v_j]) = \cosh\gamma_i \cosh\gamma_j + \sinh\gamma_i \sinh\gamma_j \cos\phi_{ij}.$$

The discrete Gaussian curvature of a vertex is calculated in a similar way as the Euclidean Ricci flow,

$$K_i = \begin{cases} 2\pi - \sum_{f=[v_i,v_j,v_k]} \theta_i^{jk}, & v_i \notin \partial\Sigma \\ \pi - \sum_{f=[v_i,v_j,v_k]} \theta_i^{jk}, & v_i \in \partial\Sigma \end{cases}$$

The discrete conformal factor u_i is defined differently for a hyperbolic or a spherical mesh M with a circle packing metric (M, Γ, Φ),

$$u_i = \begin{cases} \log\gamma_i & \mathbb{E}^2 \\ \log\tanh\frac{\gamma_i}{2} & \mathbb{H}^2 \\ \log\tan\frac{\gamma_i}{2} & \mathbb{S}^2. \end{cases}$$

The discrete Ricci flow for hyperbolic or spherical mesh is in the same form

$$\frac{du_i(t)}{dt} = \bar{K}_i - K_i.$$

Set the target Gaussian curvature $\bar{K}_i$ to zero for all vertices. The discrete Ricci flow is a negative gradient flow of discrete Ricci energy with the same form as the discrete Euclidean Ricci flow. Denote $\{v_1, v_2, \ldots, v_n\}$ the vertex set and $\mathbf{u} = \{u_1, u_2, \ldots, u_n\}$ the vector of discrete conformal factor. The following symmetry holds for both hyperbolic and spherical meshes

$$\frac{\partial K_i}{\partial u_j} = \frac{\partial K_j}{\partial u_i}.$$

Therefore the discrete hyperbolic, Euclidean or spherical energy is

$$E(\mathbf{u}) = \int_{\mathbf{u}_0}^{\mathbf{u}} \sum_{i=1}^{n} (\bar{K}_i - K_i) du_i.$$

The hyperbolic Ricci energy is convex if all the intersection angles ϕ_{ij} are acute. A global minimum of the energy, i.e., the metric $\bar{\mathbf{u}}$, induces the target curvature $\bar{\mathbf{k}}$. However, the spherical Ricci energy is not convex.

To compute a hyperbolic metric for a surface with negative Euler number, we can simply set the target curvatures of all vertices to zero and then minimize the discrete hyperbolic Ricci energy. The energy is convex, therefore, Newton's method leads to the unique global minimum with arbitrary starting point of $\mathbf{u}$. The formula for computing the Hessian matrix is given by the following formulae,

$$\frac{\partial^2 E(\mathbf{u})}{\partial u_j \partial u_i} = -\frac{K_i}{\partial u_j}.$$

5.5 Isometric Hyperbolic Embedding

Poincaré Disk Model

A commonly used model in hyperbolic space $\mathbb{H}^2$ is Poincaré disk model. The Poincaré disk model is a unit disk on the complex plane,

$$|z| < 1,$$

with Riemannian metric

$$ds^2 = \frac{4dzd\bar{z}}{(1-\bar{z}z)^2}, z \in \mathbb{C}.$$

The metric is conformal to the Euclidean metric $dzd\bar{z}$. All the rigid motions in hyperbolic space are Möbius transformations:

$$z \to e^{i\theta}\frac{z-z_0}{1-\bar{z}_0 z}, z_0 \in \mathbb{C}, \theta \in [0, 2\pi).$$

Suppose $p, q \in \mathbf{H}^2$ are two points in the Poincaré disk, the hyperbolic geodesic (hyperbolic line) is a circular arc through p and q and orthogonal to the unit circle $|z| = 1$. The hyperbolic lines through the origin are also Euclidean lines. The hyperbolic distance between p and q is

$$d(pq) = |\log \frac{|p-a|/|p-b|}{|q-a|/|q-b|}|.$$

Embedding of Fundamental Domain

Suppose M is a mesh in hyperbolic background geometry associated with a hyperbolic circle packing metric (M, Γ, Φ).

Suppose we embed one hyperbolic triangle in Poincaré disk, the positions of the three vertices are

$$v_i = (0,0), v_j = \frac{e^{l_{ij}}-1}{e^{l_{ij}}+1}, v_k = \frac{e^{l_{ik}}-1}{e^{l_{ik}}+1}e^{i\theta_i}, \tag{5.9}$$

respectively.

For general case, suppose v_i, v_j have been embedded, we need to determine the position of v_k. Denote (c, γ) a hyperbolic circle centered at c with radius γ. Then v_k locates at one of the two intersection points of (v_i, l_j) and (v_j, l_i) with the orientation of triangle $[v_i, v_j, v_k]$ counter clock wisely.

In Poincaré disk, the locus of a hyperbolic circle (c, γ) is also an Euclidean circle (C, R) with

$$C = \frac{2-2\mu^2}{1-\mu^2|c|^2}c, R^2 = |C|^2 - \frac{|c|^2-\mu^2}{1-\mu^2|c|^2},$$

where

$$\mu = \frac{e^r-1}{e^r+1}.$$

Algorithm 19 embeds the fundamental domain of a high genus surface into the Poincaré disk.

input : Fundamental domain of M with hyperbolic metric
output: Hyperbolic embedding of M into the Poincaré disk

Select a root face f, embed f using formula 5.9;
Put all faces sharing an edge with f in a queue Q;
while *Q is not empty* **do**
  Pop the queue, $f \leftarrow pop\ Q$, suppose $f = [v_i, v_j, v_k]$ where v_i, v_j have been embedded;
  Convert the hyperbolic circles (v_i, l_j) and (v_j, l_i) to Euclidean circles.;
  Compute the intersections of the two Euclidean circles;
  Choose the intersection which makes the orientation of the face counter-clock-wise, which is the coordinates of v_k;
  Put all the neighboring faces of f that haven't been embedded to the queue;
end

Algorithm 19: Hyperbolic embedding of fundamental domain

Embedding of Universal Covering Space

The whole universal covering space of a high genus surface M with hyperbolic metric, denoted as $(\bar{M}, \pi)$, can be isometrically embedded onto the Poincaré disk. A deck transformation $\rho : \bar{M} \to \bar{M}$ satisfies the following property,

$$\pi \circ \rho = \pi.$$

A deck transformation is also a Möbius transformation. It is isometric and a rigid motion of the Poincaré disk. All the deck transformations form a deck transformation group, which is isomorphic to the fundamental group of M.

Suppose D is a fundamental domain of M, $\phi : D \to \mathbb{H}^2$ is the embedding of D onto the Poincaré disk. $Deck(M)$ is the deck transformation group, $\rho \in Deck(M)$ is a Möbius transformation, then

$$\bigcup_{\rho \in Deck(M)} \rho \circ \phi(D) = \mathbb{H}^2.$$

All the copies of the embedding of a fundamental domain covers the whole Poincaré disk, namely, the embedding of universal covering space covers the whole Poincaré disk.

We can find the generators of a deck transformation group in the following manner. Suppose M is a closed genus g mesh. We first choose a base point $v \in M$ and a set of canonical fundamental group generators

$$\{a_1, b_1, a_2, b_2, \ldots, a_g, b_g\}.$$

We slice the mesh along the set of canonical fundamental group generators to form a fundamental domain D. The boundary of D is

$$\partial D = a_1 b_1 a_1^{-1} b_1^{-1} a_2 b_2 a_2^{-1} b_2^{-1} \cdots a_g b_g a_g^{-1} b_g^{-1}.$$

Then we embed D onto the Poincaré disk with Algorithm 19.

With D embedded in $\mathbb{H}^2$, a_i, b_j are curves in the $\mathbb{H}^2$. There exists a unique Möbius transformation ρ_k, which maps a_k to a_k^{-1}, namely for $p \in a_k, q \in a_k^{-1}, \pi(p) = \pi(q)$, then

$$\rho_k(p) = q, \rho_k(a_k) = a_k^{-1}.$$

Similarly, there exists a unique τ_k which maps b_k to b_k^{-1}. Then

$$\{\rho_1, \tau_1, \rho_2, \tau_2, \ldots, \rho_g, \tau_g\}$$

form the generators of the Deck transformation group of M (Fuchsian group generator).

Now, let's focus on the computation of ρ_1, the other generators can be obtained in the same way. Suppose the starting and ending points of a_1 are s, t, $\partial a_1 = t - s$, and the staring and the ending point of a_1^{-1} is t_1, s_1, $\partial a_1^{-1} = s_1 - t_1$. It is obvious that all of them are the pre-images of the base point, $\pi : \{s, t, s_1, t_1\} \to v$. We want to find a Möbius transformation ρ_1, such that

$$\rho_1(s) = s_1, \rho_1(t) = t_1.$$

Suppose the hyperbolic line through s and t is $\overline{st}$, the hyperbolic line through s_1 and t_1 is $\overline{s_1 t_1}$. Then ρ_1 maps $\overline{st}$ to $\overline{s_1 t_1}$. We first construct a Möbius transformation denoted as ϕ_1 to map s to the origin,

$$\phi_1 : z \to \frac{z - s}{1 - \bar{s}z}.$$

ϕ_1 also maps $\overline{st}$ to a radial Euclidean line. Suppose the angle from the real axis to $\phi(\overline{st})$ is θ. We construct another Möbius transformation

$$\phi_2 : z \to e^{i\theta}.$$

$\phi_2 \circ \phi_1$ maps $\overline{st}$ to the real axis with s to the origin and t to the point $\frac{e^d - 1}{e^d + 1}$, where d is the hyperbolic distance between s and t. We can use similar method to construct another Möbius transformation $\bar{\phi}_2 \circ \bar{\phi}_1$ to map s_1 to the origin and t_1 to $\frac{e^d - 1}{e^d + 1}$, then

$$\rho_1 = \bar{\phi}_1^{-1} \circ \bar{\phi}_2^{-1} \phi_2 \circ \phi_1$$

maps a_1 to a_1^{-1}.

With the computed generators of the Fuchsian group, we can transform the copies of the embedding of a fundamental domain and glue them coherently in Poincaré disk.

Figure 5.5 demonstrates the isometric hyperbolic embedding of a genus two triangular mesh. We set the target curvature to zero everywhere and compute the conformal

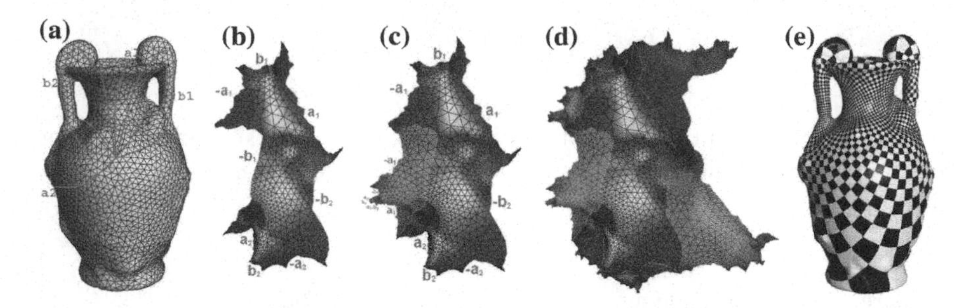

Fig. 5.5 Isometric Hyperbolic Embedding a Genus two vase model marked with a set of canonical fundamental group generators that cut the surface into a topological disk with eight sides: a_1, b_1, a_1^{-1}, b_1^{-1}, a_2, b_2, a_2^{-1}, b_2^{-1}. **b** The fundamental domain is conformally flattened onto the Poincaré disk with marked sides. **c** A Möbius transformation moves the side b_1 to b_1^{-1}. **d** Eight copies of the fundamental domain are glued coherently by eight Möbius transformations. **e** Conformal parameterization induced by the hyperbolic flattening. The corner angles of checkers are well-preserved. Image from [131]

hyperbolic metric using Newton's method of discrete hyperbolic Ricci flow. We then isometrically embed both the fundamental domain of the mesh and a portion of its universal covering space onto the Poincaré disk. Note that any pair of copies of the fundamental domain differ a Möbius transformation in Poincaré disk.

Part II
Engineering Applications

Chapter 6
Computer Graphics

Abstract This chapter introduces the applications of computational conformal geometry on computer graphics research. Specifically we focus on digital geometry processing, a subfield of computer graphics that studies 3D surfaces from a discrete differential geometry standpoint. Several research topics, including global surface parametrization, uniform remeshing, N-way rotational symmetry (N-RoSy) field and shortest homotopic cycle, together with their experimental results, are detailed in this chapter.

6.1 Introduction

Computer graphics refers to the various technologies used to create, represent, and manipulate images by a computer. Computer graphics is a very broad field. The overall methodology depends heavily on the underlying sciences of geometry, optics, and physics. Computer graphics has revolutionized movie and video game industries. Computer graphics also has a big impact in computer-aided design, computer simulation, information visualization, scientific visualization, virtual reality, and digital art.

This chapter focuses on digital geometry processing, a subfield of computer graphics that studies 3D surfaces from a discrete differential geometry standpoint. Specifically, we introduce the applications of computational conformal geometry in the following research topics in digital geometry processing field.

- *Global Surface Parametrization.* Conformal parameterization maps a surface to a canonical domain, usually a planar domain, with angles and local-shape well preserved. A conformal parameterization is global if it preserves the conformality of a surface with non-trivial topology everywhere except for a few singularity points. Conformal parameterization converts 3D geometric problems to 2D ones, significantly simplifying the computation. Conformal parametrization benefits many graphics applications including texture mapping, remeshing, morphing, and registration.
- *Uniform Remeshing.* Uniformly sampled mesh models are preferable inputs to most existing geometry processing algorithms. Remeshing, i.e., modifying the

M. Jin et al., *Conformal Geometry*, https://doi.org/10.1007/978-3-319-75332-4_6

sampling and connectivity of a geometry to generate a new mesh, is therefore a fundamental step for efficient mesh processing.

- *N-way Rotational Symmetry (N-RoSy) Field.* An N-way rotational symmetry (N-RoSy) field is invariant under rotations of an integer multiple of $\frac{2\pi}{N}$. N-RoSy field has been widely used in computer graphics to model brush strokes and hatches in non-photorealistic rendering, regular patterns in texture synthesis, and principle curvature directions in surface parameterizations and remeshing.
- *Shortest Homotopic Cycle.* Computing the shortest cycle in a given homotopic type has various applications in computer graphics, including topological segmentation and mapping of polyhedral surfaces.

6.2 Global Surface Parametrization

Surface parameterization maps a surface to a canonical domain, usually a planar one, which converts 3D geometric problems to 2D ones, thereby simplifying the computation on surface. Parametrization is important for many graphics applications including texture mapping, remeshing, morphing, registration, and so on. The main challenge of parametrization is to produce a planar triangulation that best matches the geometry of the 3D mesh, minimizing some measure of *distortion*, for example, angles or areas.

Conformal map preserves angles. Locally, conformal map introduces only a scaling factor to distance and area. Conformal map is intrinsic to the geometry of a mesh, independent of its resolution and preserving the consistency of its orientation. Conformal mapping based parametrization has been applied for texture mapping [108, 167, 189], geometry remeshing [4], and visualization [9, 100].

Unfortunately, it is impossible to map a whole closed surface with non-toroid topology to a $2D$ plane without any angle distortion. Therefore, in order to conformally parameterize a mesh with general topology, researchers have resorted to partitioning the mesh into several charts [167].

Discrete One-form introduced in [102] is the first method to achieve global conformal parameterization of surfaces with non-trivial topology. Section 6.2.1 introduces the discrete one-form based algorithm to compute optimal global conformal parameterization. Discrete Ricci flow is a more powerful and flexible tool to compute different types of automatic and seam-free global conformal parameterizations for surfaces with general topologies. Discrete Ricci flow provides the freedom to allocate the total Gaussian curvatures of a mesh in different regions: the interior vertices, the boundary vertices, or the other points. We introduce discrete Ricci flow based global conformal parameterization in Sect. 6.2.2.

6.2.1 *Optimal Global Conformal Parametrization Using Discrete One-Form*

Given a genus g surface with g equal to or greater than one, a global conformal parameterization can be obtained by integrating a holomorphic one form denoted as ω. Specifically, the surface is cut open to a topological disk, namely a *fundamental domain*. Integrating ω on the fundamental domain of the surface induces a global conformal parameterization of the whole surface to plane.

What's more, all the holomorphic one forms of the surface form a linear space, and the basis for such linear space is $2g$ dimensional. We compute a basis denoted as $\{\omega_1, \omega_2, \ldots, \omega_{2g}\}$, such that any linear combination of them is still a holomorphic 1-form, represented as $\omega = \sum_{k=1}^{2g} \lambda_k \omega_k$.

We formulate different energies to measure the quality of the global conformal parametrization. One is to measure the uniformity of the parametrization and the other is to measure the ratio of the parameter area on regions of interest in the surface. They are both denoted as $E(\omega)$, with $\omega = \sum_{i=1}^{2g} \lambda_i \omega_i$, and $E(\omega) = E(\lambda_1, \lambda_2, \ldots, \lambda_{2g})$. We need to find a set of linear combination coefficients λ_i to optimize the energy. The necessary condition for the optimal holomorphic one form is straight forward,

$$\frac{\partial E}{\partial \lambda_i} = 0, i = 1, 2, \ldots, 2g.$$

If the Hessian matrix $(\frac{\partial^2 E}{\partial \lambda_i \partial \lambda_j})$ is positive definite, E will reach the minimum; if the Hessian matrix is negative definite, E will be maximized.

Uniform Global Conformal Parametrization

Given a holomorphic one-form ω, $\omega = \sum_{k=1}^{2g} \lambda_k \omega_k$, we require the total parameter area equals the total area of the surface in R^3. The constraint function can be formulated as:

$$\sum_{[u,v,w]\in K_2} \frac{1}{2} |\omega([u, v]) \times \omega([v, w])| = \sum_{[u,v,w]\in K_2} S_{[u,v,w]}, \tag{6.1}$$

where $S_{[u,v,w]}$ is the area of face $[u, v, w]$ in R^3. The uniformity function is defined as the sum of the squared area differences of faces between parameter area and area in R^3,

$$E(\omega) = \sum_{[u,v,w]\in K_2} \left(\frac{1}{2} |\omega([u, v]) \times \omega([v, w])| - S_{[u,v,w]} \right)^2. \tag{6.2}$$

Both the constraint and the energy functions are polynomials with respect to λ_is. For example, the constraint can be reformulated as a quadratic form

$$\sum_{i,j=1}^{2g} c_{ij}\lambda_i\lambda_j = const,$$

with

$$c_{i,j} = \sum_{[u,v,w]\in K_2} \frac{1}{2}|\omega_i([u,v]) \times \omega_j([v,w])|.$$

We use the Newton's method to optimize the uniformity energy with the constraint. However, the extremal points are not unique. We randomly set initial values for λ_is. We get the most uniform parametrization by minimizing the uniformity energy, and the least uniform parametrization by maximizing the uniformity energy.

Emphasized Global Conformal Parametrization

It is also desirable to allocate more parameter areas for special regions on the surface in real applications. For example, more samples are required for regions with high Gaussian curvature or sharp features for surface remeshing. We design another function to measure the quality of parametrization for emphasized regions.

Suppose we subdivide the whole surface to two regions D_0 and D_1. The two may be disconnected with complicated topologies. If we maximize the parameter areas of D_0, we define the emphasized area energy as,

$$E(\omega) = \frac{1}{2} \sum_{[u,v,w]\in D_0} |\omega([u,v]) \times \omega([v,w])|. \tag{6.3}$$

with the same constraint in Eq. (6.1).

The function can be represented as a quadratic form directly. Let

$$c_{i,j} = \sum_{[u,v,w]\in D_0} |\omega_i([u,v]) \times \omega_j([v,w])|,$$

then the emphasized area energy is

$$E(\lambda_1, \lambda_2, \ldots, \lambda_{2g}) = \sum_{i,j=1}^{2g} c_{ij}\lambda_i\lambda_j. \tag{6.4}$$

We use conjugate gradient method for the optimization after setting initial values of λ_is. By maximizing this function, we put more samples on D_0; By minimizing it, we put more samples on D_1.

Figure 6.1 shows a genus two surface with different parameterizations. Specifically, Fig. 6.1a is the most uniform global conformal parametrization. Figure 6.1b is the least one. Figure 6.1c emphasizes the right handle that occupies the majority of the total parameter area.

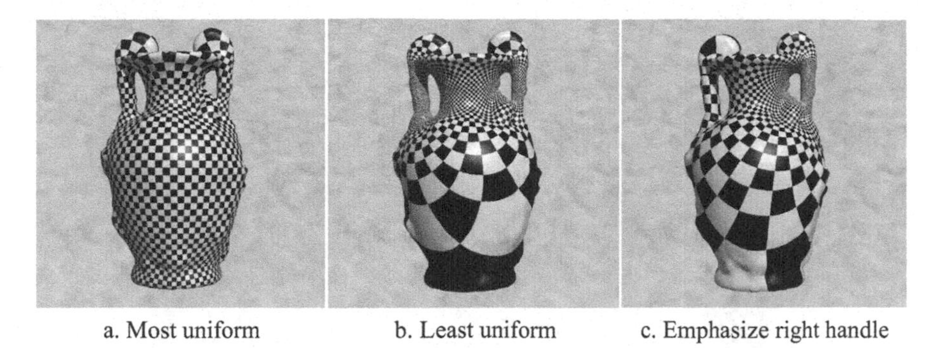

a. Most uniform b. Least uniform c. Emphasize right handle

Fig. 6.1 Amphora model. Image from [132]

Topological Optimization

For long and narrow surface regions, such as fingers and tails, the area distortion is huge with conformal parametrization. Suppose we have a long thin cylinder and we plan to conformally parameterize it. If we use polar coordinates (ρ, θ) with the center of the top mapped to the origin, the conformal factor is a function dependent only on ρ because of symmetry. The Gaussian curvature of the cylinder is zero, and

$$k(\rho, \theta) = \frac{1}{\lambda^2} \Delta \log \lambda = 0. \tag{6.5}$$

We can deduce $\lambda(\rho) = e^{a\rho+b}$, where a and b are constants. No matter which conformal parametrization method we choose, the stretching factor increases exponentially.

In order to improve the uniformity of conformal parametrization on a surface with long and narrow tubes, we can modify its topology. We first find the most uniform conformal parametrization of the surface. We then estimate the conformal factor of each vertex by the following formula:

$$\lambda(v) = \frac{1}{k} \sum_{[u,v]\in K_1} \frac{|r(u) - r(v)|^2}{|\omega([u, v])|^2}, u, v \in K_0, \tag{6.6}$$

where k is the valence of vertex v. In practice, we compute $\frac{1}{\lambda}$ instead. The inverses of conformal factors at extreme points are very close to zero. We then locate points with extremely high conformal factors and introduce small slices at the neighborhoods of those points. After double covering, i.e., gluing with a copy of the surface along the boundaries to form a closed one, we recompute the most uniform conformal parametrization of current surface. The whole process repeats until the uniformity energy is less than some threshold or converges to a limit.

Figure 6.2a shows a model with two cylinder-shaped branches where parameterizations are under-sampled. Figure 6.2b shows the new parametrization after one slice

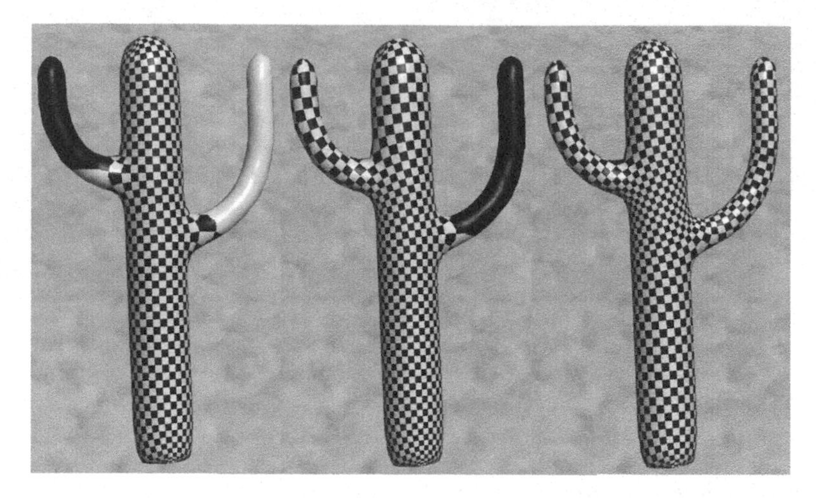

Fig. 6.2 Cactus model. **a** Most uniform parametrization on a cactus model. **b** Most uniform parametrization after one slice is introduced on the top of the left branch. **c** Most uniform parametrization after another slice is introduced on the top of the right branch. Image from [132]

a. Most uniform b. Emphasize front part

Fig. 6.3 Lion model. Image from [132]

on the top of its left branch, and (c) is the final one after adding one more slice on the top of the right branch.

The lion model shown in Fig. 6.3 has been introduced 6 slices for topological optimization. Its double covering is a genus 5 surface. Figure 6.3 shows the parameterizations of the most uniform and the emphasized front, respectively.

Zero Points Allocation

Definition 6.1 (*Zero Point*) Given a Riemann surface S with a conformal structure A, a holomorphic 1-form ω, with $\omega = f(z)dz$, where $f(z)$ is an analytic function and $z = u + iv$ is the local parameter. If $f(z)$ equals to zero at point p, p is a zero point of ω.

For a Riemann surface S with genus g, a holomorphic 1-form ω has $2g - 2$ zero points in principle. The neighborhood of a zero point will be mapped to a

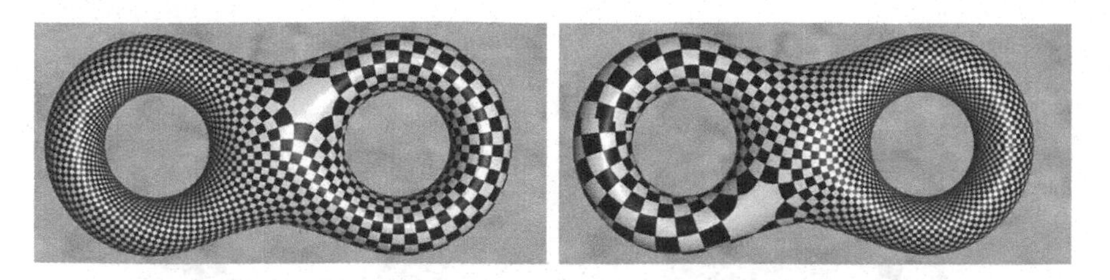

Fig. 6.4 Genus two eight model with zero points locating at different positions. Image from [132]

very small area on the parameter plane and under-sampled. Figure 6.4 shows two examples where zero points cause irregular pattern of texture mapping. Therefore, it is desirable to hide the zero points at some predetermined positions.

However, the positions of zero points are globally related. They are determined by the conformal structure of a surface. Although it is impossible to allocate all of them arbitrarily, we can still control part of them.

Suppose ω is a holomorphic 1-form, it has $p_1, p_2, \ldots, p_{2g-2}$ zero points, then $\omega(p_i) = 0, \forall i$. Let $\omega = \sum_{j=1}^{2g} \lambda_j \omega_j$, we get a linear system:

$$\sum_{j=1}^{2g} \lambda_j \omega_j(p_i) = 0, j = 1, 2, \ldots, 2g-2. \tag{6.7}$$

If $\{p_1, p_2, \ldots, p_{2g-2}\}$ is a set of zero points for some holomorphic 1-form $\omega \neq 0$, it is necessary and sufficient that the matrix $(\omega_j(p_i))$ is degenerated.

In our discrete setting, $\omega = \sum_{i=1}^{2g} \lambda_i \omega_i$, and we use the following to approximate $\omega(v), v \in K_0$.

$$\omega(v) = \sum_{[u,v]\in K_1} \frac{\omega([u,v])}{|r(u) - r(v)|} = \sum_{[u,v]\in K_1} \sum_{i=1}^{2g} \frac{\omega_i([u,v])}{|r(u) - r(v)|}.$$

Suppose we want to set n zero points $\{v_1, v_2, \ldots, v_n\}$, where $n < 2g - 2$, then we need to minimize the following energy

$$E(\omega) = \sum_{i=1}^{n} |\omega(v_i)|^2. \tag{6.8}$$

This function is a quadratic form of $\lambda_1, \lambda_2, \ldots, \lambda_{2g}$ and can be solved easily using conjugate gradient method. As long as n is not greater than g, we can fix the zero points to the predetermined positions.

For the model shown in Fig. 6.5, We predetermine the positions of two zero points and get the desired parametrization by minimizing the energy defined in Eq. (6.8).

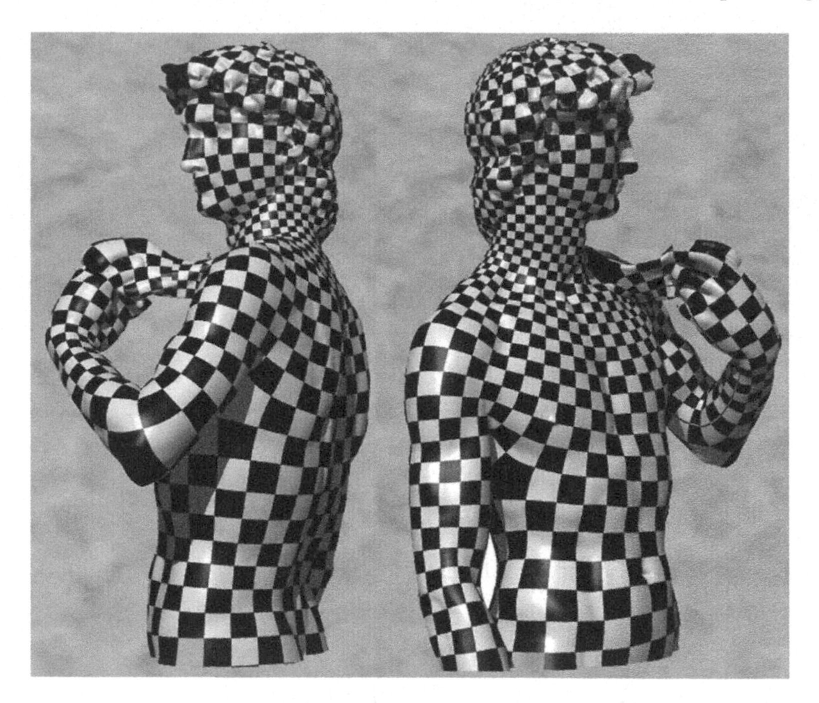

Fig. 6.5 David half body model with zero points hidden. **a** One zero point is hidden under the left armpit. **b** The other zero point is hidden under the right armpit. Image from [132]

6.2.2 Global Conformal Parameterization Using Discrete Euclidean Ricci Flow

Discrete Euclidean Ricci flow is a more powerful and flexible tool to compute surfaces' global conformal parameterizations. For surfaces with zero Euler number, like torus and annulus, we can find a special metric called flat uniformization metric, which is conformal to the original one, using discrete Euclidean Ricci flow with all target curvatures of vertices set to zero. Then surfaces with their flat uniformization metric can be conformally flattened to the plane.

For surfaces with non-zero Euler number, according to Gauss–Bonnet theorem, there must be some singularities for the parameterizations where the curvatures are not zeros. We can either concentrate all curvatures onto the singularities, while the target curvatures of all other vertices are set to zero; or if the surface is open, we can push those curvatures to the boundaries and set the target curvatures of interior vertices to zero. It is flexible to set the numbers and locations of singularities, or the boundary curvatures of open surfaces, as long as the sum of curvatures satisfy Gauss–Bonnet theorem.

Figure 6.6a, d show the two sides of a genus zero surface with 3 boundaries. We can flexibly distribute the total Gaussian curvatures of the surface to different regions.

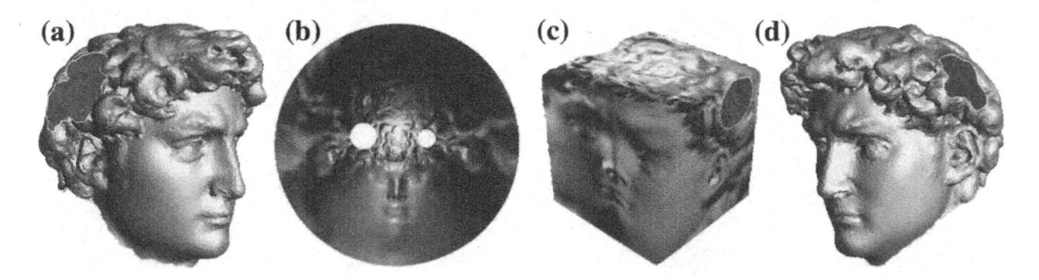

Fig. 6.6 Different global conformal parameterizations based on Ricci flow: **a** and **d** Original surface. **b** All the curvatures are pushed to the boundaries. **c** Flat metric with 4 cone singularities. Image from [131]

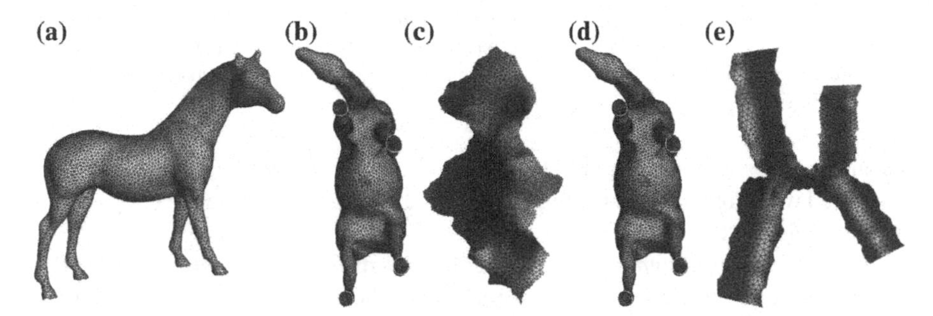

Fig. 6.7 **a**, **b** A horse model with four boundaries on its hoofs. **c** All curvatures are pushing to the four boundaries, and all interior vertices are with zero Gaussian curvature. The model is embedded onto the plane with the new metric. **d** All curvatures are concentrated on one singularity vertex, which is marked with red. **e** The horse is embedded on the plane with the new metric. Image from [131]

Specifically, we can allocate all the curvatures to boundaries and make everywhere else flat, as shown in Fig. 6.6b. Furthermore, we can make the boundaries circles. The centers and radii of the circles are determined by the geometry of the original surface. They can be used as the fingerprint of the surface valuable for shape recognition and classification. We can also set the curvatures of one boundary to zero, and the other two boundaries to be circles. All other vertex curvatures are zeros except 4 cone singularities, whose curvatures equal $\frac{\pi}{2}$, as shown in Fig. 6.6c. This can be used to construct polycube splines in geometric modeling field as we will discuss in Chap. 8.

Different parameterizations introduce various area distortions. Figures 6.7 and 6.8 compare different parameterizations and demonstrate the versatility of the Ricci flow based parametrization method.

Boundaries versus Vertices The mesh shown in Fig. 6.7a is a genus zero surface with 4 boundaries with the bottom of the four hoofs removed as shown in Fig. 6.7b. Its total Gaussian curvatures are -4π. We set the total curvatures of each boundary to $-\pi$, and all interior vertex curvatures to zero. Figure 6.7c shows the parametrization with the flat metric, where the body region has a bigger area, and the leg regions have smaller areas in the parameter domain. Then we concentrate all the curvatures

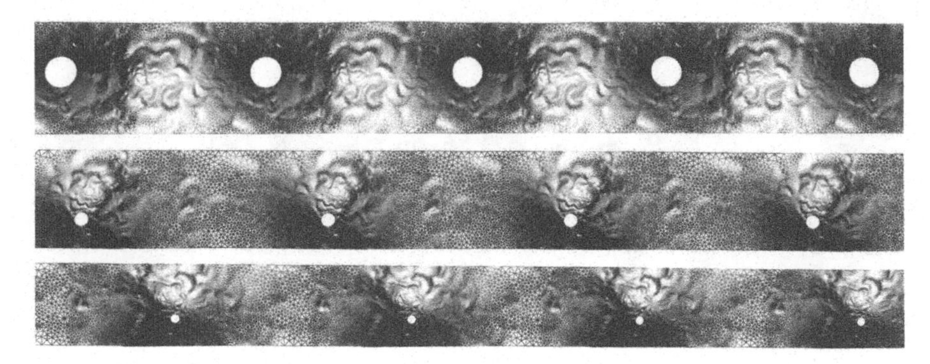

Fig. 6.8 Curvature allocation: boundary versus boundary. The original surface is illustrated in Fig. 6.6a, e. All the curvatures are allocated on one of the three boundaries. The universal covering space is flattened periodically with the corresponding metrics onto the plane. Image from [131]

at a single vertex, colored as red in Fig. 6.7d. Figure 6.7e shows the parametrization. Compared with Fig. 6.7c, the leg regions are greatly enlarged, while the body region shrinks greatly in Fig. 6.7d.

Boundary versus Boundary Figure 6.8 compares different parametrizations of the same David head model shown in Fig. 6.6a, e. Since the mesh is with three boundaries, its total Gaussian curvatures are -2π. Each row shows a periodic layout induced by a flat metric, which allocates the total curvatures on a single boundary.

6.3 Uniform Remeshing

Many applications in computer graphics require a uniform remeshing or partition of a surface. These tasks can be achieved simultaneously by computing a centroidal Voronoi tessellation (CVT) with all sites constrained on the surface.

Voronoi diagram is a well studied concept in computational geometry with a wide usage in different areas in computer graphics, geometric modeling, visualization, etc. [204]. The *centroidal Voronoi tessellation* (CVT) is a special case of Voronoi diagram, where every site coincides with the centroid of its Voronoi cell [69]. The sites in a CVT are uniformly distributed. This property is conjectured by Gersho in 1979 [92] and proved for 2D cases [75].

Such a CVT is usually known as the *constrained CVT* (CCVT) [71]. It is natural to use the geodesic distance to compute the CCVT [212], but it is difficult to compute the geodesic distance accurately. One alternative is to use the 3D Euclidean distance as an approximation [176, 226, 309], but this may lead to disconnected Voronoi cells if two regions are very close in 3D space but are far apart along the surface. A much more efficient approach is to compute a CVT in 2D parametrization domain of a surface [5] and then project the computed CVT back to the surface. By assigning appropriate density values, the computed CVT is very close to the CCVT computed using the geodesic distance. This method overcomes the shortages of both prior methods. It is more efficient since the computation is performed in a 2D Euclidean domain.

To parameterize a closed surface to 2D domain, the original surface has to be cut into a genus-0 surface. This makes the sites unable to cross the boundaries in the parametrization domain, and leads to visible artifacts along the cutting edges. A great deal of special care and delicate strategies, such as minimizing the total cutting edge length and matching the cut graph with the feature skeleton, are required in [5]. If the cut graph does not coincide much with a set of feature edges, the remeshing results become unacceptable for high-genus surfaces as indicated in [5].

Such cutting problem can be solved by computing the CVT directly on the *universal covering space* (UCS) of a surface. For closed genus-0, genus-1, and high-genus (genus>1) surfaces, we can embed the universal covering space (UCS) in 2D spherical space denoted as $\mathbb{S}^2$, Euclidean plane denoted as $\mathbb{R}^2$, and hyperbolic plane denoted as $\mathbb{H}^2$, respectively. The computation of CVT on UCS is similar as in [5] except that sites can move freely along or cross the cutting boundaries.

6.3.1 Voronoi Diagram in Different Spaces

Given n points (called *sites*) $\mathbf{s}_1, \mathbf{s}_2, \ldots, \mathbf{s}_n$ in an Euclidean domain $\Omega^{\mathbb{E}} \subset \mathbb{R}^2$, the *Voronoi cells* $\Omega_i^{\mathbb{E}}$ are the union of all points nearer to site s_i than to other sites:

$$\Omega_i^{\mathbb{E}} = \{\mathbf{p} \in \Omega^{\mathbb{E}} | d^{\mathbb{E}}(\mathbf{p}, \mathbf{s}_i) < d^{\mathbb{E}}(\mathbf{p}, \mathbf{s}_j), i \neq j\}, \tag{6.9}$$

where $d^{\mathbb{E}}(\mathbf{a}, \mathbf{b}) = \|\mathbf{a} - \mathbf{b}\| = \sqrt{(x_{\mathbf{a}} - x_{\mathbf{b}})^2 + (y_{\mathbf{a}} - y_{\mathbf{b}})^2}$ is the distance in 2D Euclidean space. The union of all the Voronoi cells is the *Voronoi diagram* of the sites.

For a 2D spherical domain $\Omega^{\mathbb{S}} \subset \mathbb{S}^2$, we can also define the Voronoi diagram on it by changing the Euclidean distance $d^{\mathbb{E}}(\cdot, \cdot)$ in (6.9) to the spherical distance $d^{\mathbb{S}}(\cdot, \cdot)$. On a sphere with radius of r, the spherical distance between two points $\mathbf{p}$ and $\mathbf{q}$ is defined as:

$$d^{\mathbb{S}}(\mathbf{p}, \mathbf{q}) = r \cos^{-1}\left(\frac{\langle \mathbf{p}, \mathbf{q} \rangle}{r^2}\right),$$

where $\langle \mathbf{p}, \mathbf{q} \rangle = x_{\mathbf{p}}x_{\mathbf{q}} + y_{\mathbf{p}}y_{\mathbf{q}} + z_{\mathbf{p}}z_{\mathbf{q}}$ is the inner product in 3D Euclidean space. Considering the spherical space on a unit sphere with radius of 1, the spherical distance is simplified to:

$$d^{\mathbb{S}}(\mathbf{p}, \mathbf{q}) = \cos^{-1}\left(\langle \mathbf{p}, \mathbf{q} \rangle\right).$$

Under this definition, the *spherical Voronoi cell* $\Omega_i^{\mathbb{S}}$ is:

$$\Omega_i^{\mathbb{S}} = \{\mathbf{p} \in \Omega^{\mathbb{S}} | d^{\mathbb{S}}(\mathbf{p}, \mathbf{s}_i) < d^{\mathbb{S}}(\mathbf{p}, \mathbf{s}_j), i \neq j\}, \tag{6.10}$$

and the spherical Voronoi diagram is thus the union of all the spherical Voronoi cells.

Because of the relationship between the spherical distance and the 3D Euclidean distance:

$$d^{\mathbb{S}}(\mathbf{p}, \mathbf{q}) = 2 \sin^{-1} \frac{d^{\mathbb{E}}(\mathbf{p}, \mathbf{q})}{2},$$

we have $d^{\mathbb{S}}(\mathbf{p}, \mathbf{s}_i) < d^{\mathbb{S}}(\mathbf{p}, \mathbf{s}_j)$ if and only if $d^{\mathbb{E}}(\mathbf{p}, \mathbf{s}_i) < d^{\mathbb{E}}(\mathbf{p}, \mathbf{s}_j)$. So the spherical Voronoi diagram is same as the constrained Voronoi diagram on the sphere, i.e. the intersection between the sphere and the 3D Euclidean Voronoi diagram [71].

Similarly, we can also define the Voronoi diagram in a 2D hyperbolic domain $\Omega^{\mathbb{H}} \subset \mathbb{H}^2$ using the hyperbolic distance $d^{\mathbb{H}}(\cdot, \cdot)$ to replace $d^{\mathbb{E}}(\cdot, \cdot)$ in (6.9). A *hyperbolic Voronoi cell* $\Omega_i^{\mathbb{H}}$ is thus:

$$\Omega_i^{\mathbb{H}} = \{\mathbf{p} \in \Omega^{\mathbb{H}} | d^{\mathbb{H}}(\mathbf{p}, \mathbf{s}_i) < d^{\mathbb{H}}(\mathbf{p}, \mathbf{s}_j), i \neq j\}. \tag{6.11}$$

The *hyperbolic Voronoi diagram* is the union of hyperbolic Voronoi cells. We compute the Voronoi diagram in Klein disk as it is easy to be both computed and visualized. The geodesic distance in Klein disk is a chord of the circle. The distance between two points $\mathbf{p}$ and $\mathbf{q}$ in the Klein disk is

$$d_K^{\mathbb{H}}(\mathbf{p}, \mathbf{q}) = \cosh^{-1} \frac{1 - \langle \mathbf{p}, \mathbf{q} \rangle}{\sqrt{(1 - \|\mathbf{p}\|^2)(1 - \|\mathbf{q}\|^2)}},$$

where $\langle \cdot, \cdot \rangle$ and $\| \cdot \|$ are inner product and vector norm computed in Euclidean space as defined above. Using this distance, it can be proved that the bisector of two points in Klein disk is a straight line in Euclidean space and, for a set of sites $\mathbf{s}_i$ in the Klein disk, the Voronoi diagram in hyperbolic space is a *power diagram* [14] in Euclidean space. More specifically, for every site $\mathbf{s}_i$ in the Klein disk, a corresponding weighted point $wp_i =< \mathbf{t}_i, w_i^2 >$ can be created in Euclidean space, where $\mathbf{t}_i = \frac{\mathbf{s}_i}{2\sqrt{1-\|\mathbf{s}_i\|^2}}$ and $w_i^2 = \frac{\|\mathbf{s}_i\|^2}{4(1-\|\mathbf{s}_i\|^2)} - \frac{1}{\sqrt{1-\|\mathbf{s}_i\|^2}}$. The power diagram of weighted points wp_i in Euclidean space is same as the Voronoi diagram of the sites $\mathbf{s}_i$ in the Klein disk. The derivation details can be found in [203].

6.3.2 *Centroid Voronoi Tessellation in Different Spaces*

The *centroidal Voronoi tessellation* (CVT) is a special Voronoi diagram where every site $\mathbf{s}_i$ locates exactly at the centroid $\mathbf{c}_i$ of its Voronoi cell [69, 70]. To define the CVT in Euclidean, spherical, and hyperbolic spaces, we first define the centroid in these spaces. Combined with the definitions of the Voronoi diagram above, we can well define the CVT in these spaces.

In Euclidean space, the centroid of a region $V_i^{\mathbb{E}} \subset \mathbb{R}^2$ is defined as:

$$\mathbf{c}_i^{\mathbb{E}} = \frac{\int_{V_i^{\mathbb{E}}} \rho(\mathbf{p})\mathbf{p}\,\mathrm{d}\sigma}{\int_{V_i^{\mathbb{E}}} \rho(\mathbf{p})\,\mathrm{d}\sigma}, \tag{6.12}$$

where $\rho(\mathbf{p}) \geq 0$ is a given density function.

We extend the idea of *model centroid* proposed by Galperin [88] from discrete points to a continuous region. The model centroid unifies the definition of the in spaces with constant curvature – Euclidean space, spherical space, and hyperbolic space.

Given n discrete points $\mathbf{p}_i$ in a k-dimensional space, and n mass values m_i corresponding to these points, the position of the centroid of these points can be located as follows: We find a "model" of the k-dimensional space in $(k+1)$-dimensional Euclidean space. For every point $\mathbf{p}_i$, a vector is built from the origin pointing to the point. The vectors are first scaled by the corresponding mass values, and then are added up. The intersection between the sum vector and the model is defined as the position of the centroid of these points. This definition is proved to be well-defined for any number of discrete points and satisfied the axioms given in [88].

For 2D Euclidean space, the model is the plane $z = 1$ in 3D Euclidean space. Figure 6.9 illustrates the side view of the computation of the centroid of two points in 2D Euclidean space. For this case, the sum vector $\mathbf{q} = m_1\mathbf{p}_1 + m_2\mathbf{p}_2$. To compute the intersection between any vector and the model (the plane $z = 1$), we can divide it by its z-component. So we have

$$\mathbf{c} = \frac{\mathbf{q}}{z_{\mathbf{q}}} = \frac{m_1\mathbf{p}_1 + m_2\mathbf{p}_2}{m_1 z_{\mathbf{p}_1} + m_2 z_{\mathbf{p}_2}} = \frac{m_1\mathbf{p}_1 + m_2\mathbf{p}_2}{m_1 + m_2}$$

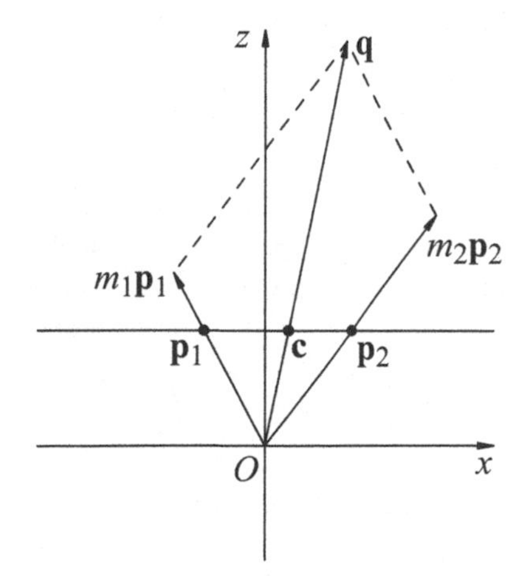

Fig. 6.9 Side view of the computation for the centroid of two discrete points in Euclidean space. Image from [225]

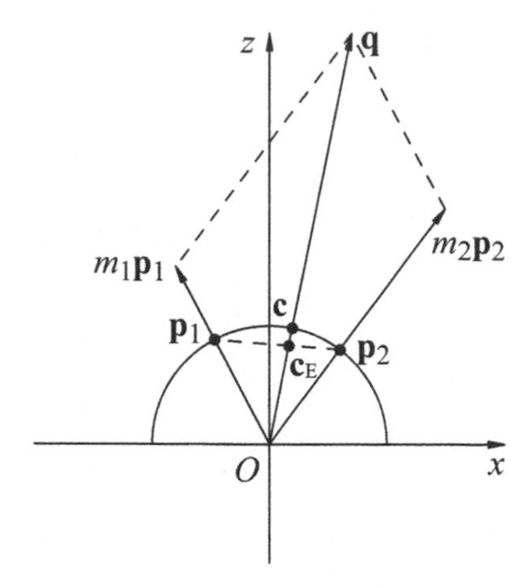

Fig. 6.10 Side view of the computation for the centroid of two discrete points in spherical space. Image from [225]

which is same as the traditional definition of the centroid of two points in Euclidean space.

When we replace the summation of the n vectors with the integral over a continuous region, we can compute the centroid of the region. Using the computation similar to above, it is easy to verify that the model centroid defined in this way is same as the centroid $\mathbf{c}_i^{\mathbb{E}}$ defined in Eq. (6.12).

For 2D spherical space, the model is the unit sphere $x^2 + y^2 + z^2 = 1$. Figure 6.10 illustrates the side view of the computation of the centroid of two points in spherical space. The sum vector is same as above, i.e. $\mathbf{q} = m_1\mathbf{p}_1 + m_2\mathbf{p}_2$. To compute the intersection between any vector and the sphere, we normalize the vector using its Euclidean norm:

$$\mathbf{c} = \frac{\mathbf{q}}{\|\mathbf{q}\|} = \frac{m_1\mathbf{p}_1 + m_2\mathbf{p}_2}{\|m_1\mathbf{p}_1 + m_2\mathbf{p}_2\|}.$$

Compare this with the Euclidean centroid of these two points

$$\mathbf{c}_E = \frac{\mathbf{q}}{\|\mathbf{q}\|} = \frac{m_1\mathbf{p}_1 + m_2\mathbf{p}_2}{m_1 + m_2},$$

we note their numerators are same, which indicates these two vectors are collinear. In other words, the spherical centroid defined by model centroid is the central projection with respect to the origin point of the 3D Euclidean centroid onto the sphere.

Similarly, we can extend the definition of the spherical centroid from discrete points to a continuous region $V_i^{\mathbb{S}} \subset \mathbb{S}^2$ by replacing the summation by integral:

$$\mathbf{c}_i^{\mathbb{S}} = \frac{\int_{V_i^{\mathbb{S}}} \rho(\mathbf{p})\mathbf{p}\,\mathrm{d}\sigma}{\|\int_{V_i^{\mathbb{S}}} \rho(\mathbf{p})\mathbf{p}\,\mathrm{d}\sigma\|}, \tag{6.13}$$

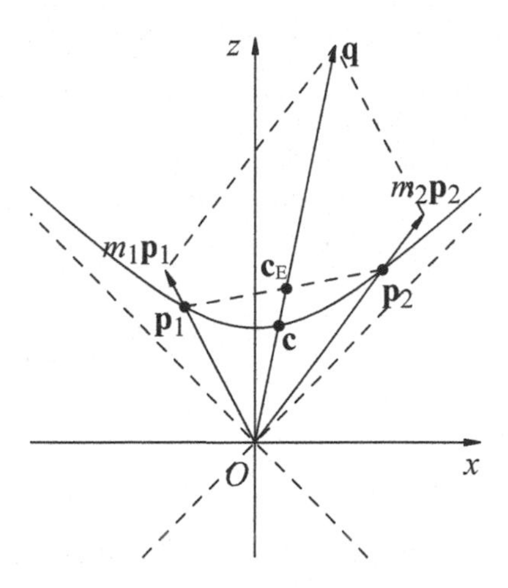

Fig. 6.11 Side view of the computation for the centroid of two discrete points in hyperbolic space. Image from [225]

and the projection property still holds.

For 2D hyperbolic space, the model is the Minkowski model, which is the upper sheet of a two-sheeted hyperboloid $-x^2 - y^2 + z^2 = 1$ in 3D Euclidean space. Figure 6.11 illustrates the side view of the computation of the centroid of two points in hyperbolic space. Again, the sum vector here is $\mathbf{q} = m_1\mathbf{p}_1 + m_2\mathbf{p}_2$. To compute the intersection between any vector with the hyperboloid, we have to normalize the vector using its *Minkowski norm* defined as $\|\mathbf{p}\|_M = \sqrt{-x_\mathbf{p}^2 - y_\mathbf{p}^2 + z_\mathbf{p}^2}$. So, in Fig. 6.11, we have:

$$\mathbf{c} = \frac{\mathbf{q}}{\|\mathbf{q}\|} = \frac{m_1\mathbf{p}_1 + m_2\mathbf{p}_2}{\|m_1\mathbf{p}_1 + m_2\mathbf{p}_2\|_M}.$$

It also has the same numerator with the 3D Euclidean centroid $\mathbf{c}_E$, and thus these two vectors are collinear.

When we extend this definition to continuous regions using integral, we have the hyperbolic centroid of a region $V_i^{\mathbb{H}} \subset \mathbb{H}^2$ as:

$$\mathbf{c}_i^{\mathbb{H}} = \frac{\int_{V_i^{\mathbb{H}}} \rho(\mathbf{p})\mathbf{p}\, d\sigma}{\| \int_{V_i^{\mathbb{H}}} \rho(\mathbf{p})\mathbf{p}\, d\sigma \|_M}. \tag{6.14}$$

It is the central projection with respect to the origin point of the 3D Euclidean centroid onto the hyperboloid.

6.3.3 CVT Energy in Different Spaces

In Euclidean space, suppose we have an ordered set of sites $\mathbf{S} = (\mathbf{s_1}, \mathbf{s_2}, \ldots, \mathbf{s_n})$ in a Euclidean domain $\Omega^{\mathbb{E}} \subset \mathbb{R}^2$, and a tessellation $V = (V_1^{\mathbb{E}}, V_2^{\mathbb{E}}, \ldots, V_n^{\mathbb{E}})$, where part V_i correspond to site $\mathbf{s}_i$, $V_i^{\mathbb{E}} \cap V_j^{\mathbb{E}} = \emptyset$ if $i \neq j$, and $\bigcup_{i=1}^{n} V_i^{\mathbb{E}} = \Omega^{\mathbb{E}}$. The CVT energy of $\mathbf{S}$ and V is defined as:

$$\begin{aligned} F^{\mathbb{E}}(\mathbf{S}, V) &= \sum_{i=1}^{n} \int_{V_i^{\mathbb{E}}} \rho(\mathbf{p}) \left(d^{\mathbb{E}}(\mathbf{p}, \mathbf{s}_i)\right)^2 \mathrm{d}\sigma \\ &= \sum_{i=1}^{n} \int_{V_i^{\mathbb{E}}} \rho(\mathbf{p}) \|\mathbf{p} - \mathbf{s}_i\|^2 \mathrm{d}\sigma. \end{aligned} \tag{6.15}$$

If the sites are fixed and the tessellation varies, we have the following conclusion:

Lemma 6.2 *When the sites are fixed, the CVT energy $F^{\mathbb{E}}(\mathbf{S}, V)$ is minimized when the tessellation V is the Voronoi diagram of the sites.*

If the tessellation is fixed and the sites move, for simplicity, we ignore the tessellation and write the energy as $F^{\mathbb{E}}(\mathbf{S})$.

Lemma 6.3 *When the tessellation is fixed, the CVT energy $F^{\mathbb{E}}(\mathbf{S})$ is minimized when the sites locate at centroids of their corresponding parts.*

Recall that the distance in spherical space is $d^{\mathbb{S}}(\mathbf{a}, \mathbf{b}) = \cos^{-1}(\langle \mathbf{a}, \mathbf{b} \rangle)$. So $\cos\left(d^{\mathbb{S}}(\mathbf{a}, \mathbf{b})\right) = \langle \mathbf{a}, \mathbf{b} \rangle$. Since $\langle \mathbf{a}, \mathbf{a} \rangle = \|\mathbf{a}\|^2$, $\cos\left(d^{\mathbb{S}}(\mathbf{a}, \mathbf{b})\right)$ has the same dimension as $\left(d^{\mathbb{E}}(\mathbf{a}, \mathbf{b})\right)^2$. Based on this observation, we define the CVT energy in spherical space as:

$$\begin{aligned} F^{\mathbb{S}}(\mathbf{S}, V) &= \sum_{i=1}^{n} \int_{V_i^{\mathbb{S}}} \rho(\mathbf{p}) \cos\left(d^{\mathbb{S}}(\mathbf{p}, \mathbf{s}_i)\right) \mathrm{d}\sigma \\ &= \sum_{i=1}^{n} \int_{V_i^{\mathbb{S}}} \rho(\mathbf{p}) \langle \mathbf{p}, \mathbf{s}_i \rangle \mathrm{d}\sigma. \end{aligned} \tag{6.16}$$

Similar to Euclidean space, if the sites are fixed and the tessellation varies, we have the following conclusion in spherical space:

Lemma 6.4 *When the sites are fixed, the CVT energy $F^{\mathbb{S}}(\mathbf{S}, V)$ is maximized when the tessellation V is the Voronoi diagram of the sites.*

This lemma shows that the CVT energy has a maximum value in spherical space, while the CVT energy in Euclidean space and hyperbolic space (defined later) has no upper bound. This is because the distance between any two points in spherical space has the maximum value – π, while it has no upper bound in the other two spaces. To be consistent with the other two spaces, we can slightly modify the definition of the CVT energy as:

$$\begin{aligned} F^{\mathbb{S}}(\mathbf{S}, V) &= \sum_{i=1}^{n} \int_{V_i^{\mathbb{S}}} \rho(\mathbf{p}) \cos\left(\pi - d^{\mathbb{S}}(\mathbf{p}, \mathbf{s}_i)\right) \mathrm{d}\sigma \\ &= -\sum_{i=1}^{n} \int_{V_i^{\mathbb{S}}} \rho(\mathbf{p}) \cos\left(d^{\mathbb{S}}(\mathbf{p}, \mathbf{s}_i)\right) \mathrm{d}\sigma \\ &= -\sum_{i=1}^{n} \int_{V_i^{\mathbb{S}}} \rho(\mathbf{p}) \langle \mathbf{p}, \mathbf{s}_i \rangle \, \mathrm{d}\sigma. \end{aligned} \tag{6.17}$$

Obviously, for the CVT energy under this definition, we have:

Lemma 6.5 *When the sites are fixed, the CVT energy $F^{\mathbb{S}}(\mathbf{S}, V)$ is minimized when the tessellation V is the Voronoi diagram of the sites.*

To distinguish these two definitions, we denote the two CVT energy functions defined by Eqs. (6.16) and (6.17) by $F^{\mathbb{S}}_{max}(\mathbf{S}, V)$ and $F^{\mathbb{S}}_{min}(\mathbf{S}, V)$ respectively.

Same as in Euclidean space, when the tessellation is fixed, we ignore it and write the energy functions as $F^{\mathbb{S}}_{max}(\mathbf{S})$ and $F^{\mathbb{S}}_{min}(\mathbf{S})$. For this situation, we have:

Lemma 6.6 *When the tessellation is fixed, the CVT energy $F^{\mathbb{S}}_{min}(\mathbf{S})$ is minimized when the sites locate at centroids of their corresponding parts, and the CVT energy $F^{\mathbb{S}}_{max}(\mathbf{S})$ is maximized when the sites locate at centroids of their corresponding parts.*

To be consistent with Euclidean space and hyperbolic space, from now on, when we mention the CVT energy in spherical space, we always mean $F^{\mathbb{S}}_{min}(\mathbf{S})$ and thus write it as $F^{\mathbb{S}}(\mathbf{S})$.

In Sect. 6.3.1, we have proved that the Voronoi diagram in spherical space is same as the constrained Voronoi diagram on the sphere. We can also define the CVT energy for this constrained Voronoi diagram in 3D Euclidean space:

$$\begin{aligned} F^{\mathbb{E}}_{c}(\mathbf{S}) &= \sum_{i=1}^{n} \int_{V_{c,i}} \rho(\mathbf{p}) \left(d^{\mathbb{E}}(\mathbf{p}, \mathbf{s}_i)\right)^2 \mathrm{d}\sigma \\ &= \sum_{i=1}^{n} \int_{V_{c,i}} \rho(\mathbf{p}) \|\mathbf{p} - \mathbf{s}_i\|^2 \, \mathrm{d}\sigma, \end{aligned} \tag{6.18}$$

where $V_{c,i}$ is the constrained Voronoi cell of site $\mathbf{s}_i$, i.e. the intersection between the sphere and the 3D Voronoi cell $\Omega^{\mathbb{E}}_i$. The point $\mathbf{c}^*_i \in V_{c,i}$ which minimizes the partial CVT energy $F^{\mathbb{E}}_{c}(\mathbf{s}_i)$ is defined as the *constrained centroid* of V_i [71]. It is proved in [71] that if $\mathbf{c}_E$ is the centroid of $V_{c,i}$ in 3D Euclidean space, the vector $\mathbf{c}_E\mathbf{c}^*$ is collinear with the normal vector at $\mathbf{c}^*$. For sphere, this indicates that $\mathbf{c}^*$ is the central projection of $\mathbf{c}_E$ onto the sphere. So the constrained centroid $\mathbf{c}^*$ is same as the spherical centroid $\mathbf{c}_i^{\mathbb{S}}$ defined by Eq. (6.13), and the spherical CVT is thus identical to the CCVT on the sphere.

We define the hyperbolic CVT energy on Minkowski model, since the hyperbolic distance on this model has a simple form. The distance between two points $\mathbf{p}$ and $\mathbf{q}$ on Minkowski model is:

$$d_M^{\mathbb{H}}(\mathbf{p}, \mathbf{q}) = \cosh^{-1}\left(\langle \mathbf{p}, \mathbf{q} \rangle_M\right),$$

where $\langle \mathbf{p}, \mathbf{q} \rangle_M = -x_{\mathbf{p}} x_{\mathbf{q}} - y_{\mathbf{p}} y_{\mathbf{q}} + z_{\mathbf{p}} z_{\mathbf{q}}$ is the Minkowski inner product defined in 3D Euclidean space.

Similar to spherical space, we define the CVT energy in hyperbolic space as:

$$\begin{aligned} F^{\mathbb{H}}(\mathbf{S}, V) &= \sum_{i=1}^{n} \int_{V_i^{\mathbb{H}}} \rho(\mathbf{p}) \cosh\left(d^{\mathbb{H}}(\mathbf{p}, \mathbf{s}_i)\right) \, \mathrm{d}\sigma \\ &= \sum_{i=1}^{n} \int_{V_i^{\mathbb{H}}} \rho(\mathbf{p}) \, \langle \mathbf{p}, \mathbf{s}_i \rangle_M \, \mathrm{d}\sigma. \end{aligned} \quad (6.19)$$

We have similar conclusions in hyperbolic space as follows:

Lemma 6.7 *When the sites are fixed, the CVT energy $F^{\mathbb{H}}(\mathbf{S}, V)$ is minimized when the tessellation V is the Voronoi diagram of the sites.*

Again, when the tessellation is fixed, we ignore it and write the hyperbolic energy as $F^{\mathbb{H}}(\mathbf{S})$. For this situation, we have:

Lemma 6.8 *When the tessellation is fixed, the CVT energy $F^{\mathbb{H}}(\mathbf{S})$ is minimized when the sites locate at centroids of their corresponding parts.*

6.3.4 Computing Centroidal Voronoi Tessellations

Lloyd's algorithm is an iterative algorithm to minimize the CVT energy. It starts with an arbitrary set of initial sites. In each iteration, the algorithm computes first the Voronoi diagram of current sites, and then the centroid of each Voronoi cell. These centroids are set as new sites for next iteration. The procedure is repeated until certain stopping condition is satisfied (e.g. the moving distance of every site is smaller than a threshold). Lloyd's algorithm converges in Euclidean, hyperbolic, and spherical spaces.

Since a spherical CVT is identical to a CCVT on the sphere, we can directly compute a CCVT as in [71].

Hyperbolic CVT is more complicated. Since the hyperbolic Voronoi diagram is computed in Klein disk and the hyperbolic centroid is defined in Minkowski model, we need to convert positions between these two models. From now on, we will use normal letters to represent points in the Minkowski model, and letters with bars for

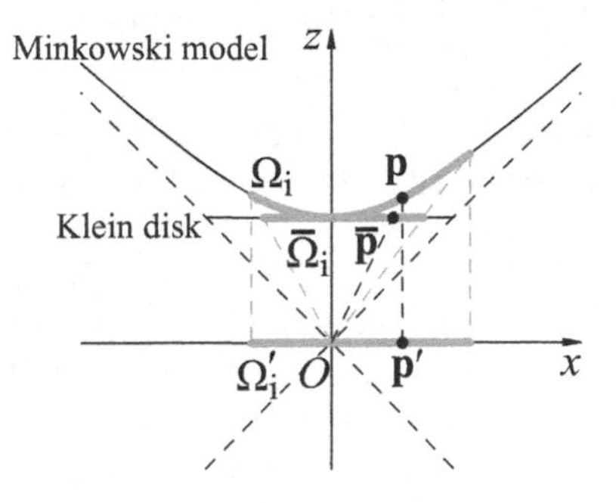

Fig. 6.12 Illustration of mapping functions φ and ψ. Image from [225]

points in the Klein disk. Furthermore, we use letters with primes to represent points on the plane $z = 0$.

When embedded in the plane $z = 1$ with the center on z-axis, the Klein disk is the central projection of the Minkowski model with respect to the origin point. The correspondence between a point $\mathbf{p}(x_\mathbf{p}, y_\mathbf{p}, z_\mathbf{p})$ in the Minkowski model and a point $\overline{\mathbf{p}}(\overline{x}_\mathbf{p}, \overline{y}_\mathbf{p})$ in the Klein disk is given by the following formulas:

$$\begin{cases} \overline{x}_\mathbf{p} = x_\mathbf{p}/z_\mathbf{p} \\ \overline{y}_\mathbf{p} = y_\mathbf{p}/z_\mathbf{p} \end{cases} \tag{6.20}$$

$$\begin{cases} x_\mathbf{p} = \overline{x}_\mathbf{p}/\sqrt{1-(\overline{x}_\mathbf{p}^2+\overline{y}_\mathbf{p}^2)} \\ y_\mathbf{p} = \overline{y}_\mathbf{p}/\sqrt{1-(\overline{x}_\mathbf{p}^2+\overline{y}_\mathbf{p}^2)} \\ z_\mathbf{p} = 1/\sqrt{1-(\overline{x}_\mathbf{p}^2+\overline{y}_\mathbf{p}^2)} \end{cases} . \tag{6.21}$$

Formula (6.21) defines a mapping function φ from the Klein disk to the Minkowski model, where $\varphi(\overline{\mathbf{p}}) = \mathbf{p}$. We define another mapping function ψ which orthogonally projects a point $\mathbf{p}$ in the Minkowski model to the plane $z = 0$ to get the point $\mathbf{p}'$, i.e. $\psi(\mathbf{p}) = \mathbf{p}'$. Note that these two mapping functions can be naturally extended to Voronoi cells:

$$\varphi(\overline{\Omega}_i^{\mathbb{H}}) = \bigcup_{\overline{\mathbf{p}}\in\overline{\Omega}_i^{\mathbb{H}}} \varphi(\overline{\mathbf{p}}) = \Omega_i^{\mathbb{H}},\ \psi(\Omega_i^{\mathbb{H}}) = \bigcup_{\mathbf{p}\in\Omega_i^{\mathbb{H}}} \psi(\mathbf{p}) = \Omega_i'^{\mathbb{H}}.$$

Figure 6.12 illustrates the two mapping functions.

Note the mapping functions are only for the computation of centroids. The hyperbolic Voronoi diagram is directly computed as a power diagram in Euclidean space.

6.3.5 Uniform Remeshing and Partition

Given a general topology surface with a set of initial sites **S**, we apply Lloyd's algorithm to compute the CVT of the sites in the embedded UCS of the surface.

Specifically, we compute the uniformization metric of the surface and isometrically embed a partial of the UCS of the surface into space with constant Gaussian curvature: genus zero surfaces into spherical space, genus one surfaces into Euclidean space, and genus larger than one surfaces into hyperbolic space. For surfaces embedded in Euclidean and hyperbolic spaces, we apply the rigid motion on sites in **S** to compute the corresponding points in neighbor domains of the fundamental domain (those sharing an edge or a vertex with the fundamental domain) and denote the union of them by $\mathbf{S}'$ ($\mathbf{S}' = \emptyset$ for spherical space). We compute the Voronoi diagram of $\mathbf{S} \cup \mathbf{S}'$ and locate the centroids of Voronoi cells of sites in **S**. If a centroid is outside of the fundamental domain by one side, say a_1, we move it to the "opposite" side a_1^{-1} of the fundamental domain by performing a corresponding rigid motion. The adjusted centroids are then inside the fundamental domain and used as the new sites in the next iteration.

The usage of the UCS allows sites to move freely everywhere on the surface: a site can cross the boundary of the fundamental domain, and come into the fundamental domain from the "opposite" side of the boundary. So there is no artifact along the cutting edges any more. This is also the major advantage of the method over Alliez et al.'s algorithm [5].

Since uniformization metric introduces area distortion only, we compensate such distortion by assigning an appropriate density value at each point of the embedding domain. Specifically, we define *conformal factor* denoted as cf to measure the local scaling of a conformal mapping. For each vertex **v** on the surface, cf can be computed as the ratio of the sums of the areas of its incident triangles in 3D space and in 2D parametrization domain, i.e. $cf(\mathbf{v}) = \frac{A_{3D}(\mathbf{v})}{A_{2D}(\mathbf{v})}$. For a non-vertex point on the surface, cf can be computed by linearly interpolating the computed conformal factors of the three vertices of the triangle containing that point.

We then define a *sizing field* where every point **p** on the surface has a desired size $\mu(\mathbf{p})$. For every triangle t, its sizing ratio is computed as $sr(t) = \frac{longest\_edge(t)}{\mu(centroid(t))}$. For a given sizing field, we say a triangle mesh satisfies it if the sizing ratio of every triangle is less than or equal to 1. For the ideal case, the sizing ratios of all triangles should be 1 to minimize the number of triangles used. So the length of the longest edge of every triangle is approximately equal to the sizing at its centroid. The area of the triangle is thus proportional to the square of the sizing at its centroid, i.e. $A(t) \sim \mu(centroid(t))^2$. Since we want the sites uniformly distributed on the surface, the sizing field on the surface should be a constant value everywhere. After a proper normalization, we have $A_{3D}(\mathbf{p}) \sim 1$ and $A_{2D}(\mathbf{p}) \sim \mu(\mathbf{p})^2$, thus the conformal factor $cf(\mathbf{p}) = \frac{1}{\mu(\mathbf{p})^2}$. It is pointed out by Du and Wang [72] that the dual mesh of a CCVT with density values $\rho(\mathbf{p}) = \frac{1}{\mu(\mathbf{p})^{d+2}}$ will satisfy the given sizing field, where d is the dimension of the problem. In 2D, we assign the density value at point **p** as:

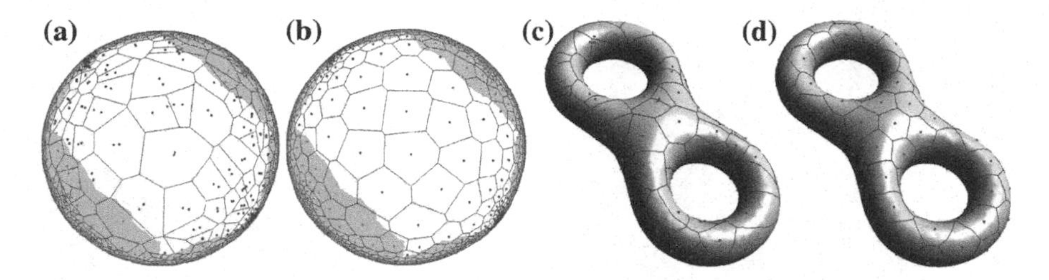

Fig. 6.13 **a** Voronoi diagram of 100 random initial sites in the UCS, and the centroids for Voronoi cells of sites in **S**. Red dots are sites in **S**, green dots are sites in **S**′, and blue dots are centroids. **b** CVT result generated from **a**. **c** Initial Voronoi diagram on surface. **d** CVT result on surface. Image from [225]

$$\rho(\mathbf{p}) = \frac{1}{\mu(\mathbf{p})^4} = cf(\mathbf{p})^2. \tag{6.22}$$

The uniformity of the sites in the CVT result is compensated by the distortion, and thus uniformly distributed on the surface.

Figure 6.13a shows the Voronoi diagram of 100 random initial sites marked with red inside the fundamental domain of a genus two surface and their corresponding points marked with green in neighboring domains in UCS of the surface. The adjusted centroids for Voronoi cells of sites in **S** are all inside the fundamental domain, marked with blue. Figure 6.13b shows the CVT result generated by 100 Lloyd's iterations from initial sites with the modulated density values. The Voronoi diagram of initial sites and the CVT results on the surface are shown in Fig. 6.13c and d respectively. As we can see the final sites are uniformly distributed on the surface.

Similarly, Fig. 6.14 shows the initial Voronoi diagram and the corresponding CVT results both in Euclidean parametrization domain and on a genus one surface. Figure 6.15 shows the results both in spherical parametrization domain and on a genus zero surface.

6.4 Metric-Driven RoSy Fields Design

N-way rotational symmetry (N-RoSy) field has been widely used in computer graphics and digital geometry processing to model brush strokes and hatches in non-photorealistic rendering, regular patterns in texture synthesis, and principle curvature directions in surface parameterizations and remeshing. An important requirement for an N-RoSy field design system is to allow a user to fully control the topology of the field, including the number, positions and indices of the singularities, and the turning numbers of the loops. Automatic generation of N-RoSy fields with user prescribed topologies is highly desired.

We introduce a conformal geometry based method that allows the user to design N-RoSy fields with full control of the topology and without inputting any initial

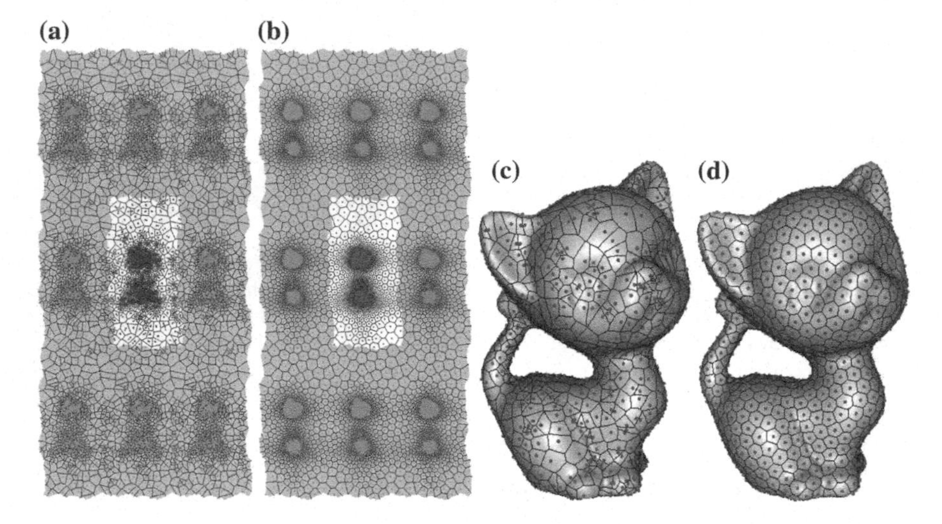

Fig. 6.14 **a** Voronoi diagram of 1,000 random initial sites in the UCS, and the centroids for Voronoi cells of sites in **S**. Red dots are sites in **S**, green dots are sites in **S**′, and blue dots are centroids. **b** CVT result generated from **a**. **c** Initial Voronoi diagram on surface. **d** CVT result on surface. Image from [225]

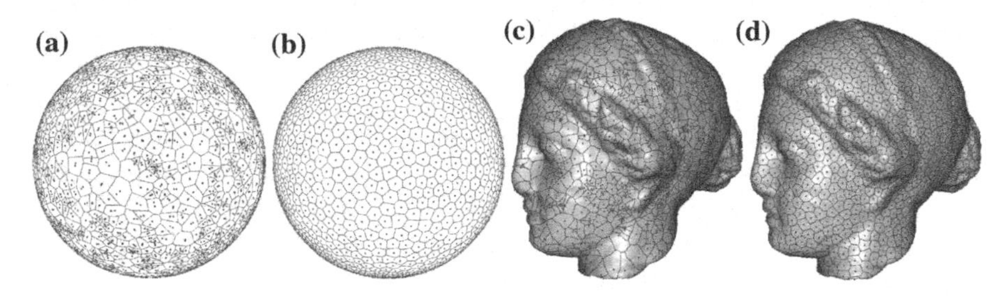

Fig. 6.15 **a** Voronoi diagram of 2,000 random initial sites in the UCS, and the centroids for Voronoi cells of sites in **S**. Red dots are sites in **S** and blue dots are centroids. **b** CVT result generated from **a**. **c** Initial Voronoi diagram on surface. **d** CVT result on surface. Image from [225]

field. Furthermore, the algorithm can automatically generate a smooth field with the desired topology and allow the user to further modify it interactively. The method is inspired by the work in [222]. An N-RoSy field has *local symmetry* that is invariant under rotations of an integer multiple of $\frac{2\pi}{N}$. A surface has *global symmetry* that is intrinsically determined by the Riemannian metric. If the global symmetry is *compatible* with the N-RoSy symmetry, then smooth N-RoSy fields can be constructed on the surface directly. Roughly speaking, if a surface admits an N-RoSy field, then for any loop on the surface, the total turning angle of the tangent vectors along the loop cancels the total turning angle of the N-RoSy field along the loop.

6.4.1 Theory of Compatibility

Let M be a triangular mesh in $\mathbb{R}^3$. A *metric* of M is a configuration of edge lengths, such that the triangular inequality holds on all faces. The *vertex curvature* is the angle deficit, i.e., 2π-the total angle around the vertex. A *flat cone metric* is a metric such that the curvatures are zero for almost all the vertices, except at a few ones. The vertices with non-zero curvatures are called the *cone singularities*, denoted as $S = \{s_1, s_2, \ldots, s_n\}$. Denote $\bar{M}$ as the mesh obtained by removing all the cone singularities from M, $\bar{M} = M/S$.

Note that a metric determines curvatures. Reversely, the curvatures uniquely determine the metric. However, the total curvature of a surface is determined by the topology of the mesh, which is equal to $2\pi\chi(M)$, where $\chi(M)$ is the Euler characteristic number.

Parallel Transport

Parallel transportation is the direct generalization of planar translation. Let γ be a path consisting of a sequence of consecutive edges on $\bar{M}$. The sorted vertices of γ are $\{v_0, v_1, \ldots, v_n\}$. Let N_i denote the one-ring neighborhood of v_i (the union of all the faces adjacent to v_i), then the one-ring neighborhood of γ is defined as the union of all N_i's: $N(\gamma) = \bigcup_{i=0}^{n} N_i$.

The *development* of $N(\gamma)$ refers to the following process: first we flatten N_0 on the plane, and then we extend the flattening to N_1, such that the common faces in both N_0 and N_1 coincide on the plane. This process is repeated until N_n is flattened. In this way, we develop $N(\gamma)$ to the plane. We denote the development map as $\phi : N(\gamma) \to \mathbb{R}^2$. Note that the restriction of the development map on each triangle is a planar rigid motion. *Parallel transportation* on the mesh along γ is defined as the translation on the development of $N(\gamma)$. See Fig. 6.16 for the illustration of parallel transportation.

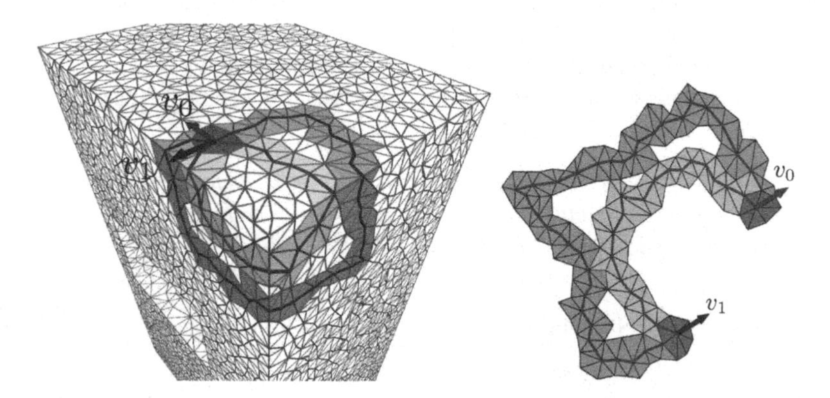

Fig. 6.16 Discrete parallel transportation and holonomy. Homotopic loops sharing the base vertex have the same holonomy

Holonomy

In practice, we are more interested in the *loop* case, i.e. $v_0 = v_n$. When parallel transporting a tangent vector at v_0 along γ to v_n, the resulting vector differs from the original vector by a rotation, which is the *holonomy* of the loop, denoted as $h(\gamma)$. Given a vector field $\mathbf{v}$ along γ, we parallel transport the vector at the starting point. The vector at the ending point differs from the transported vector, which is the *absolute rotation* of the field along γ, denoted as $R_{\mathbf{v}}(\gamma)$.

Two loops γ_1, γ_2 sharing a base point p are *homotopic*, if one can deform to the other. The concatenation of γ_1, γ_2 through p is still a loop, which is the product of them. All homotopy classes of loops form a group, the so-called homotopy group $\pi(\bar{M})$.

Homotopic loops share the same holonomy if the underlying surface has a flat cone metric. In this case, we can define the *holonomy map*, $h : \pi(\bar{M}) \to SO(2)$, where $SO(2)$ is the rotation group in the plane. Its image $h(\pi(\bar{M}))$ is the *holonomy group* of M, denoted as $holo(\bar{M})$.

Compatibility

The *relative rotation* of a vector $\mathbf{v}$ along γ is defined as the difference of the absolute rotation of $\mathbf{v}$ and the holonomy of γ, $T_{\mathbf{v}}(\gamma) = R_{\mathbf{v}}(\gamma) - h(\gamma)$. The relative rotation is equivalent to the *turning number* defined in [222]. Ray et al. proved that for a smooth N-RoSy field, the turning number along any loop must be integer times of $\frac{2\pi}{N}$.

$$T_{\mathbf{v}}(\gamma) = R_{\mathbf{v}}(\gamma) - h(\gamma) \equiv 0, mod \frac{2\pi}{N}. \tag{6.23}$$

Furthermore, the turning numbers on a basis of the homotopy group $\pi(\bar{M})$

$$\{T_{\mathbf{v}}(a_1), T_{\mathbf{v}}(b_1), \ldots, T_{\mathbf{v}}(a_g), T_{\mathbf{v}}(b_g), T_{\mathbf{v}}(c_1), \ldots, T_{\mathbf{v}}(c_n)\} \tag{6.24}$$

determine the homotopy class of the N-RoSy field. We develop our theoretical results based on these fundamental facts. All the proofs are given in the appendix.

The following theorems lay down the theoretical foundation of the metric-driven method, which claims that the topological properties of a vector field are preserved by metric deformation.

Theorem 6.9 *Suppose* $\mathbf{v}$ *is a smooth vector field on a surface* M. $\mathbf{g}(t)$ *is a one parameter family of Riemannian metric tensors. Then for any closed loop* γ *on* M, *the relative rotation* $T_{\mathbf{v}}(\gamma)$ *on* $(M, \mathbf{g}(t))$, *i.e.* M *with the metric* $g(t)$, *is constant for any* t.

The simplest N-RoSy field is a parallel field. The following theory leads to the design of the algorithm.

Theorem 6.10 *Suppose* M *is a surface with a flat cone metric. A parallel N-RoSy field exists on the surface, if and only if all the holonomic rotation angles of the metric are integer times of* $\frac{2\pi}{n}$.

For genus zero closed surfaces, the curvatures of cone singularities determine the holonomy.

Corollary 6.11 *Suppose M is a genus zero closed surface with finite cone singularities. M has a parallel N-RoSy field, if and only if the curvature for each cone singularity is $\frac{2k\pi}{N}$.*

According to this corollary, it is easy to verify the symmetry of a platonic solid. If a platonic solid has N vertices, then a vertex curvature is $\frac{4\pi}{N}$. A $\frac{N}{2}$-RoSy field exists on the platonic solid with a rotational homology group generated by the rotation of angle $\frac{4\pi}{N}$. For examples, an octahedron with 6 vertices exists a 3-RoSy field, and a dodecahedron with 20 vertices exists a 10-RoSy field.

The following existence theorem guarantees the existence of N-RoSy fields on surfaces with arbitrary flat cone metrics.

Theorem 6.12 *Suppose M is a surface with flat cone metric, then there exists a smooth N-RoSy field.*

Suppose $\tilde{M}$ is a branch covering of M (defined in [144]), then the holonomy group of $\tilde{M}$ is a subgroup of that of M. $\tilde{M}$ may have more N-RoSy fields with lower N. For example, in [144], M has a parallel 4-RoSy field, its 4-layer branch covering $\tilde{M}$ allows a parallel 1-RoSy field, namely, a vector field.

If the symmetry of the metric on a surface is compatible with the symmetry of the planar tessellation of the surface, we can re-mesh the surface based on the planar tessellation.

We generalize holonomy to include both translation and rotation. Figure 6.16 shows the concept. Given a loop γ, the starting vertex v_1 coincides with the ending vertex v_n, we develop its neighborhood $N(r)$ onto the plane, then the development of N_1 and that of N_n differs by a planar rigid motion, which is defined as the *general holonomy* along γ. Two loops sharing the common base vertex share the same general holonomy. Therefore, general holonomy maps the homotopy group to a subgroup of planar rigid motion $E(2)$. We denote the image as $Holo(\bar{M})$, and call it the *general holonomy group* of $\bar{M}$.

Suppose T is a tessellation of the plane $\mathbb{R}^2$, τ is a rigid motion preserving T, $\tau(T) = T$. The *symmetry group* of T is defined as

$$G_T = \{\tau \in E(2) | \tau(T) = T\}.$$

Theorem 6.13 *Suppose M is with a flat cone metric, the holonomy group of $\bar{M}$ is $Holo(\bar{M})$, if $Holo(\bar{M})$ is a subgroup of G_T, then T can be defined on M.*

The proof of Theorems 6.9–6.13 can be found in [158].

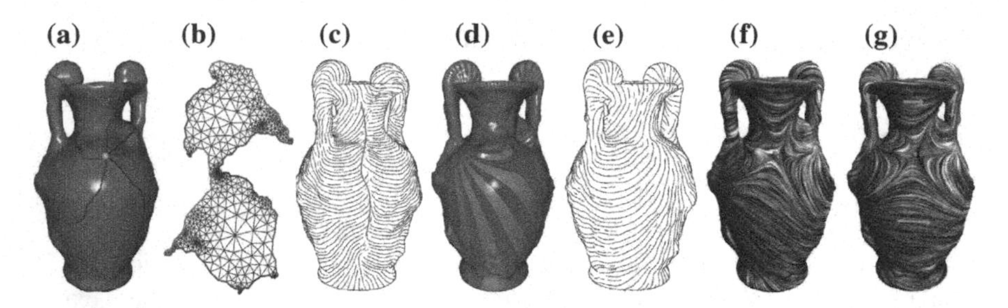

Fig. 6.17 **Algorithm pipeline.** **a** Users specify the desired singularities with both positions and indices. Here we show one singularity point marked with blue and with the index −2. The curves are homotopy group basis. **b** We compute a flat metric, the curvature at the singularity is -4π, everywhere else 0. The surface is cut along the base curves and flatten to the plane. Note that boundaries with the same color can match each other by a rigid motion. **c** Parallel vector field. The field has discontinuities along the red curve. **d** Compute a harmonic 1-form to compensate the holonomy. **e** A smooth vector field after rotation compensation. **f**, **g** Users input geometric constraints (red arrows) to guide the direction of the field, then the field is modified from **f** to **g**

6.4.2 *Computing N-RoSy Fields*

Figure 6.17 illustrates the pipeline of the algorithm. A user first specifies the desired singularities with both positions and indices. The algorithm then computes a flat cone metric, such that all the cone singularities coincide with those specified by the user. The algorithm parallel transports a tangent vector at the base point to construct a parallel vector field. If the parallel field has jumps when it goes around handles or circulates singularities, the jumps are eliminated by two methods: *rotation-based compensation* adjusts the rotation of the vector field; *metric-based compensation* modifies the rotation of the loops by deforming the surface. In the editing stage, the user can interactively edit the rotation and magnitude of the vector field to incorporate further constraints.

Computing a Flat Cone Metric

The cone singularities are fully determined by the singularities on the desired N-RoSy field. Let v be a cone singularity, then its curvature and index are closely related by the formula $Ind(v) = \frac{k(v)}{2\pi}$, where $Ind(v)$ is the index of v. If the index is non-positive, then it is easy to define the curvature of v. For vertex with a positive index, it is more complicated to find the curvature. We handle this situation in the following way. We punch a small hole at the cone singularity. Suppose the boundary vertices of the small hole are $\{v_1, v_2, \ldots, v_m\}$, then the index of the singularity and the total curvature of the boundary are related by $Ind(v) = \frac{\sum_{i=1}^{m} k_i}{2\pi} + 1$. Given the desired target Gaussian curvatures, we compute a flat metric using discrete Euclidean Ricci flow. Figure 6.18 illustrates a vector field constructed on Michelangelo's David head surface model, a closed genus zero surface with one singularity of index +2. According to Corollary 6.11, the flat cone metrics on a closed genus zero mesh satisfy the compatible condition automatically.

Fig. 6.18 A vector field on a genus zero closed surface with a single singularity with index +2

Computing Holonomy

For closed genus zero meshes, if the cone singularity curvatures satisfy the compatibility condition Eq. (6.23), then a flat cone metric of the surface satisfies the same condition. For high genus meshes, the cone singularity curvatures do not guarantee the holonomy compatibility. Explicit computation is required. Specifically, we compute a basis of the homotopy group $\pi(\bar{M})$. Then we compute the development of each base loop γ to obtain the holonomy $h(\gamma)$. The holonomies of all the base loops form the generators of the holonomy group.

Compensating Holonomy

We can compensate holonomy using either rotation-based or metric-based methods. Rotation-based compensation is to adjust the absolute rotation of the direction field $R_{\mathbf{v}}(\gamma)$. Metric-based compensation modifies the metric to change the holonomy $h(\gamma)$ such that the relative rotation is equal to $\frac{2k\pi}{N}$ along arbitrary loops.

Rotation-Based Compensation

The homotopy class of an N-RoSy field is determined by the relative rotations on the basis of homotopy group in Eq. (6.24). We use a conventional method to compute a harmonic 1-form ω on $\bar{M}$, such that for any homotopy group generator γ, the following condition holds: for N-RoSy field design,

$$\int_{\gamma} \omega = T_{\mathbf{v}}([\gamma]) - h([\gamma]),$$

such a harmonic 1-form exists and is unique. Conceptually, the tangent field corresponding to the 1-form ω is constructed in the following way. We select a tangent vector w_0 at the base vertex. Suppose v is another vertex, the shortest path on $\bar{M}$ from v_0 to v is γ, then we parallel transport w_0 to v along γ to obtain w, then we rotate w clock-wisely about the normal by an angle $\theta = \int_{\gamma} \omega$. By this way, we propagate the tangent vector w_0 to cover the whole mesh.

In practice, we use an equivalent fast marching method to propagate the vector field.
1. Select a tangent vector w_0 at v_0, put v_0 in a queue.
2. If the queue is empty, stop. Otherwise, pop the head vertex v_i of the queue. Go through all the neighbors of v_i. For each neighboring vertex v_j, which hasn't been accessed, parallel transport w_i from v_i to v_j, rotate it counter-clock-wisely by angle $\omega(v_i, v_j)$. Enqueue v_j.
3. Repeat step 2, until all the vertices have been processed.

Metric-Based Compensation

Rotation-based compensation is enough to design a smooth N-RoSy field. Metric-based compensation is good for remeshing purpose. In contrast to rotation-based compensation, metric-based compensation modifies a flat cone metric to achieve the desired holonomy that satisfies the compatibility condition in Theorem 6.13.

Specifically, a flat cone metric on a polycube in [265] satisfies the compatibility condition in Eq. (6.23) for a 4-RoSy field. A flat metric on a mesh with all faces equilateral triangles is compatible with a 6-RoSy field.

Using genus zero surfaces as example, the following algorithm computes the desired flat cone metric.
1. First, the user specifies the singularities of the N-RoSy field for both positions and indices, such that the curvatures satisfy the holonomy condition in Eq. (6.23) and are positive. Furthermore, the user specifies the connectivity of a polyhedron P, whose vertices are the cone singularities, and faces are either quadrilaterals or triangles.
2. We use the discrete Ricci flow method to compute a flat cone metric. If $\{s_i, s_j\}$ is an edge in P, we compute the shortest path connecting s_i, s_j under the flat metric. P becomes a convex polyhedron.
3. We use Alexandrov embedding method [27] to embed P in $\mathbb{R}^3$.
4. We adjust the positions of all the vertices of P to make the general holonomy of P satisfy the condition in Theorem 6.13.

Editing N-RoSy Field

Suppose users add some geometric constraints to an N-Rosy field, we can incorporate them into the introduced algorithms easily. We decompose the constraints as orientation and length constraints. Suppose users specify the directions of the vectors at special point set $\omega \subset M$. Let $p \in \omega$, the angle between $\mathbf{w}(p)$ and the desired direction is $\psi(p)$. Then we compute a harmonic function using the method described in [202], $\psi : M \to \mathbb{R}$ with the boundary condition on Ω. Then at each point $q \in M$, we rotate $\mathbf{w}(q)$ by an angle $\psi(q)$. The length constraint can be satisfied using the similar harmonic function method. Figure 6.19 demonstrates a vector field editing process on a genus one kitten surface. A vector field is modified interactively to follow the specified directions marked with red arrows.

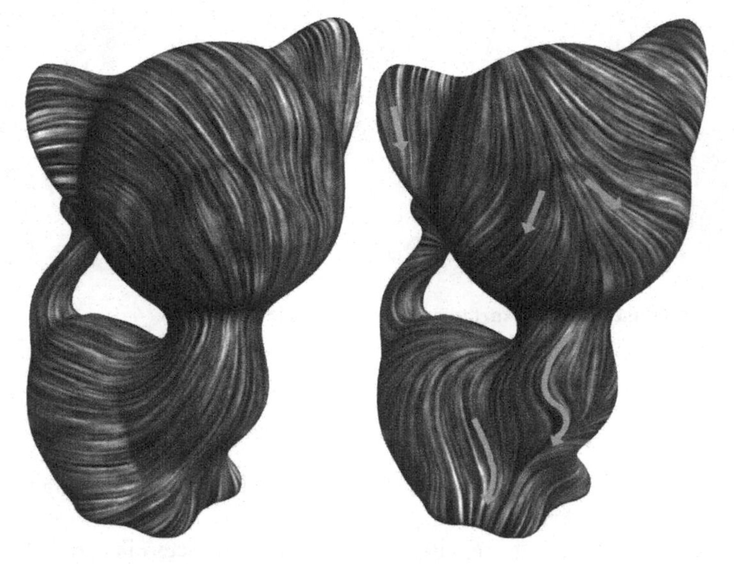

Fig. 6.19 Vector field editing

Fig. 6.20 Metric-driven N-RoSy field design. From left to right, a 3-RoSy field, a 4-RoSy field, a flat cone metric visualized as an obelisk, triangle-quad mixed remeshing based on the metric, quad-remeshing, woven Celtic knot design over the surface based on the quad-remeshing. Image from [158]

6.4.3 Remeshing

Metric-based compensation method can adjust the metric to satisfy the tessellation compatibility condition in Theorem 6.13 to induce a desired tessellation.

Figure 6.20 demonstrates the results of N-RoSy fields on a buddha model. Figure 6.20a, b show a 3-RoSy and a 4-RoSy field on the buddha model, respectively. In Fig. 6.20c, a flat cone metric deforms the mesh in the shape of an obelisk to induce a mixed 4-RoSy and 3-RoSy field on the mesh. Figure 6.20d shows a mixed quadrilateral and triangle tessellation based on the flat cone metric. Figure 6.20f shows a Celtic knot buddha surface constructed based on the quad-remeshing given in Fig. 6.20e.

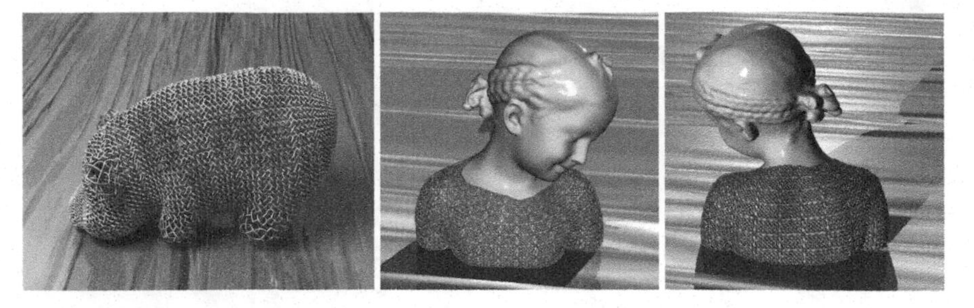

Fig. 6.21 Celtic knots designed surfaces. Image from [158]

6.4.4 *Celtic Knot on Surfaces*

Celtic knot refers to a variety of endless knots, which in most cases contain delicate symmetries and entangled structures.

The local symmetry and the quality of remeshing of surfaces play crucial roles for the knotwork on surfaces. Those uniform quads and triangles provide a perfect canvas for Celtic knot design. Similar to the method in [146], control points can be directly set on surfaces, connected with polynomials. Compared with traditional geometric texture synthesis approaches, there is no need of shell mapping from planar domains to surfaces. Figure 6.21 shows the Celtic knots synthesis results on different surfaces. The knotwork has complicated structures and rich symmetries.

6.4.5 *Pen-and-Ink Sketching of Surfaces*

Pen-and-ink sketching of surfaces is a non-photo-realistic style of shape visualization. The introduced method enables a user to fully control the number, positions, and indices of singularities, and edit a field interactively. These merits make the method rather desirable for NPR applications. Figure 6.22 shows a pen-and-ink sketching on a genus zero model with the hatching quality improved by an editing process.

6.5 Computing Shortest Homotopic Cycles on Polyhedral Surfaces

Two closed curves on a surface are homotopic if they can deform to each other without leaving the surface. A loop is contractible if it can shrink to a point on the surface. Computing the shortest cycle in a given homotopy type has various applications in computer graphics. Most of the previous works [54–56, 87, 96, 120, 121, 163, 187, 193, 304, 317] consider the shortest homotopic cycle problem on either combinatorial

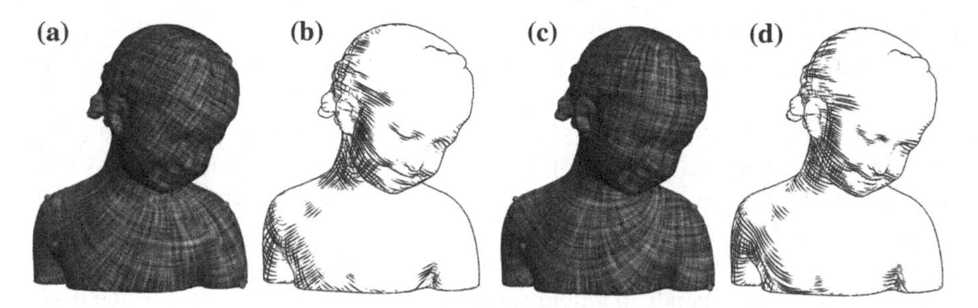

Fig. 6.22 Pen-and-ink sketching of the bimba mesh model before (the left two) and after editing (the right two). The hatch directions follow the natural directions better (e.g. neck, arm). Image from [158]

or piecewise-linear surfaces. The best result, to our knowledge, is given by Colin de Verdière and Erickson [54] with time complexity $O(gnk \log(nk))$, where g and n are the genus and complexity of the combinatorial surface, respectively, and k is the number of edges of the input cycle. Their computation of the shortest cycle homotopic to a given one with complexity k on a surface is based on the preprocessing of the surface: a tight octagonal decomposition of the surface in $O(n^2 \log n)$ time.

In general, a closed geodesic in a given homotopy type is not unique. Local shortening algorithms easily get stuck in local minima, as illustrated in Fig. 6.23. Therefore, complicated global processing algorithms are required. However, if we consider a topologically non-trivial surface (i.e., $genus > 1$), there exists a hyperbolic uniformization metric such that the induced Gaussian curvatures are constantly negative everywhere on the surface according to the Uniformization theorem. It is

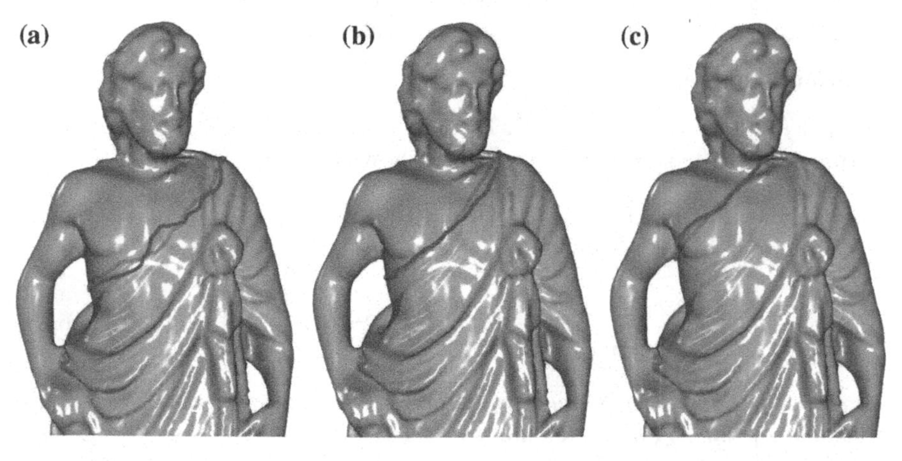

Fig. 6.23 A local loop shortening process may become stuck at a local minimum. **a** An input loop on a model. **b** Shorten the input loop after 2 iterations. **c** Shorten the input loop after 5 iterations. Image from [134]

not difficult to prove that there exists a unique closed geodesic, a cycle with a global minimal length in each homotopy class for surfaces with negative Gaussian curvatures.

We can find the unique closed geodesic in a homotopy class using the duality between the fundamental and Fuchsian groups. Specifically, the universal covering space of a surface with non-trivial topology can be isometrically embedded onto the hyperbolic space $\mathbb{H}^2$. All the deck transformations of the universal covering space are Möbius transformations. They form the so called Fuchsian group such that a homotopy class corresponds to a unique deck transformation. The axis of the deck transformation corresponds to the unique geodesic in the homotopy class.

We can also find the unique closed geodesic in a homotopy class using local shortening methods. In contrast to the stuck of a local loop shortening process at a local minimum as shown in Fig. 6.23, a simple local shortening algorithm based on hyperbolic uniformization metric simplifies the algorithm design to find out the closed geodesic homotopic to a given one on a surface with non-trivial topology with local minimum free.

6.5.1 Geodesic Uniqueness

Homotopy

Two closed curves γ and $\bar{\gamma}$ on a surface M are *homotopic* to each other if there is a continuous map $h : [0, 1] \times S^1 \to M$ such that $h(0, t) = \gamma(t)$ and $h(1, t) = \bar{\gamma}(t)$ for all t. A loop is contractible if it is homotopic to a constant point. In simple words, two loops are homotopic if they can deform to each other without leaving the surface. A loop is contractible if it can shrink to a point on the surface. So all the closed loops on a surface can be classified by homotopy equivalence.

Uniformization Metric

Let M be a surface embedded in R^3. The total Gaussian curvatures of M is solely determined by the topology of the surface, as shown below:

Theorem 6.14 (Gauss–Bonnet Theorem [219]) *The total Gaussian curvature of a compact surface M is given by*

$$\int_M K dA + \int_{\partial M} k_g ds = 2\pi\chi(M), \tag{6.25}$$

where K is the Gaussian curvature on interior points, K_g is the geodesic curvature on boundary points, $\chi(M)$ is the Euler characteristic number of M.

Let g be the Riemannian metric of M induced from its Euclidean metric in R^3. Uniformization theorem claims that:

Theorem 6.15 (Uniformization Theorem [20]) *Let* $(M, \mathbf{g})$ *be a compact 2-dimensional Riemannian manifold with Riemannian metric* $\mathbf{g}$ *and negative Euler characteristic* $\chi < 0$. *There exists a Riemannian metric* $\bar{\mathbf{g}}$ *such that* $\bar{\mathbf{g}}$ *induces constant* -1 *Gaussian curvature everywhere of* M *and is conformal to* $\mathbf{g}$.

According to the Gauss–Bonnet theorem, the sign of the constant Gaussian curvatures is determined by the Euler characteristic of the surface. So surfaces with negative Euler characteristics (i.e., $\chi < 0$) have hyperbolic uniformization metric with -1 Gaussian curvature everywhere.

The geodesics are the locally shortest curves on surfaces. They are defined closely related to metric. The geodesic lengths reflect the global information of the surface. For general surfaces with Euclidean metric, there may be multiple globally shortest and many locally shortest geodesics in each homotopy class. For surfaces with hyperbolic uniformization metric, there exists only one unique globally shortest geodesic and no any other locally shortest geodesics in each homotopy class, which can be directly deduced from the following theorems:

Theorem 6.16 (Geodesic Uniqueness) *Suppose* $(M, \mathbf{g})$ *is a closed compact surface with metric* $\mathbf{g}$. *If Gaussian curvature is negative everywhere, then each homotopy class has a unique globally shortest geodesic and no other locally shortest geodesics.*

The proof of Theorem 6.16 can be found in [134]. For surfaces with boundaries, proof of geodesic uniqueness can be found in [38].

Corollary 6.17 *Each homotopy class has a unique globally shortest geodesic and no other locally shortest geodesics for surface* ($\chi < 0$) *with hyperbolic uniformization metric.*

The proof is straightforward based on the Geodesic Uniqueness theorem.

6.5.2 Fuchsian Group Method

A *covering space* of S is a space $\tilde{S}$ together with a continuous surjective map $h : \tilde{S} \rightarrow S$, such that for every $p \in S$ there exists an open neighborhood U of p such that $h^{-1}(U)$ (the inverse image of U under h) is a disjoint union of open sets in $\tilde{S}$ each of which is mapped homeomorphically onto U by h. The map h is called the *covering map*. A simply connected covering space is a *universal cover*.

Suppose $\gamma \subset S$ is a loop through the base point p on S. Let $\tilde{p}_0 \in \tilde{S}$ be a preimage of the base point p, $\tilde{p}_0 \in h^{-1}(p)$, then there exists a unique path $\tilde{\gamma} \subset \tilde{S}$ lying over γ (i.e. $h(\tilde{\gamma}) = \gamma$) and $\tilde{\gamma}(0) = \tilde{p}_0$. $\tilde{\gamma}$ is a *lift* of γ.

A *deck transformation* of a cover $h : \tilde{S} \rightarrow S$ is a homeomorphism $f : \tilde{S} \rightarrow \tilde{S}$ such that $h \circ f = h$. All deck transformations form a group, the so-called *deck transformation group*. A *fundamental domain* of S is a simply connected domain, which intersects each orbit of the deck transformation group only once.

The deck transformation group $Deck(S)$ is isomorphic to the fundamental group $\pi_1(S, p)$. Let $\tilde{p}_0 \in h^{-1}(p)$, $\phi \in Deck(S)$, $\tilde{\gamma}$ is a path in the universal cover connecting $\tilde{p}_0$ and $\phi(\tilde{p}_0)$, then the projection of $\tilde{\gamma}$ is a loop on S, ϕ corresponds to the homotopy class of the loop, $\phi \to [h(\tilde{\gamma})]$. This gives the isomorphism between $Deck(S)$ and $\pi_1(S, p)$.

Suppose S is a high genus closed surface with the hyperbolic uniformization metric $\tilde{\mathbf{g}}$. Then its universal covering space $(\tilde{S}, \tilde{\mathbf{g}})$ can be isometrically embedded into $\mathbb{H}^2$. Any deck transformation of $\tilde{S}$ is a Möbius transformation, and called a *Fuchsian transformation*. The deck transformation group is called the *Fuchsian group* of S.

Let ϕ be a Fuchsian transformation and $z \in \mathbb{H}^2$. The *attractor* and *repulsor* of ϕ are $\lim_{n\to\infty} \phi^n(z)$ and $\lim_{n\to\infty} \phi^{-n}(z)$, respectively. The *axis* of ϕ is the unique geodesic through its attractor and repulsor.

Figure 6.24 illustrates the major steps to find the unique closed geodesic in a homotopy class using the Fuchsian group based method.

Specifically, given a mesh denoted as M with a negative Euler number and genus $g > 1$ as shown in Fig. 6.24a, we first compute a set of canonical fundamental group generators through a base vertex denoted as $\{a_1, b_1, a_2, b_2, \ldots, a_g, b_g\}$ shown in Fig. 6.24b. We slice M along the set of loops to get an open mesh denoted as $\bar{M}$. The boundary of $\bar{M}$ is

$$\partial \bar{M} = a_1 b_1 a_1^{-1} b_1^{-1} \cdots a_g b_g a_g^{-1} b_g^{-1}.$$

We compute the hyperbolic uniformization metric of M and isometrically embed $\bar{M}$ onto the Poincaré disk shown in Fig. 6.24c. We then compute the Fuchsian group generators, i.e., a set of Möbius transformations corresponding to the fundamental group generators. We denote the Fuchsian transformations as α_i corresponding to a_i, β_j corresponding to b_j.

Suppose a loop is given on the surface, denoted as γ shown in Fig. 6.24d marked with red. We start from a randomly chosen vertex $v \in \gamma$ to trace γ. Each time γ crosses a fundamental group generator, we record the ID. When tracing back to the starting vertex v, we obtain a word denoted as $w = \sigma_1\sigma_2 \cdots \sigma_k$ where $\sigma_i (1 \le i \le k)$ is in $\{a_1, b_1, a_1^{-1}, b_1^{-1}, \ldots, a_g, b_g, a_g^{-1}, b_g^{-1}\}$. We then convert w to

$$\phi = \phi_k \circ \phi_{k-1} \cdots \phi_2 \circ \phi_1,$$

where $\phi_i (1 \le i \le k)$ is in $\{\alpha_1, \beta_1, \alpha_1^{-1}, \beta_1^{-1}, \ldots, \alpha_g, \beta_g, \alpha_g^{-1}, \beta_g^{-1}\}$, the Fuchsian transformation corresponding to σ_i.

Denote the axis of ϕ as $\tilde{\Gamma}$. We start the trace of $\tilde{\Gamma}$ from the fundamental domain $\bar{M}$, called **geodesic tracing**. Once $\tilde{\Gamma}$ crosses the boundary of the current fundamental domain, we transform a copy of the current fundamental domain by ϕ_i, the Fuchsian transformation corresponding to the boundary segment. We continue the tracing on the new fundamental domain. At the same time, project $\tilde{\Gamma}$ to the original surface, denoted as Γ. When Γ forms a closed loop, we can stop tracing $\tilde{\Gamma}$ in universal covering space. Then Γ is the unique geodesic for the homotopy class $[\gamma]$. Figure 6.24e, f illustrate part of the tracing process. Figure 6.24g shows the tracing result of the geodesic

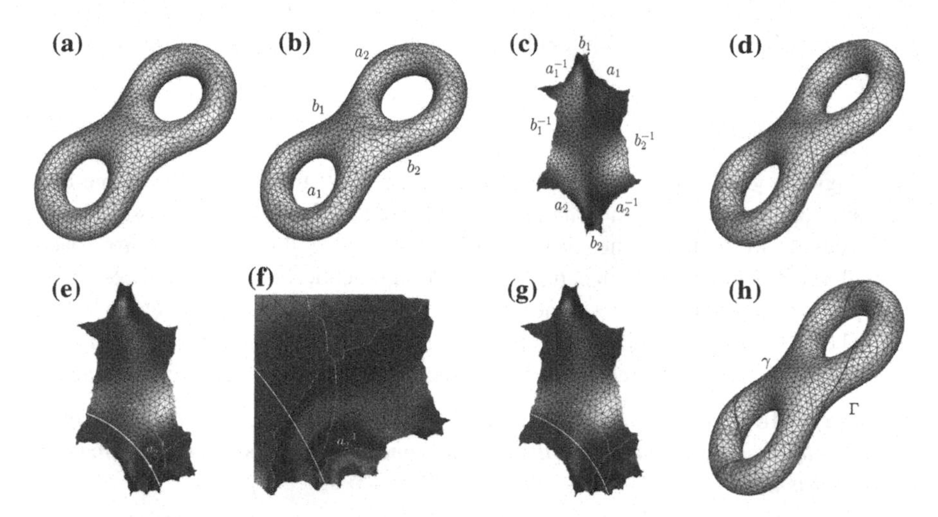

Fig. 6.24 Algorithm pipeline: **a** A genus two triangular mesh; **b** A set of canonical fundamental group basis $\{a_1, b_1, a_2, b_2\}$; **c** The fundamental domain is isometrically embedded onto $\mathbb{H}^2$ with the hyperbolic uniformization metric. **d** A given loop marked with red on the surface; **e**, **f** illustrate part of the tracing process. **g** The tracing result of the geodesic in a portion of the universal covering space; **h** The geodesic marked with blue on the original surface is the canonical representative of the homotopy class of the given loop marked with red. Image from [320]

in a portion of the universal covering space. Figure 6.24h shows the geodesic, the canonical representative of the homotopy class $[\gamma]$ marked with blue on the original surface.

6.5.3 Birkhoff Curve Shortening Method

Injective radius of a point p on a smooth manifold is the largest radius for which the exponential map at p is a diffeomorphism. Injective radius of a point p on a discrete surface denoted as M can be defined as well. For each direction denoted as v in the tangent plane at point $p \in M$, we can trace a geodesic on the surface from p in v to a point at distance t, denoted $q_v(t)$. The points at distance t from all directions form a geodesic circle. The map from the circle on the surface to the one centered at p with radius t on the tangent plane is called the *exponential map*. As the geodesic circle grows, it may touch itself. We define the injective radius of p where the exponential map is no longer 1-to-1 at the touching point. The minimum of the injective radii among all points, denoted as r, is called the injective radius of M. Any two points $p, q \in M$ with $d(p, q) < r$ can be joined by a locally unique geodesic.

Let γ^0 be a closed curve in M. We divide γ^0 into n segments with division points $p_1, p_2, \dots, p_n$. Such subdivision is fine enough that ensures the distance of nearby division points within the distance of injective radius r. We then replace each arc

$\widetilde{p_i p_{i+1}}$ with the geodesic $\overline{p_i p_{i+1}}$ connecting p_i to p_{i+1} and update γ^0 to:

$$\gamma^1 = \overline{p_1 p_2} \bigcup \overline{p_2 p_3} \bigcup \cdots \bigcup \overline{p_{n-1} p_n}.$$

The length of γ^1 is strictly smaller than the length of γ^0 unless $\gamma^1 = \gamma^0$. Note that the homotopy type of γ^1 won't change because each arc is within distance r.

Successive midpoints of the n segments of γ^1 are also within distance r from each other. The arc connecting them can be replaced with geodesics. This produces a new loop γ^2. By iterating this process, each new loop is always homotopic to the previous one. The following theorem proves that the algorithm converges for surfaces with hyperbolic uniformization metric with the final loop the unique homotopy geodesic.

Corollary 6.18 *For a closed compact surface M with hyperbolic uniformization metric, a loop on M converges to the unique homotopy geodesic with the Birkhoff curve shortening.*

Chart Size Estimation

Birkhoff curve shortening process requires the maximal segment length no longer than the injective radius of a surface. While for a given loop, the convergence speed with the curve shortening process is related to the number of segments. In general, the more segments, the more time is needed for each iteration, unless they are computed in a parallel way. Also, the smaller each segment, the slower the overall convergence speed is. A trade-off is to compute an approximation of the injective radius and set as the maximal length of each segment.

The following algorithm approximates the injective radius of a triangular surface. The basic idea is to grow a disk as big as possible with the center at a vertex of the surface before the boundary of the disk meets from different directions. The key is to distinguish whether the disk meets from different directions or one direction due to the discrete structure of the surface. We associate each edge with two directions. $[v_i, v_j]$ indicates the direction from v_i to v_j and $[v_j, v_i]$ with the reverse direction. Each face $[v_i, v_j, v_k]$ has a CCW orientation indicated by its vertex sequence.

1. For each vertex v_i, initialize an empty boundary edge list L_i associated with v_i.
2. Randomly mark one face $f = [v_i, v_j, v_k]$ that contains vertex v_i, and update L_i to $\{[v_i, v_j], [v_j, v_k], [v_k, v_i]\}$. Note that the sequence and directions of the edges are consistent with the CCW orientation of $[v_i, v_j, v_k]$.
3. Mark the neighboring faces of $[v_i, v_j, v_k]$ if they have not been marked by v_i. For example, $[v_i, v_j, v_k]$ has not been marked by v_i. Then mark $[v_j, v_i, v_l]$ and update L_i by replacing $[v_i, v_j]$ with $[v_i, v_l], [v_l, v_j]$. The boundary list L_i is $\{[v_i, v_l], [v_l, v_j], [v_j, v_k], [v_k, v_i]\}$.
4. Continue to grow the disk centered at v_i. In each iteration, mark those non-marked faces along the current boundary of disk and update the boundary list L_i. L_i also needs be checked to remove those edges that appear successive but with opposite directions, like $[v_i, v_j], [v_j, v_i]$. Keep checking L_i until no more edges can be removed. Note that the first edge in the list is considered successive with the last one in L_i.

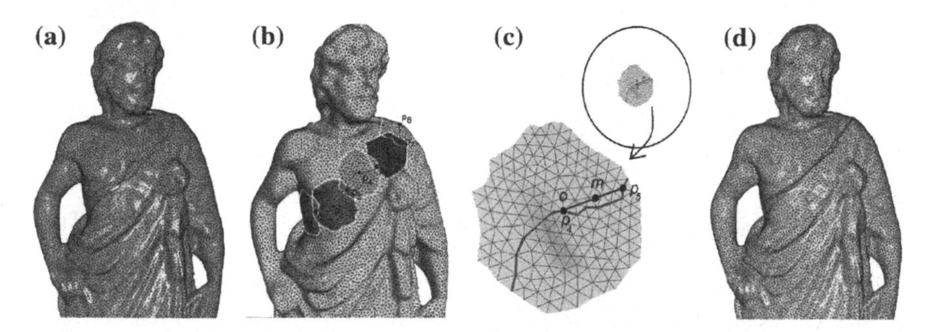

Fig. 6.25 Algorithm illustration: **a** An input loop denoted as γ is marked with red on a triangular manifold M; **b** Loop γ is divided by points $p_0, p_1, \ldots, p_{n-1}$ into n segments such that each segment is shorter than the injective radius of the surface. One chart covers one segment with marked color. **c** One chart is isometrically embedded into the Poincaré disk with one of the two division points (it is p_4 in this example) in the center of the disk. Therefore the geodesic path between p_4 and p_5 in the Poincaré disk coincides with an Euclidean straight line, marked with blue. m is their middle point. **d** Connecting the n midpoints of segments with geodesic paths forms a new closed loop $\bar{\gamma}$, shorter than and homotopic to γ. Image from [134]

5. The disk stops growing as soon as there exists one edge that appears twice in L_i but can't be removed.
6. Each vertex computes such a disk. The smallest radius r among all the radii is set as the threshold that no segment of a divided loop can be longer than r.

The above algorithm is efficient but provides only an approximation based on the combinatorial structure of the surface, which depends on the triangulation quality of the surface and prefers a uniform one. If the triangulation is extremely irregular, the disk can grow based on geodesic offsets with the price of high computational cost. The algorithm proposed by [308] can be adapted by replacing the Euclidean metric with the hyperbolic uniformization metric.

Loop Shortening

Given a triangular manifold M with $\chi < 0$, we compute its hyperbolic uniformization metric using discrete surface Ricci flow method [131]. Since it is the most time consuming step, we compute the metric only once and then stored - a positive real number - at each edge.

We divide a given loop γ on M into n segments with the length of each segment less than the injective radius r, as one example given in Fig. 6.25a. For a segment with two ending vertices p_i and p_{i+1}, a local chart, denoted as C, fully covers the segment as shown in Fig. 6.25b and is isometrically embedded to Poincaré disk based on the pre-computed hyperbolic uniformization metric as shown in Fig. 6.25c. Denote the embedding function as $uv : V \rightarrow \mathbb{H}^2$. The following algorithm gives the details.

1. Start from vertex p_i (v_i). Randomly pick one of its neighboring triangles $f = [v_i, v_j, v_k]$ to embed to Poincaré disk, with

$$uv(v_i) = (0,0), uv(v_j) = \left(\tanh\frac{l_{ij}}{2}, 0\right),$$

$$uv(v_k) = \left(\tanh\frac{l_{ik}}{2}\cos\theta_0^{jk}, \tanh\frac{l_{ik}}{2}\sin\theta_0^{jk}\right).$$

Here, l_{ij} is the hyperbolic uniformization metric of edge $e = [v_i, v_j]$. Then all the neighboring triangles of $f = [v_i, v_j, v_k]$ is put to a queue Q.

2. Pop out the first triangle $f = [v_j, v_i, v_l]$ from Q. Suppose except v_l, v_i and v_j have been embedded to the Poincaré disk with $uv(v_i)$ and $uv(v_j)$ respectively. The coordinates of v_l can be computed as one of the intersection points of the two hyperbolic circles
$$(uv(v_i), l_{il}) \cap (uv(v_j), l_{jl}),$$
with centers $uv(v_i)$ and $uv(v_j)$ and radii l_{il} and l_{jl} respectively, and satisfying
$$(uv(v_j) - uv(v_i)) \times (uv(v_l) - uv(v_i)) > 0.$$
Then the algorithm puts all the neighboring triangles of $f = [v_j, v_i, v_l]$ to Q.
3. We keep growing and embedding the chart with such a breadth first search, until all the neighboring triangles of vertex p_{i+1} have been embedded to the Poincaré disk.

Since p_i is embedded in the center of the Poincaré disk, the hyperbolic geodesic connecting p_i and p_{i+1} coincides with the Euclidean straight line connecting p_i and p_{i+1}, as illustrated in Fig. 6.25c. The algorithm can explicitly compute the geodesic by splitting crossing edges embedded in the Poincaré disk with a straight line and then project it back to the surface. By replacing each segment with its geodesic on surface, the new loop $\bar{\gamma}$ is shorter than the original γ.

For further shortening, the middle point of a segment with ending vertices p_i and p_{i+1} can be computed directly from the following formula on the embedded chart containing the segment:
$$m = \frac{1+\mu^2}{2} uv(p_{i+1}),$$
where
$$\mu = tanh\left(\frac{r}{2}\right), r = \frac{1}{2} log\frac{1 + |uv(p_{i+1})|}{1 - |uv(p_{i+1})|}.$$

The geodesic connecting two successive middle points can be constructed in the similar way of computing the geodesic connecting two successive subinterval points. Joining every two successive middle points with a geodesic produces a new loop $\bar{\bar{\gamma}}$ shorter than $\bar{\gamma}$ as illustrated in Fig. 6.25d. The algorithm records the length of loop in current iteration. The iteration stops if the difference of lengths of the loop between two successive iterations is less than a threshold.

Note that it is unnecessary to explicitly compute and project each geodesic segment back to the original surface. For each iteration, the computation needs only a list of middle points in sequence. Since the embedded charts on Poincaré disk differ only by Möbius transformations for overlapping parts, cross ratio is invariant. The algorithm stores cross ratios instead of barycentric coordinates of the positions of middle points. For a middle point denoted as P located inside face $[v_i, v_j, v_k]$, its cross ratio is computed as

$$(uv(v_i), uv(v_j); uv(v_k), uv(P)) = \frac{(uv(v_i) - uv(v_k))(uv(v_j) - uv(P))}{(uv(v_j) - uv(v_k))(uv(v_i) - uv(P))}.$$

When a loop has shrunk such that the distances between neighboring subdivision points are much less than the injective radius, we merge those short neighboring subdivisions to reduce the number of segments and improve the convergence speed, especially for loops homotopic to zero.

Figure 6.26 visualizes the shortening processes of multiple loops on the same model based on the stored hyperbolic uniformization metric.

The Birkhoff curve shortening process itself is not affected by various triangulations, as shown by examples in Fig. 6.27. Loops with the same homotopy type shrink to the unique homotopy geodesic of a mesh model with different triangulations.

6.5.4 Extremal Quasiconformal Mapping

Constructing a smooth and one-to-one mapping with the least global distortion between surfaces provides a useful tool for numerous applications in computer graphics, geometric modeling, and visualization, where mapping quality is largely measured by the introduced angular distortion. However, it is a challenging problem especially for surfaces with complicated topology. In general, there does not exist an angle preserved mapping between two randomly chosen topologically equivalent surfaces, but among all possible diffeomorphisms between the two surfaces, the extremal quasiconformal mapping, the one minimizing the angular distortion the most, exists.

Given a topologically nontrivial surface, its unique closed geodesics - shortest homotopic cycles - can be sorted according to their lengths under the measurement of surface hyperbolic uniformization metric. They form a geodesic spectrum and represent the signature of the surface conformal structure. If the conformal structures of two given surfaces are close, their geodesic spectra are also close. More explicitly, not only the lengths of the geodesics, but also their distribution on the surface and the skeleton formed by them are also similar. Therefore, the closed geodesics under the hyperbolic uniformization metric are the major feature curves of the surface. They capture the conformal structure of the shapes. We can use geodesic spectrum to guide us to approximate the extremal quasiconformal mapping between surfaces with similar shapes and the same non-trivial topology to achieve less angle distortions.

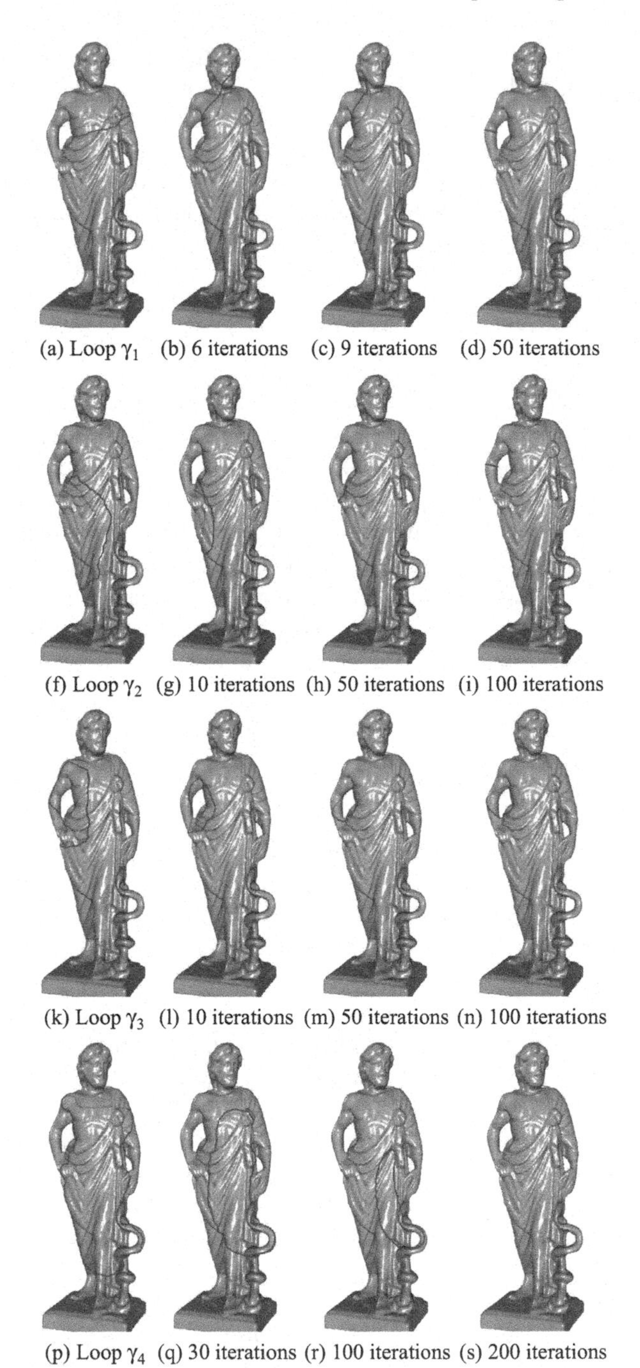

Fig. 6.26 Visualization of loop shortening on Greek model. Image from [134]

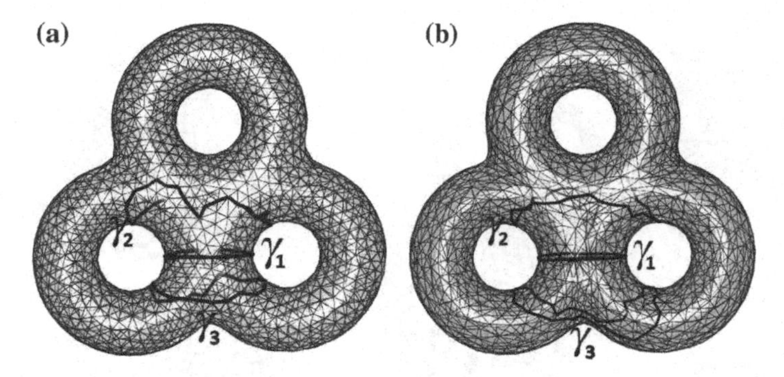

Fig. 6.27 Loops shrink to the unique geodesic γ_1 of the same homotopy type with the same mesh model approximated by different triangulations. Image from [134]

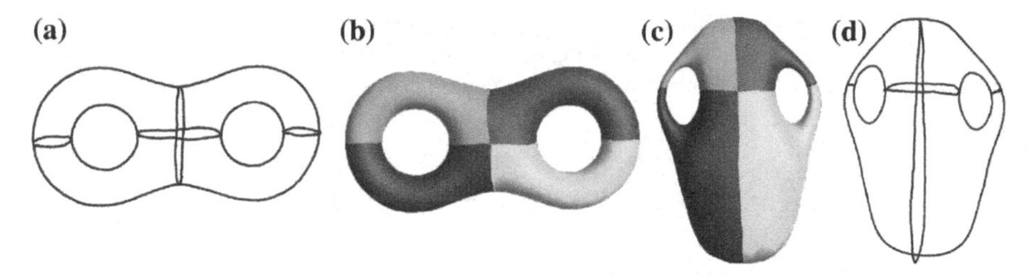

Fig. 6.28 **a**, **d** Geodesic spectra of two mesh models; **b**, **c** Surfaces are segmented to a set of patches, same color marked for corresponding pair of patches. Image from [134]

Although there are infinite number of homotopy types in a general non-trivial topology surface, a partial geodesic spectrum is enough to guide surfaces mapping. Geodesic spectra segment a surface into a set of patches. A global mapping is reduced to a set of mappings between pairs of corresponding surface patches. The mapping between a pair of surface patches can be induced from a conformal map between parameterized patches in 2D plane [131]. Figure 6.28 gives one example.

Texture transfer is the best way to intuitively visualize the globally least angle distortion property of the computed map between surfaces. Figure 6.29 shows the blackboard texture transferred from the amphora model to the vase model based on the computed extremal quasiconformal mapping between the two models. Texture transfer from the top teapot model to the bottom one is in a similar way. Surface morphing is another way to visualize the mapping result. Figure 6.30 shows the surface mapping results by conducting linear interpolation between source and target surfaces to generate the morphing process. For the last two morphing examples, the original models are genus zero surfaces. By topological surgery, adding cuts on the hands and feet for the girl models, cuts on the tips of fingers for the hand models, these models are converted to closed high genus surfaces after double covering (gluing two copies of a same open surface along their boundaries).

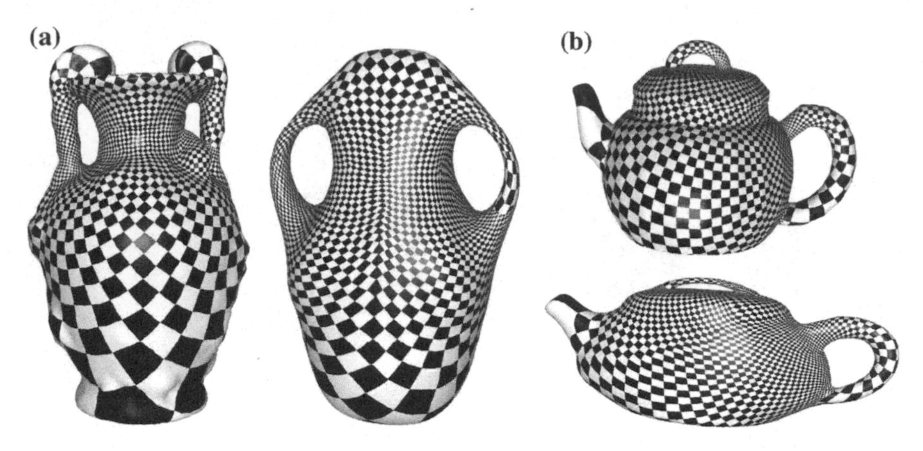

Fig. 6.29 Texture transfer based on surface mapping: **a** An amphora model transfers its checkboard texture to a vase model based on the mapping between them; **c** the top teapot model transfers its checkboard texture to the bottom one based on the mapping between them. Image from [134]

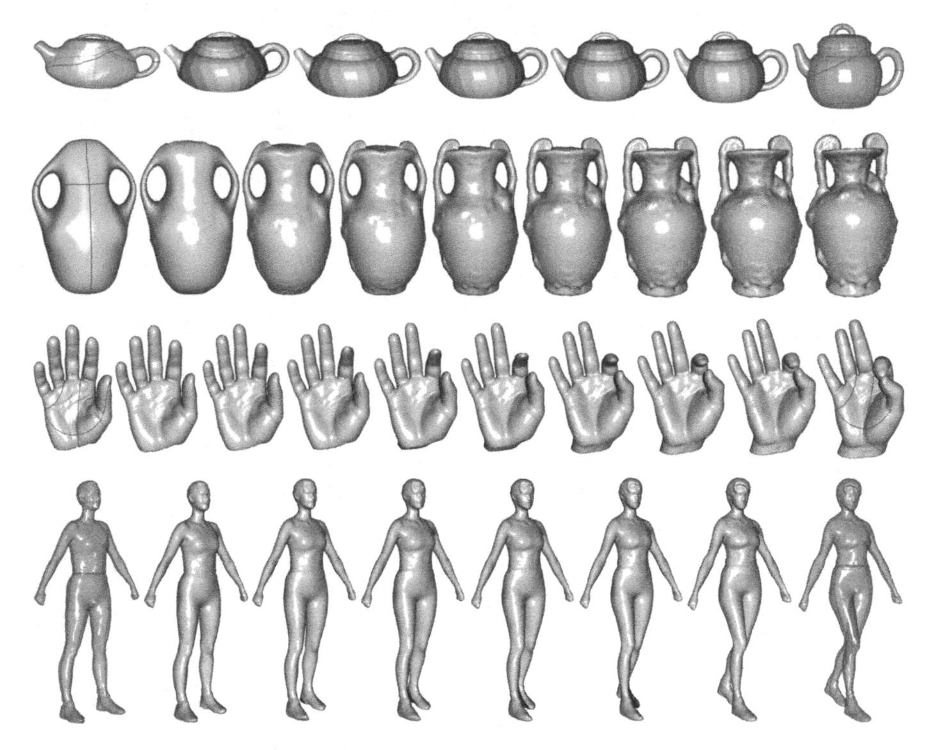

Fig. 6.30 Geodesic spectrum based surface mapping visualized by surface morphing. Red loops on source and target surfaces are computed geodesic spectra. Image from [134]

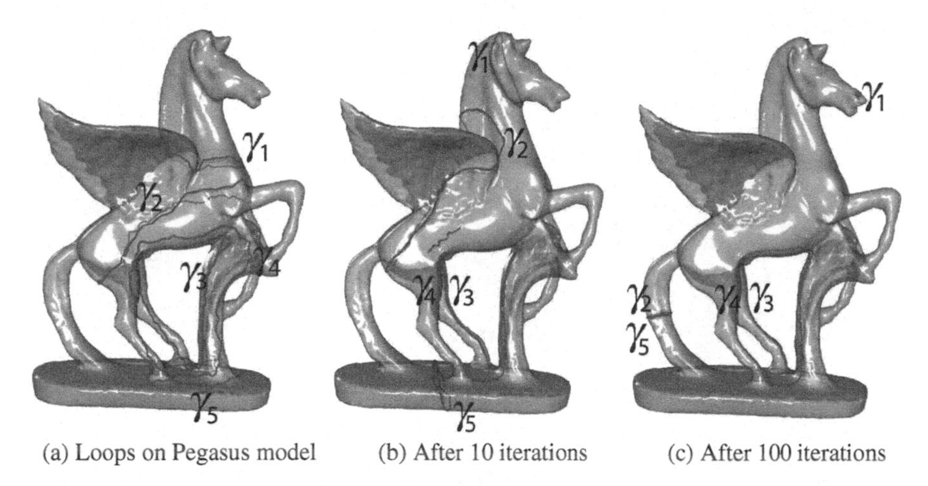

Fig. 6.31 Homotopy detection: **a** Five different loops $\gamma_1, \gamma_2, \gamma_3, \gamma_4, \gamma_5$ on Pegasus model; **b** After 10 iterations; **c** After 100 iterations. Image from [134]

6.5.5 Homotopy Detection

A basic topological problem is to determine whether two cycles γ_i and γ_j or paths l_i and l_j that start and end at the same points are homotopic to each other or not. We pick the unique geodesic under surface hyperbolic uniformization metric in each homotopy class as a homotopy representative. If two cycles γ_i and γ_j are homotopic to each other, they will shrink to the same homotopy representative γ. If two paths l_i and l_j are homotopic to each other, the closed loop $l_i l_j^{-1}$ which concatenates l_i with the reverse of l_j, can shrink to a point on the surface. One example is given on the Pegasus model in Fig. 6.31. After 150 iterations, five different loops γ_1, γ_2, γ_3, γ_4, and γ_5 on the surface as shown in Fig. 6.31a, shrink to three geodesics and one point as shown in Fig. 6.31c, which indicates that γ_2 and γ_5 are homotopic to each other, and γ_1 is topologically trivial.

6.6 Summary and Further Reading

Geometry plays a big role in computer graphics, esp. digital geometry processing field. We refer readers with interests in digital geometry processing to [30] that covers the general pipeline of geometry processing with discussion of techniques in each process based on triangular meshes.

Chapter 7
Computer Vision

Abstract This chapter introduces the applications of computational conformal geometry on computer vision research. Specifically, we focus on algorithms utilizing the topology and geometry information to effective index, classify, and register 3D shapes. Several research topics, including Teichmüller shape space, 3D facial shape index and signatures of 2D Shapes, together with their experimental results, are detailed in this chapter.

7.1 Introduction

Computer vision is an interdisciplinary field. It requires knowledge from computer science, electrical engineering, mathematics, physiology, biology, and cognitive science. The ultimate goal of computer vision is to model, replicate, and more importantly exceed human vision using computer software and hardware at different levels. The Low-level computer vision refers to the process of images for feature extraction including edges, corners, and optical flows. The middle-level computer vision refers to object recognition, motion analysis, and 3D reconstruction using features obtained from the low-level one. The high-level computer vision refers to interpretation of the evolving information provided by the middle level one, for example, conceptual description of a scene like activity and intention.

Effective index and classification of 3D objects are becoming demanding with the dramatically increasing of 3D geometric models in online repositories, while also challenging. This chapter focuses on the algorithms utilizing the topology and geometry information to index, classify, and register 3D shapes.

- *Teichmüller Shape Space.* Teichmüller shape space provides a theoretically sound foundation for surface classification, where surfaces are classied based on conformal equivalence relation. Specifically, in this space, each point represents a class of conformal equivalent surfaces where there exist conformal deformations (i.e., angle preserved deformations) among them. A geodesic between any two points represents an extremal quasiconformal deformation, the one minimizing the angular distortion among all possible dieomorphisms between the two confor-

M. Jin et al., *Conformal Geometry*, https://doi.org/10.1007/978-3-319-75332-4_7

mally inequivalent classes of compact Riemann surfaces. We introduce algorithms to compute coordinates of surfaces in Teichmüller Shape Space.
- *3D Facial Shape Index*. 3D human faces can also be classified by conformal equivalence relation. Facial shapes belonging to the same conformal equivalent class share the same conformal invariants. We introduce algorithms to compute the conformal invariants, the so called conformal module of 3D facial shapes, i.e., genus zero surfaces with multiple boundaries using generalized discrete Ricci flow.
- *Signatures of 2D Shapes*. Shape analysis of planar objects benefits applications in computer vision including image classification, recognition, and retrieval. Effective shape analysis of planar objects requires effective representations of the observed silhouettes and a robust metric measuring their dissimilarity. We introduce a representation to model planar objects with arbitrary topologies using conformal welding. The signature uniquely represents 2D shapes with arbitrary topologies up to scaling and translation. A metric can be defined in the representation space to measure the dissimilarities between objects.

7.2 Teichmuller Shape Space

A *conformal map* preserves angles between two surfaces. Surfaces with the same topology are *conformally equivalent* or belong to the same conformal class if there exists a bijective conformal map between them. All conformal classes form a space called Teichmüller space [38], which can be modeled as a finite dimensional manifold. A surface has a set of unique coordinates in the space. The dimension of coordinates is determined by the topology of the surface. Two surfaces share the same set of coordinates in Teichmüller space if and only if they belong to the same conformal class. Therefore, surfaces can be easily classified and differentiated by conformal equivalence.

Figure 7.1 provides intuitive examples to visualize coordinates of scanned humans faces in Teichmüller space. We conformally map each face shown in Fig. 7.1a–c to two congruent right-angled hyperbolic polygons in Poincarédisk as shown in Fig. 7.1d–f, respectively. The dimension of a human face with three boundaries in Teichmüller space is three, equivalent to the number of boundaries. The values of coordinates are the geodesic lengths of boundaries under hyperbolic uniformization metric. Since the edge lengths are not equal, they do not belong to the same conformal class.

Figure 7.2 gives another example to visualize Teichmuller coordinates. The Teichmuller coordinates of a toy face (with different view points in Fig. 7.2a, b) are visualized as the radii of three inner circles in Fig. 7.2c. After we isometrically deform the toy face (with different view points in Fig. 7.2d, e), the values of Teichmüller coordinates (visualized as the three inner circles radii in Fig. 7.2f) do not change. The example demonstrates that conformal equivalence is invariant under conformal deformation including isometric deformation, rigid motion, and scaling.

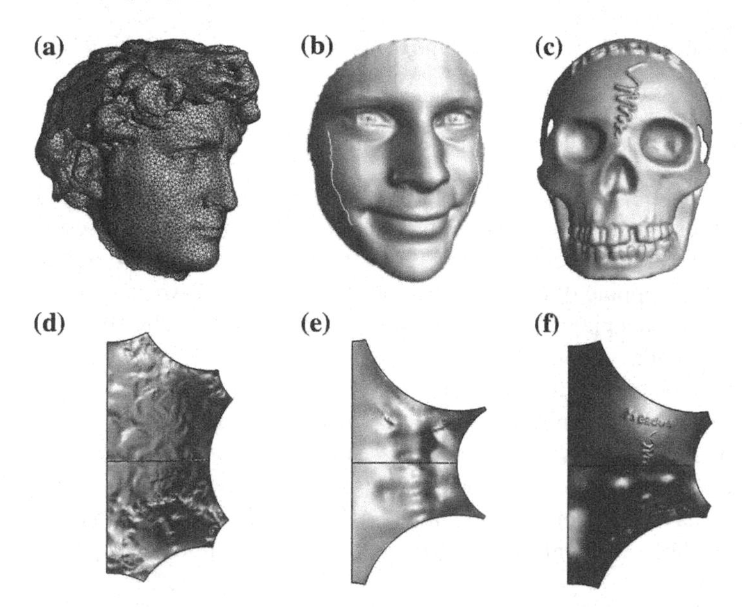

Fig. 7.1 Three human faces with the same topology (two holes annulus) are not conformally equivalent. There doesn't exist a conformal mapping between them, which can be verified by comparing their coordinates in Teichmüller space: the edge lengths of conformally mapped hyperbolic hexagons under hyperbolic uniformization metric. Image from [133]

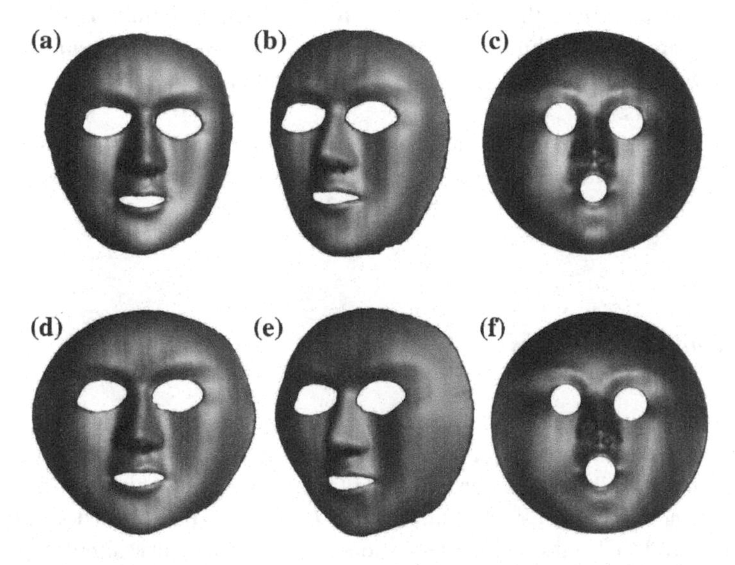

Fig. 7.2 Conformal equivalence is invariant under isometric deformation. The first row shows different views of the original surface and its conformal image in plane; the second row shows different views of the isometrically deformed surface and its conformal image in plane. Their coordinates in Teichmüller space are visualized as the inner circles radii. It is obvious that the conformal images are identical under isometric deformation, which means their coordinates in Teichmüller space are same. Image from [133]

Denote (g, r) as the topological type of an oriented surface, where g is the number of handles (genus), and r is the number of boundaries. The Euler number of a surface with topological type (g, r) is $2 - 2g - r$. We briefly summarize the Teichmüller spaces for surfaces with different Euler numbers.

- The Teichmüller space for (0, 0) type surfaces, namely closed genus zero surfaces, has only one point, which means that all closed genus zero surfaces are conformally equivalent. We can conformally map a closed genus zero surface to a unit sphere. By mapping different surfaces to the unit sphere, we can easily construct a conformal mapping between any two of them. The area distortion induced by the conformal mapping is called the *conformal factor*. In [99], we proved that the conformal factor and the mean curvature determine the surface uniquely up to a rigid rotation of the sphere. We use area distortion and mean curvature as shape descriptors for shape comparison purposes in [99].
- The Teichmüller space for (0, 1) type surfaces, namely genus zero surface with a single boundary, has only one point too. All such surfaces can be mapped to a unit disk. Similarly, the conformal factor and mean curvature can be applied as shape descriptors.
- The Teichmüller space for $(1, 0)$ type surfaces, namely tori, is two dimensional. The Teichmüller coordinates of a torus can be computed using global surface conformal parameterization method in [102]. Basically, we can compute a holomorphic 1-form. By integrating the 1-form, we can map the universal covering space of the surface to the plane $\mathbb{R}^2$. Each fundamental domain is mapped to a parallelogram. The Teichmüller coordinates of a torus are the angle and the length ratio between the two adjacent edges of the parallelogram. We refer readers to [102] for details.
- There are different coordinates defined in Teichmüller space for surfaces with negative Euler numbers. We introduce the algorithms to compute Luo's and Fenchel-Nielsen coordinates in Sects. 7.2.1 and 7.2.2, respectively.

7.2.1 Luo Coordinates

Given a surface with topological type $\Sigma_{(g,r)}$ and negative Euler number, we can decompose the surface into three types of building blocks, as shown in Fig. 7.3a–c. The procedure to build Σ from the building blocks is illustrated by Fig. 7.3d. We use $I \bigcap II$ to denote the process of gluing blocks I and II. The gluing does not mean combining two blocks along their corresponding boundary curves, but by merging their overlapping regions. For example, in the first gluing step of the figure, the overlapping region of I and II is a two-holed annulus. From the left to right, we use basic building blocks I and II, so that $I \bigcap II$ is homeomorphic to $\Sigma_{(1,2)}$, a genus one surface with two boundaries; smoothly joining building block III, so that $\Sigma_{(1,2)} \bigcap III$ is homeomorphic to $\Sigma_{(2,1)}$, a genus two surface with one boundary; then

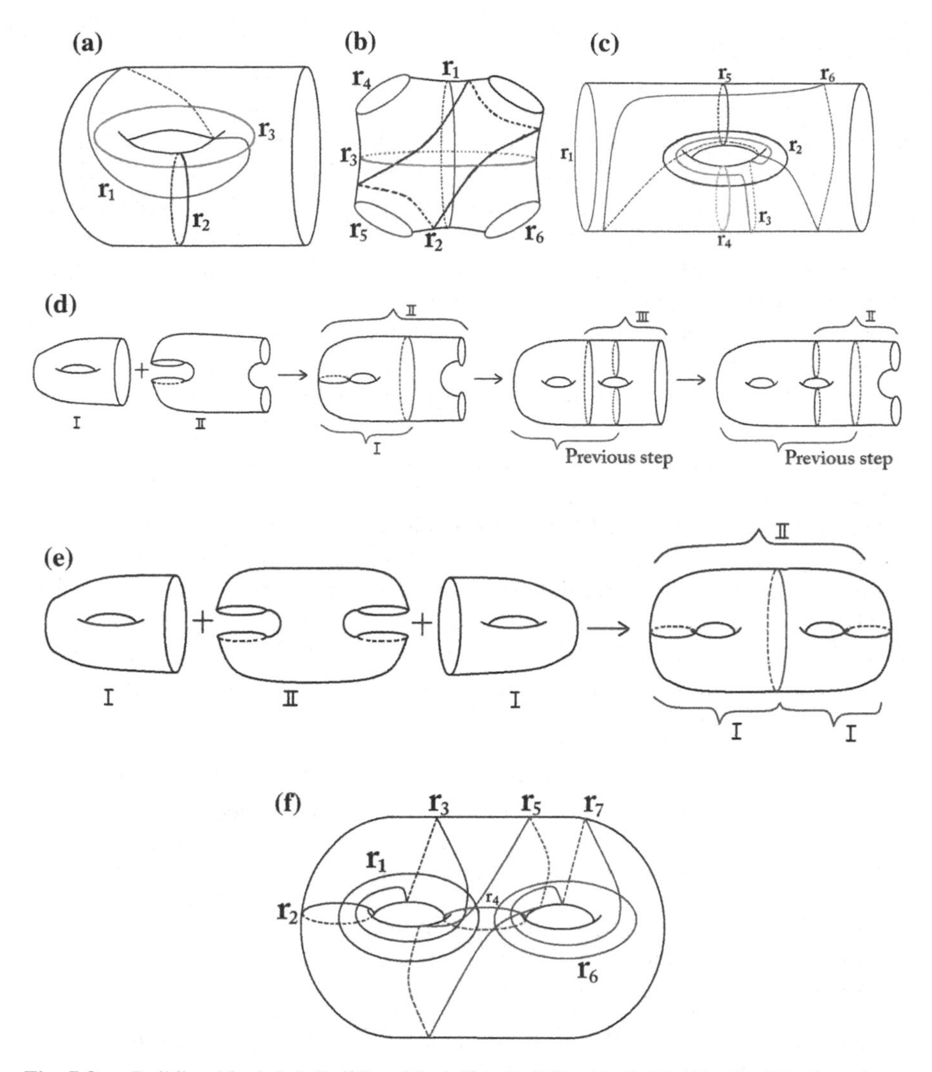

Fig. 7.3 **a** Building block I, **b** Building block II, **c** Building block III. For all of the three basic building blocks, the lengths of geodesics homotopic to the labeled curves determine the building block's metric. **d** Using building blocks I, II and III to build all surfaces: from left to right, using building block I and II to build genus one surface with two boundaries. Then adding building block III to build genus two surface with one boundary. Then adding building block II to build genus two surface with two boundaries. Repeating to get all surfaces. Note that marked curves on surface indicate the boundaries of overlapping part where two building blocks are glued together, and red and blue colors are used to distinguish boundaries coming from different building blocks. **e** The construction of genus two surface. **f** The geodesic lengths of the set of color labeled curves determine the metric of a genus two surface. Blue curves and green curves come from the first and the second Building blocks with type I; red curves come from building block with type II. Note that two of the curves for building block with type II and one for the second building block with type I are redundant and have been canceled off. Image from [133]

joining building block II, so that $\Sigma_{(2,1)} \bigcap II$ is homeomorphic to $\Sigma_{(2,2)}$, a genus two surface with two boundaries; repeating this procedure, we can generate all types of surfaces with negative Euler surfaces. By this construction, a simple method is provided to define Luo's coordinates in Techmüller space for general surfaces.

For each building block, its conformal structure is determined by the lengths of geodesics homotopic to those red loops under the hyperbolic uniformization metric.

For general surfaces, there may exist multiple geodesics in each homotopy class. They are the locally shortest curves on surfaces. However, for surfaces with hyperbolic uniformization metric, the geodesic is unique in each homotopy class. It can be proved by Gauss-Bonnet theorem. We refer readers to [240] for details.

When two building blocks are glued together to form a new surface, non-homotopic loops on the original blocks may become homotopic on the resulting surface. After canceling off the redundant loops, the lengths of geodesics homotopic to remaining loops determine the conformal structure of the resulting surface, which are the coordinates of this surface in Techmüller space. For example, for a closed genus two surface, constructed from two building blocks of type I and one building block of type II as shown in Fig. 7.3e, its Techmüller coordinates are the lengths of geodesics homotopic to those loops marked with different colors as shown in Fig. 7.3f. Loops with the same color indicate that they come from the same building block. In general, for a surface $\Sigma_{(g>1,r)}$ with a negative Euler number, their Teichmüller coordinates are determined by the lengths of $6g + 3r - 5$ closed geodesics.

Algorithms to Compute Length Coordinates

The Luo coordinates of a surface with negative Euler number in Teichmüller space are the lengths of a special set of geodesics under surface hyperbolic uniformization metric. We can compute the lengths of the set of geodesics symbolically from Fuchsian group generators. The pipeline of the whole algorithm is as follows:

1. Compute surface hyperbolic uniformization metric.
2. Compute Fuchsian group generators.
3. Compute Luo coordinates in Teichmüller space.

Since we have already discussed the algorithm to compute surface hyperbolic uniformization metric using discrete surface Ricci flow, we focus on the algorithms to compute Fuchsian group generators and the Luo coordinates in Teichmüller space.

Compute Fuchsian Group Generators

We first enumerate the handles of a surface with $h_1, h_2, \ldots, h_g$, where g is the genus. We then pick a point on the surface as the base point that can be any vertex on the surface. For each handle h_i, we can uniquely decide a tunnel loop a_i that goes around the circle, a handle loop b_i that goes around the handle, and both of them go through the base point. By doing in this way, we get a set of canonical fundamental group generators denoted as $\{a_1, b_1, a_2, b_2, \ldots, a_g, b_g\}$ shown in Fig. 7.4a, where a set of canonical fundamental group generators is marked with different colors on a vase model. We refer readers for more details of computing canonical fundamental group generators in [41].

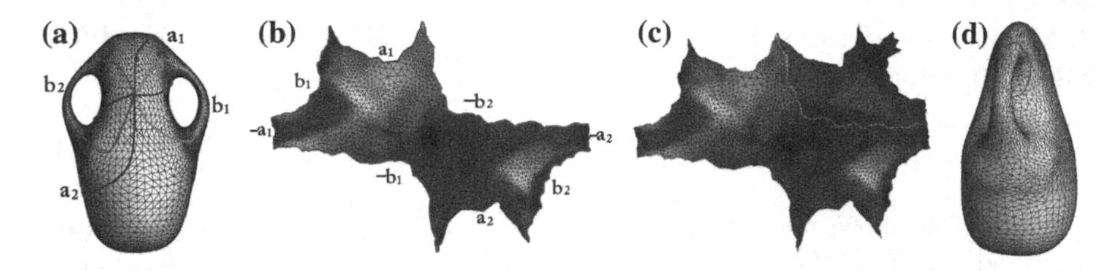

Fig. 7.4 **a** A vase model with a set of canonical fundamental group generators marked with red. **b** The fundamental domain of the vase model embedded in the Poincaré disk with the hyperbolic uniformization metric. **c** One deck transformation maps the left period to the right one. **d** Two closed loops homotopic to the red one on the vase model are lift as two blue paths in the universal covering space. Image from [133]

We then slice the surface open along the fundamental group generators to form a topological disk D, called fundamental domain, and isometrically embed D onto the Poincarédisk based on the hyperbolic uniformization metric. The boundary of D takes the form $\partial D = a_1 b_1 a_1^{-1} b_1^{-1} a_2 b_2 a_2^{-1} b_2^{-1} \cdots a_g b_g a_g^{-1} b_g^{-1}$. Let $\tau : D \rightarrow \mathbb{H}^2$ denote the isometric embedding. The embedding of D in hyperbolic space has $4g$ sides, $\tau(a_1), \tau(b_1), \tau(a_1^{-1}), \tau(b_1^{-1}), \ldots, \tau(a_g), \tau(b_g), \tau(a_g^{-1}), \tau(b_g^{-1})$. Figure 7.4b shows the isometric embedding of the fundamental domain of the vase model onto the Poincaré disk with its hyperbolic uniformization metric.

For each pair of the sides of the embedded D, there exist unique Möbius transformations α_k, β_k that map $\tau(a_k)$ and $\tau(b_k)$ to $\tau(a_k^{-1})$ and $\tau(b_k^{-1})$, respectively. To compute the set of Möbius transformations $\{\alpha_1, \beta_1, \alpha_2, \beta_2, \ldots, \alpha_g, \beta_g\}$, we use β_1 as an example to explain the algorithm. Let the starting and ending vertices of b_1 and b_1^{-1} be: $\partial\tau(b_1) = q_0 - p_0$ and $\partial\tau(b_1^{-1}) = p_1 - q_1$, respectively. Since the geodesic distance from p_0 to q_0 equals to the geodesic distance from p_1 to q_1 in the Poincarédisk, we can align them. We first construct a Möbius transformation τ_0, which maps p_0 to the origin, and q_0 to a positive real number, with

$$\tau_0 = e^{-i\theta_0} \frac{z - p_0}{1 - \bar{p}_0 z}, \theta_0 = arg \frac{q_0 - p_0}{1 - \bar{p}_0 q_0}.$$

Similarly, we construct another Möbius transformation τ_1, which maps p_1 to the origin, and q_1 to a real number, with $\tau_1(q_1)$ equivalent to $\tau_0(q_0)$. By composing the two Möbius transformation, we have Möbius transformation $\beta_1 = \tau_1^{-1} \circ \tau_0$, which satisfies $p_1 = \beta_1(p_0)$ and $q_1 = \beta_1(q_0)$, and aligns the two sides well.

The set of the Möbius transformations $\{\alpha_1, \beta_1, \alpha_2, \beta_2, \ldots, \alpha_g, \beta_g\}$ forms the Fuchsian group generators of the surface. Then we convert the Fuchsian group generators from the Poincarédisk model to the upper half plane model for the purpose of the later computation of the geodesics' lengths. Specifically, the conformal transformation that maps the upper half plane to the Poincarédisk is

$$T = \frac{i - z}{i + z}.$$

A Möbius transformation denoted as ϕ on the Poincarédisk can be converted to a Möbius transformation on the upper half plane as

$$T^{-1} \circ \phi \circ T. \tag{7.1}$$

Figure 7.4c shows one Fuchsian group generator acting on one copy of the fundamental domain of the vase model, which maps $\tau(b_k)$ to $\tau(b_k^{-1})$. The two ending points on the embedded fundamental domain are the pre-images of the same point on the vase model. Paths connecting them are projected to closed loops homotopic to the red loop on the vase model, see Fig. 7.4b.

Compute Luo Coordinates

Surface with enumerated handles has fixed decomposition to building blocks since such decomposition is purely based on topology. We remove redundant loops that share the same homotopy class while belonging to different building blocks. We then compute the lengths of geodesics homotopic to the remaining loops. Since hyperbolic uniformization metric induces a constant and negative Gauss curvature everywhere, geodesics are unique in each homotopy class under hyperbolic uniformization metric. The lengths of the geodesics form the Luo coordinates of the surface in Teichmüller space.

With the computed fundamental group generators $\{a_1, b_1, a_2, b_2, \ldots, a_g, b_g\}$ and the corresponding Fuchsian group generators $\{\alpha_1, \beta_1, \alpha_2, \beta_2, \ldots, \alpha_g, \beta_g\}$, we compute the length of the geodesic homotopic to a loop, for example: γ on surface. We first determine its homotopy class, which can be symbolically represented, for example: $\gamma = a_1 b_1 a_1^{-1} b_1^{-1}$. Then by mapping each a_i to α_i and b_j to β_j, we get its representation using corresponding Fuchsian transformations, still the previous example: $\phi_\gamma = \alpha_1 \beta_1 \alpha_1^{-1} \beta_1^{-1}$. Let the length of γ denoted as l_γ, and we use the matrix representation of ϕ_γ on the upper half plane. l_γ can be easily computed from the following relation:

$$|tr(\phi_\gamma)| = 2\cosh\left(\frac{l_\gamma}{2}\right).$$

7.2.2 Fenchel-Nielsen Coordinates

A closed high genus surface with hyperbolic metric can be decomposed to $2g-2$ pairs of pants, i.e., genus zero surfaces with three boundaries, by cutting the surface along $3g-3$ geodesic loops. Two adjacent pairs of pants are glued together along the cutting geodesic loop with an angle, called twisting angle. The lengths of the cutting loops and the twisting angles give the coordinates of the surface in Teichmüller space, which are called *Fenchel-Nielsen coordinates*. The Fenchel-Nielsen coordinates uniquely determine the conformal structure of a closed high genus surface. They can be treated as the fingerprint of the surfaces and applied for shape comparison and classification.

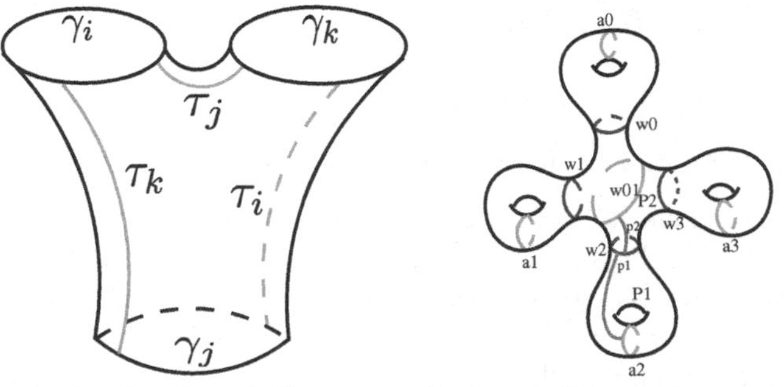

(a) A pair of hyperbolic pants (b) Hyperbolic pants decomposition

Fig. 7.5 **a** A pair of hyperbolic pants with three geodesic boundaries. **b** A genus g surface with hyperbolic metric is decomposed to $2g-2$ pairs of hyperbolic pants by $3g-3$ geodesic cutting loops. The twisting angles and lengths of cutting loops give the Fenchel-Nielsen coordinates in the shape space. Here we visualize the twisting angle on w_2, which equals to the ratio between the hyperbolic distance of $|p_1, P_2|$ and the geodesic length of w_2. Image from [130]

Let S be a closed surface of genus $g > 1$, then its Fenchel-Nielsen coordinates in Teichmüller space can be constructed in the following way.

Definition 7.1 (*Pants*) A pair of topological pants is a genus zero surface with three boundaries.

Given a genus g surface, it can be decomposed to $2g-2$ pairs of pants. Figure 7.5 illustrates one example. Assume all the cutting loops are geodesics $\{\gamma_1, \gamma_2, \ldots, \gamma_{3g-3}\}$, then each pair of pants is pair of hyperbolic pants.

Definition 7.2 (*Hyperbolic Pants*) A pair of pants is called a pair of hyperbolic pants, if it is with a hyperbolic metric, and all boundaries are geodesics.

For each pair of hyperbolic pants P with three boundaries $\gamma_i, \gamma_j, \gamma_k$, there are three shortest paths connecting each pair of boundaries, e.g. τ_i connects γ_j, γ_k, and intersects γ_j and γ_k with right corner angles.

Suppose two pairs of hyperbolic pants P_1 and P_2 are glued together along γ. The shortest path τ_1 on P_1 intersects γ at p_1, and the shortest path τ_2 on P_2 intersects γ at p_2, then the *twisting angle* on γ is given by

$$\theta = 2\pi \frac{d(p_1, p_2)}{|\gamma|}$$

where $d(p_1, p_2)$ is the geodesic distance between p_1 and p_2, and $|\gamma|$ is the length of γ.

Definition 7.3 (*Fenchel-Nielsen Coordinates*) Suppose S is a genus $g > 1$ closed surface with hyperbolic metric. S is decomposed to pairs of pants $\{P_1, P_2, \ldots, P_{2g-2}\}$ by closed geodesics $\{\gamma_1, \gamma_2, \ldots, \gamma_{3g-3}\}$. Then Fenchel-Nielsen coordinates of S in the Teichmüller space T_g are given by

$$\{(l_1, \theta_1), (l_2, \theta_2), \ldots, (l_{3g-3}, \theta_{3g-3})\},$$

where (l_k, θ_k) are the length and twisting angle of γ_k.

Algorithms to Compute Fenchel-Nielsen Coordinates

The key to compute Fenchel-Nielsen coordinates is the hyperbolic pants decomposition of a given closed high genus surface. The geodesic lengths of cutting loops that segment the surface into pairs of hyperbolic pants and the angles of gluing pairs of pants together form the Fenchel-Nielsen coordinates of the surface. The pipeline of the whole algorithms is as follows:

1. Compute topological pants decomposition.
2. Compute surface hyperbolic uniformization metric.
3. Compute hyperbolic pants decomposition.
4. Compute Fenchel-Nielsen coordinates in Teichmüller space.

Compute Topological Pants Decomposition

To decompose a surface with hyperbolic pants, we need to decompose the surface with topological pants first.

For a closed $g > 1$ surface, the size of cutting loops that decompose the surface to topological pants is $3g - 3$. Surface topological pants decomposition has been studied [112, 279]. Figure 7.6 illustrates the algorithm to consistently decompose surfaces with the same topology to a set of corresponding pants as introduced in [66, 171]. The following introduces briefly the algorithm.

1. Compute tunnel and handle loops marked as a_i and b_i, respectively for a given surface as shown in Fig. 7.6a.
2. Slice each surface handle h_i open along its tunnel loop a_i and handle loop b_i. The boundary curve is $c_i = a_i \dot{b}_i \dot{a}_i^{-1} \dot{b}_i^{-1}$ as shown in Fig. 7.6b.

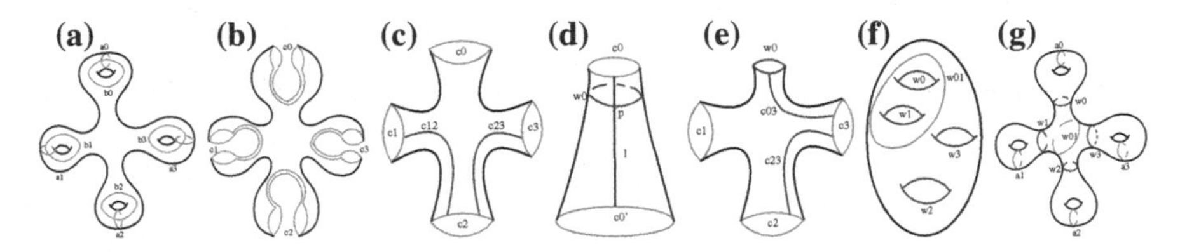

Fig. 7.6 Topological pants decomposition for surface with g handles: **a** Compute surface handle loops and tunnel loops; **b** Slice surface open along tunnel and handle loops; **c** Connect all other boundaries except c_0 to form a big boundary and get a topological cylinder; **d** Find a locally shortest loop w_0 along the path l connecting boundaries c_0 and $\acute{c}_0$, which is the waist of the handle; **e** Repeat the process to find waists for each handle; **f** Cutting handles out along each waist, we get a topological sphere with g holes. Repeat this process as long as the total number of boundaries is less than 4: a locally shortest loop w_{ij} which is homotopic to $w_i \circ w_j$ is computed, and surface patch bounded by w_i, w_j, and w_{ij} is cut out. **g** The set of cutting loops are tunnel loops computed in **a**, waists computed in **e**, and loops computed in **f**. They decompose the surface to topological pants. Image from [130]

3. A waist denoted as w_i is the shortest loop homotopic to c_i. To compute w_i, we connect all other c_js with $i \neq j$ to form a large boundary loop $\acute{c_i}$. We have a topological cylinder as shown in Fig. 7.6c. We compute a shortest path denoted as l connecting c_i and $\acute{c_i}$. The waist w_i is then the shortest loop along l as shown in Fig. 7.6d. We cut off the handle bounded by c_i and w_i and replace w_i with c_i as shown in Fig. 7.6e. We repeat this process until we compute waists for all handles. The surface is now a topological sphere with g holes (g is the genus).
4. If $g > 3$, then for each pair of w_i and w_j (from the increasing number of indexes), we compute the shortest loop $\acute{w_{ij}}$ that bounds w_i and w_j and remove the pant with boundaries $\acute{w_{ij}}$, w_i and w_j. We repeat this process until the number of boundaries is less than or equal to 3 as shown in Fig. 7.6f.

All the tunnel loops computed in the first step, waists computed in the third step, and loops computed in the fourth step form the set of cutting loops that segment the given surface to topological pants as shown in Fig. 7.6g. Since we have indexed surfaces' handles, the ordered sets of topological pants are consistent with surfaces of the same topology.

Compute Hyperbolic Pants Decomposition

The key to decompose a surface to hyperbolic pants is to compute geodesics homotopic to the set of cutting loops that decompose the given surface to topological pants under surface hyperbolic metric. The main idea is to embed the universal cover of the given surface to hyperbolic space and map the set of cutting loops to a set of paths in hyperbolic space. For each path, its two ending points will be projected to the same point on the surface, while in the universal cover, the two ending points for each path induce a Möbiustransformation. The axis of each Möbiustransformation, when projected from universal cover to the surface, is a geodesic loop homotopic to the original cutting loop. We introduce the details of the algorithm in the following.

Similar to we compute the Luo coordinates in Teichmüller space, we slice a surface open along a set of fundamental group generators to form a topological disk D. The boundary of D takes the form $\partial D = a_1 b_1 a_1^{-1} b_1^{-1} a_2 b_2 a_2^{-1} b_2^{-1} \cdots a_g b_g a_g^{-1} b_g^{-1}$. We then isometrically embed D onto the Poincarédisk based on the hyperbolic uniformization metric, and compute a set of Möbius transformations $\{\alpha_1, \beta_1, \alpha_2, \beta_2, \ldots, \alpha_g, \beta_g\}$, i.e., the Fuchsian group generators of the surface. The Fuchsian group generators transform one copy of $\tau(D)$ in Poincarédisk and match with the original copy along the mate boundaries, a_i and a_i^{-1}, or b_i and b_i^{-1}. We repeat this process and glue copies of $\tau(D)$ along their mate boundaries to form a portion of the universal cover of the surface embedded in the Poincaré disk.

Based on the constructed universal cover of the surface, we compute a set of geodesics that are homotopic to the set of topological cutting loops of the surface and decomposes the surface to hyperbolic pants.

For each of the topological cutting loops denoted as η, we perform a "lifting" process that lifts the loop to the universal cover. In practice, to save space, the lifting is only needed to perform in a finite portion of the universal cover, which contains $\tilde{\eta}$. The portion is constructed during the lifting process "on the fly", which means we glue one more copy of the fundamental domain only if we have to. The steps of the "lifting" can be summarized as:

1. For one cutting loop η on surface M, we choose one point p (can be arbitrary point in η) as the base point of the loop.
2. To lift η to universal cover, We first lift the base point p to the center fundamental domain $\bar{M}$.
3. Then we lift next vertex connecting p through edge e_p along the loop under CCW direction. we extend the lifting vertex by vertex. Whenever the lifted loop intersects the boundary segment of the fundamental domain, we compute a Möbius transformation to glue a new copy of the fundamental domain along that boundary segment, then we continue the extension of the lifting. If the lifted path goes through a corner point of the fundamental domain, we need to compute $4g - 1$ Möbius transformations and glue $4g - 1$ copies at that corner.
4. When the lifting process comes back to the edge e_p, we have lifted the cutting loop η in M to a path $\tilde{\eta}$ in universal cover, with the base point lifted to the two points $\tilde{p}_0$ and $\tilde{p}_1$, and edge e_p lifted to two edges $\tilde{e_{p0}}$ and $\tilde{e_{p1}}$ of $\tilde{\eta}$.
5. Similarly, We can construct a deck transformation τ, such that $\tau(\tilde{e_{p0}}) = \tilde{e_{p1}}$.
6. Since τ is a Möbius transformation, its two fixed points can be computed as

$$s = \lim_{n\to\infty} \tau^n(z), t = \lim_{n\to\infty} \tau^{-n}(z),$$

   where z is an arbitrary point in the unit disk.
7. A unique geodesic $\tilde{\gamma}$ in Poincarédisk passing through s and t can be computed, which is the axis of τ.

Then the projection of $\tilde{\gamma}$, $\gamma = h(\tilde{\gamma})$, from universal cover back to the original surface, is the geodesic homotopic to η.

Compute the Fenchel-Nielsen Coordinates

Figure 7.7 visualizes the computation of Fenchel-Nielsen coordinates on a genus two model.

We first decompose surface M to hyperbolic pants based on a set of geodesic cutting loops $\{\gamma_1, \gamma_2, \ldots, \gamma_{3g-3}\}$. For a pair of hyperbolic pants S, the three boundaries $\partial S = \gamma_i + \gamma_j + \gamma_k$ are geodesics in hyperbolic space. Since we have indexed each handle at the step of topological pants decomposition, we can consistently assign a number to each boundary of the pant.

To compute the Fenchel-Nielsen coordinates, we first compute the length of each geodesic cutting loop. They can be easily computed using hyperbolic geometry. Here are the steps:

1. For each geodesic cutting loop γ_k on M, similar to we lift a topological cutting loop to universal cover, we first choose one base point p on that loop, then lift that base point to universal cover. We extend the lifting vertex by vertex along this loop until we are back to the base point.

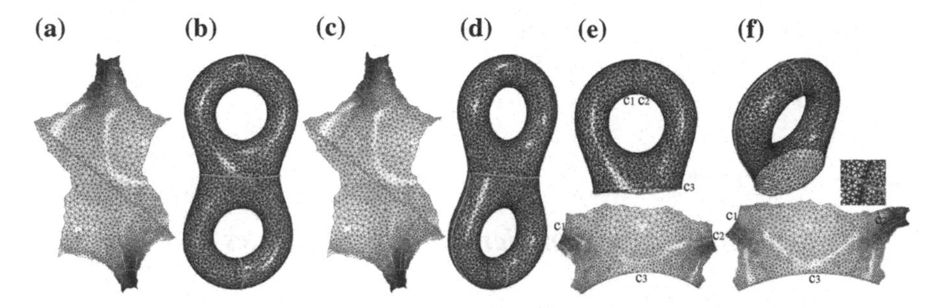

Fig. 7.7 Pipeline of computing Fenchel-Nielsen coordinates: The fundamental domain of the eight surface is embedded in hyperbolic disk as shown in **a**. The three geodesics on its domain, when lifted on to the surface as shown in **b**, will decompose the eight surface to two hyperbolic pants. One pant is shown in **e**, and its boundaries, c_1, c_2, and c_3 are geodesics in hyperbolic disk. We compute the geodesics perpendicular to the boundaries of the two pants, and get intersection points. One is shown in **d**, and the other is shown in **f**. The twisting angle can be computed from the distance of the two intersection points along the same cut loop. For this eight model, since it is very symmetric, all its twisting angles are close to zero. As visualized in **f**, the distance between the two intersection points is very small, almost coincide. Both the three geodesic lengths in **a** and the twisting angles are Fenchel-Nielsen coordinates. Image from [130]

2. Then γ_k is lifted to universal cover as part of a geodesic hyperbolic line, with p lifted to $\tilde{p}_0$ and $\tilde{p}_1$. The geodesic hyperbolic line will intersect the unit circle at q_0 and q_1, then the length of γ_k is given by the logarithm of the cross ratio of $\{q_0, \tilde{p}_0, \tilde{p}_1, q_1\}$.

To compute the twisting angle associated with each geodesic cutting loop, the algorithm is:

1. Suppose geodesic cutting loop γ_k glues the two pairs of pants P_1 and P_2 together. The lifting of γ_k and other boundaries of pants P_1 and P_2 (other geodesic cutting loops) are geodesic hyperbolic lines in Poincarédisk.
2. The geodesic ζ_1 between $\tilde{\gamma_k}$ and $\tilde{\gamma_1}$ (let $\tilde{\gamma_1}$ be one of the other two lifted boundaries of pant P_1, with the smallest assigned number) is also a hyperbolic line in Poincarédisk, which is not only perpendicular to $\tilde{\gamma_k}$ and $\tilde{\gamma_1}$, but also perpendicular to the unit circle. Namely, we compute a circular arc, orthogonal to three circles, the unit circle, $\tilde{\gamma_k}$, and $\tilde{\gamma_1}$. ζ_1 is unique.
3. The same we compute the geodesic Υ_2 between $\tilde{\gamma_k}$ and $\tilde{\gamma_2}$ (the lifted boundary of pant P_2).
4. Suppose ζ_1 intersects γ_k with point q_1, and ζ_2 intersects γ_k with point q_2, hyperbolic distance between q_1 and q_2 is $|q_1q_2|$, then the twisting angle is given by

$$\theta_k = 2\pi\frac{|q_1q_2|}{l_k},$$

where l_k is the length of $\tilde{\gamma_k}$ in Poincarédisk.

Then the Fenchel-Nielsen coordinates are given by

$$\{(l_1, \theta_1), (l_2, \theta_2), \ldots, (l_{3g-3}, \theta_{3g-3})\}.$$

7.2.3 Robustness of Teichmuller Space Coordinates

Teichmuller space coordinates are intrinsic properties of surfaces, independent of translation, rotation, scaling, and insensitive to the resolutions and local noise of surfaces. Figure 7.8 illustrates the robustness of Luo's coordinates in Teichmüller space. The vase model shown in Fig. 7.8 is tessellated using different resolutions, with the number of faces 5, 10, 20 and 40k, respectively. Table 7.1 lists the computed coordinates including the mean average and standard deviation with relative error less than 0.3%.

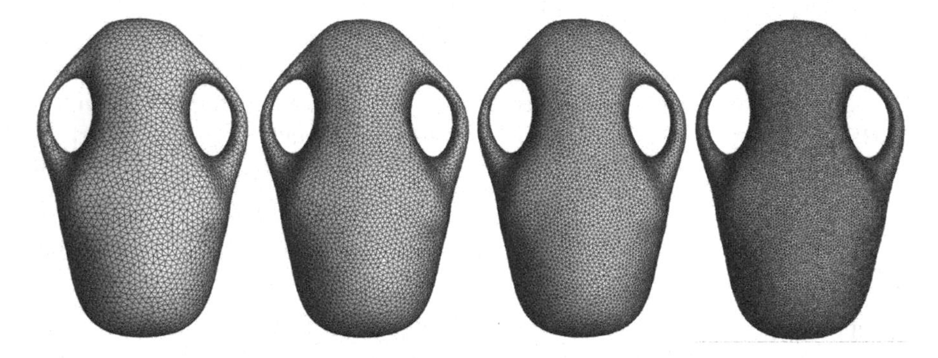

Fig. 7.8 Same model with different triangulation density: 5, 10, 20 and 40k. Comparison of Teichmuller space coordinates with different densities is listed in Table 7.1. Image from [133]

Table 7.1 Comparison of coordinates of vase model with different densities. The dimension of Teichmüller space coordinates for closed genus two surfaces is seven

Vase Model	Coordinates of Vase Model						
	1st	2nd	3rd	4th	5th	6th	7th
Face #: 5k	3.55027	0.99990	3.88055	5.55885	6.11438	3.33029	3.66071
Face #: 10k	3.55700	0.99832	3.88144	5.55611	6.11180	3.33369	3.66703
Face #: 20k	3.55805	0.99759	3.88316	5.55517	6.11112	3.33357	3.66713
Face #: 40k	3.55905	0.99559	3.88416	5.55417	6.11012	3.33367	3.66813
Average	3.55609	0.99785	3.88232	5.55607	6.11185	3.33280	3.66575
Std. dev.	0.00343	0.00154	0.00141	0.00174	0.00157	0.00145	0.00294

7.2.4 *Surface Indexing and Classification*

Teichmuller coordinates can be directly applied for indexing and classification of surfaces with the same topology. We approximate the distance of surfaces in Teichmüller space using the Euclidean distances of their Teichmüller space coordinates.

Figure 7.9 shows the clustering of a randomly chosen set of genus two surfaces, where the x-coordinate and y-coordinate represent the twisting angles and geodesic lengths, respectively. We first classify them into three big groups based on the twisting angles, and then get more refined groups with marked circles when taking geodesic lengths into consideration.

Table 7.2 compares the Fenchel-Nielsen and Luo coordinates distances between a selected teapot model (the left one) and others. The sorting result based on Fenchel-Nielsen coordinates is the same with Luo coordinates.

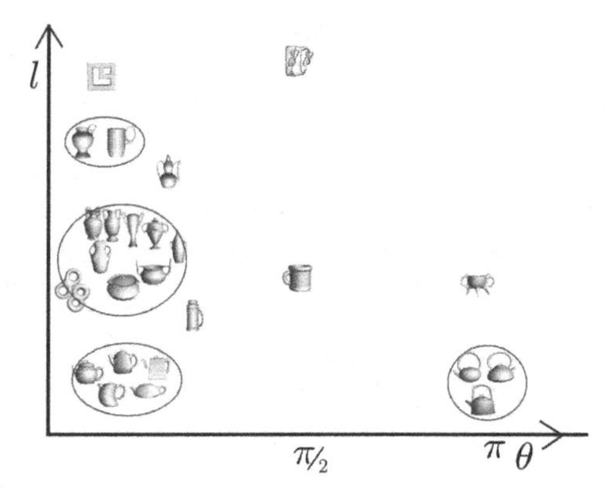

Fig. 7.9 Clustering of surfaces based on their Fenchel-Nielsen coordinates. The x-coordinate indicates the twisting angle, and the y-coordinate indicates the geodesic length. Surfaces are clustered based on both their twisting angle and geodesic lengths, with different groups marked with circles. Image from [130]

Table 7.2 Comparison: sorted distances between the selected teapot model (the left one) and others using their Fenchel-Nielsen and Luo coordinates. Image from [130]

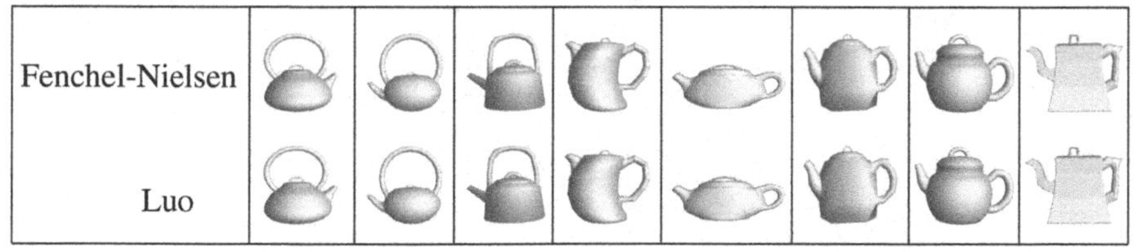

7.3 3D Facial Shape Index

Discrete Ricci flow uses conventional circle packing metric to approximate discrete metric with applications in 3D facial shape analysis [104] and 3D face matching and registration [327]. However, discrete Ricci flow based methods [104, 327] can handle only the case shown in Fig. 7.10a instead of the one shown in Fig. 7.10b. Generalized discrete Ricci flow uses both conventional and inversive circle packing metric to approximate discrete metric to handle both cases shown in Fig. 7.10.

We briefly introduce the application of generalized discrete Ricci flow in 3D facial shape index. Specifically, surfaces can be classified by conformal equivalence relation. Surfaces belonging to the same conformal equivalent class share the same conformal invariants. We introduce algorithms to compute the conformal invariants, the so called conformal module, of 3D facial shapes, i.e., genus zero surfaces with multiple boundaries using generalized discrete Ricci flow.

7.3.1 *Generalized Discrete Ricci Flow*

A *circle packing* associates each vertex with a circle. The circle at vertex v_i is denoted as c_i. The two circles c_i and c_j on an edge $[v_i, v_j]$ are disjoint as shown in Fig. 7.10b, or intersect each other at acute angle $\theta_{ij} < \frac{\pi}{2}$ as shown in the left frame.

Definition 7.4 (*Inversive distance*) Suppose the length of $[v_i, v_j]$ is l_{ij}, the radii of c_i and c_j are γ_i and γ_j respectively, then the inversive distance between c_i and c_j is given by

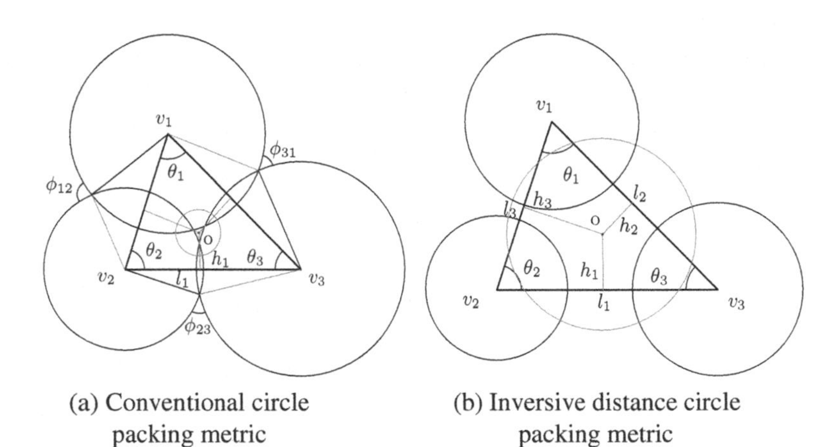

(a) Conventional circle packing metric

(b) Inversive distance circle packing metric

Fig. 7.10 Circle packing metric is generalized from conventional case (**a**) to include the inversive distance metric (**b**). This greatly improves the generality and flexibility of discrete Ricci flow. Image from [324]

$$I(c_i, c_j) = \begin{cases} \frac{l_{ij}^2 - \gamma_i^2 - \gamma_j^2}{2\gamma_i\gamma_j} & \mathbb{E}^2 \\ \frac{\cosh l_{ij} - \cosh\gamma_i \cosh\gamma_j}{\sinh\gamma_i \sinh\gamma_j} & \mathbb{H}^2 \end{cases} \tag{7.2}$$

The generalized circle packing metric is defined as

Definition 7.5 (*Generalized Circle Packing Metric*) A generalized circle packing metric on a mesh M is to associate each vertex v_i with a circle c_i, whose radius is γ_i, associate each edge $[v_i, v_j]$ with a non-negative number I_{ij}. The edge length is given by

$$l_{ij} = \begin{cases} \sqrt{\gamma_i^2 + \gamma_j^2 + 2I_{ij}\gamma_i\gamma_j} & \mathbb{E}^2 \\ \cosh^{-1}(\cosh\gamma_i \cosh\gamma_j + I_{ij}\sinh\gamma_i \sinh\gamma_j) & \mathbb{H}^2 \end{cases} \tag{7.3}$$

The circle packing metric is denoted as (Γ, I, M), where $\Gamma = \{\gamma_i\}$, $I = \{I_{ij}\}$.

A *discrete conformal deformation* is to change radii γ_i's only, and preserve inverse distance I_{ij}'s. The discrete Ricci flow is defined as follows. Let

$$u_i = \begin{cases} \log\gamma_i & \mathbb{E}^2 \\ \log\tanh\frac{\gamma_i}{2} & \mathbb{H}^2 \end{cases}$$

Definition 7.6 (*Discrete Ricci Flow*) Given a circle packing metric (Γ, I, M), the discrete Ricci flow is

$$\frac{du_i}{dt} = \bar{K}_i - K_i, \tag{7.4}$$

where $\bar{K}_i$ is the user defined curvature at vertex v_i.

Given a triangle $[v_i, v_j, v_k]$ with a circle packing, there exists a unique circle c, which is orthogonal to c_i, c_j, c_k, shown as the red circles in Fig. 7.10. The center of c is O. The distance from O to edge $[v_i, v_j]$ is denoted as h_k, the edge length of $[v_i, v_j]$ is denoted as l_k.

Lemma 7.7 *The following symmetric relation holds for Euclidean Ricci flow:*

$$\frac{\partial\theta_i}{\partial u_j} = \frac{\partial\theta_j}{\partial u_i} = \frac{h_k}{l_k}. \tag{7.5}$$

and

$$\frac{\partial\theta_i}{\partial u_i} = -\frac{\partial\theta_i}{\partial u_j} - \frac{\partial\theta_i}{\partial u_k} \tag{7.6}$$

For the hyperbolic Ricci flow, similar symmetry holds, albeit with a more complex formula. On one face $[v_1, v_2, v_3]$,

$$\begin{pmatrix} d\theta_1 \\ d\theta_2 \\ d\theta_3 \end{pmatrix} = \frac{-1}{\sin\theta_1 \sinh l_2 \sinh l_3} M \begin{pmatrix} du_1 \\ du_2 \\ du_3 \end{pmatrix} \tag{7.7}$$

$$M = \begin{pmatrix} 1-a^2 & ab-c & ca-b \\ ab-c & 1-b^2 & bc-a \\ ca-b & bc-a & 1-c^2 \end{pmatrix} \Lambda \begin{pmatrix} 0 & ay-z & az-y \\ bx-z & 0 & bz-x \\ cx-y & cy-x & 0 \end{pmatrix}$$

$$\Lambda = \begin{pmatrix} \frac{1}{a^2-1} & 0 & 0 \\ 0 & \frac{1}{b^2-1} & 0 \\ 0 & 0 & \frac{1}{c^2-1} \end{pmatrix}$$

where

$$(a, b, c) = (\cosh l_1, \cosh l_2, \cosh l_3),$$

and

$$(x, y, z) = (\cosh \gamma_1, \cosh \gamma_2, \cosh \gamma_3).$$

Lemma 7.8 *The following symmetric relation holds for hyperbolic Ricci flow:* $\frac{\partial \theta_i}{\partial u_j} = \frac{\partial \theta_j}{\partial u_i}$.

Let $\mathbf{u}$ represent the vector $(u_1, u_2, \ldots u_n)$, $\mathbf{K}$ represent the vector $(K_1, K_2, \ldots, K_n)$. where $n = |V|$. Fixing the inversive distances, all possible $\mathbf{u}$s that ensures the triangle inequality on each face form the *admissible metric space* of M, which is a simply connected domain in R^n. The above lemma proves that the differential 1-form $\omega = \sum_i (\bar{K}_i - K_i) du_i$ is a closed 1-form. The discrete Euclidean Ricci energy and hyperbolic Ricci energy have the same formula.

Definition 7.9 (*Discrete Ricci Energy*) The *discrete Euclidean and Hyperbolic Ricci energy* is defined as

$$E(\mathbf{u}) = \int_{\mathbf{u}_0}^{\mathbf{u}} \sum_i (\bar{K}_i - K_i) du_i,$$

where $\mathbf{u}_0 = (0, 0, \ldots, 0)$.

The discrete Ricci flow in Eq. (7.4) is the negative gradient flow of the Ricci energy. The following convexity theorem lays down the theoretic foundation of the algorithm.

Theorem 7.10 (Convexity of Ricci Energy) *The discrete Euclidean Ricci energy is convex on the hyper plane* $\sum_i u_i = 0$ *in the admissible metric space. The discrete hyperbolic Ricci energy is convex in the admissible metric space.*

Detailed proof for the inversive distance circle packing metric can be found in [106]. The metric inducing the target curvature is the unique global optimum of the Ricci energy. Therefore, the discrete Ricci flow will converge to the global optimum.

Algorithm 20 computes the generalized circle packing metric on meshes for both Euclidean and hyperbolic Ricci flow.

Require: A triangular mesh M, embedded in $\mathbb{R}^3$.
1: **for all** face $[v_i, v_j, v_k] \in M$ **do**
2: Compute $\gamma_i^{jk} = \frac{l_{ij}+l_{ki}-l_{jk}}{2}$.
3: **end for**
4: **for all** vertex $v_i \in M$ **do**
5: Compute the radius $\gamma_i = \min_{jk} \gamma_i^{jk}$.
6: **end for**
7: **for all** edge $[v_i, v_j] \in M$ **do**
8: Compute the inversive distance using Eq. (7.2).
9: **end for**

Algorithm 20: Initial Circle Packing Metric

Algorithm 21 gives the generalized discrete Euclidean and Hyperbolic Ricci flow with user defined target curvatures $\bar{K}$.

Require: A triangular mesh M, the target curvature $\bar{K}$.
1: Compute the initial circle packing metric using Algorithm 20.
2: **repeat**
3: For each edge, compute the edge length using Formula (7.3).
4: For each face $[v_i, v_j, v_k]$, compute the corner angles θ_i, θ_j and θ_k.
5: For each face $[v_i, v_j, v_k]$, compute $\frac{\partial\theta_i}{\partial u_j}$, $\frac{\partial\theta_j}{\partial u_k}$, and $\frac{\partial\theta_k}{\partial u_i}$ using Eq. (7.5) for $\mathbb{E}^2$ case and (7.7) for $\mathbb{H}^2$ case.
6: For each face $[v_i, v_j, v_k]$, compute $\frac{\partial\theta_i}{\partial u_i}$, $\frac{\partial\theta_j}{\partial u_j}$ and $\frac{\partial\theta_k}{\partial u_k}$ using Eq. (7.6) for $\mathbb{E}^2$ case, and (7.7) for $\mathbb{H}^2$ case.
7: Construct the Hessian matrix H.
8: Solve linear system $H\delta\mathbf{u} = \bar{K} - K$.
9: Update discrete conformal factor $\mathbf{u} \leftarrow \mathbf{u} + \delta\mathbf{u}$.
10: For each vertex v_i, compute the Gaussian curvature K_i.
11: **until** $\max_{v_i \in M} |\bar{K}_i - K_i| < \epsilon$

Algorithm 21: Generalized Euclidean and Hyperbolic Ricci Flow

With the computed discrete metric of a mesh, we embed the mesh onto $\mathbb{R}^2$ or $\mathbb{H}^2$, isometrically.

7.3.2 Doubly Connected Domain

A genus zero surface with two boundaries is called a *doubly connected domain*, as shown in Fig. 7.11. The following theorem postulates that the conformal module of a doubly connected domain is one dimensional, given by $\frac{1}{2\pi}\log\frac{R}{r}$. Algorithm 22 computes the conformal module of a doubly connected domain. Figure 7.11 shows the computed conformal module for a doubly connected domain.

Theorem 7.11 (Doubly Connected Domain [90]) *Suppose* $(S, \mathbf{g})$ *is a doubly connected domain, then it can be conformally mapped to a planar annulus, with two concentric circular boundaries. Suppose the radii of the outer boundary and inner boundary are R and r respectively, then the conformal module of S is* $\frac{1}{2\pi}\log\frac{R}{r}$.

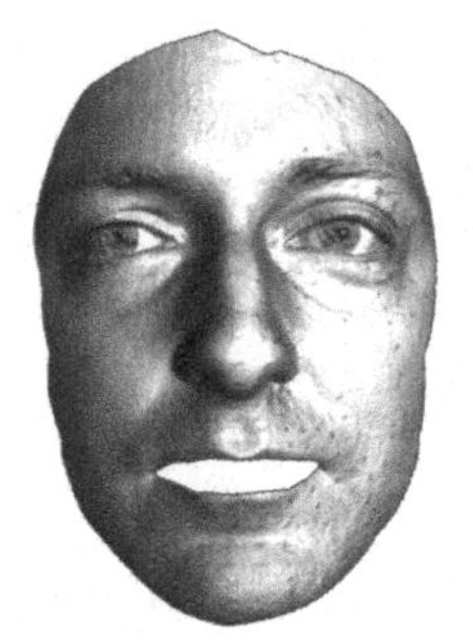
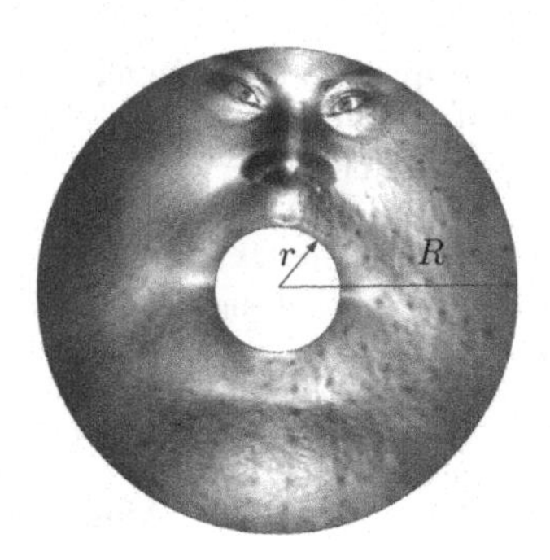

Fig. 7.11 A doubly connected domain is conformally mapped onto a planar annulus, the conformal module (fingerprint) of the shape is $\frac{1}{2\pi}\log\frac{R}{r}$. Image from [324]

Require: A triangular mesh M, which is a doubly connected domain.
1: Set target curvature equal to zero everywhere.
2: Compute a flat metric using Euclidean Ricci flow Algorithm 21.
3: Compute the shortest path γ between two boundaries.
4: Slice the M along γ to get $\bar{M}$.
5: Flatten $\bar{M}$ onto plane based on the computed flat metric.
6: Scale and translate the planar image of $\bar{M}$, such that the image of the outer boundary is aligned with the imaginary axis, the length of outer boundary is 2π.
7: Use the exponential map $z \rightarrow e^z$ to map the planar image of $\bar{M}$ to an annulus.
8: Measure the radii of inner and outer boundary circles r and R respectively. The conformal module is given by $\frac{1}{2\pi}\log\frac{R}{r}$.

Algorithm 22: Conformal Module for Doubly Connected Domains

7.3.3 Multiply Connected Domain

Genus zero surfaces with multiple holes denoted as $n \geq 2$ are called *multiply connected domains*. Multiply connected domains can be conformally mapped to a planar disk with circular holes, and all such mappings differ by Möbius transformations.

The following theorem postulates that the conformal module of a multiply connected domain is $3n - 3$ dimensional. Algorithm 23 computes the conformal module

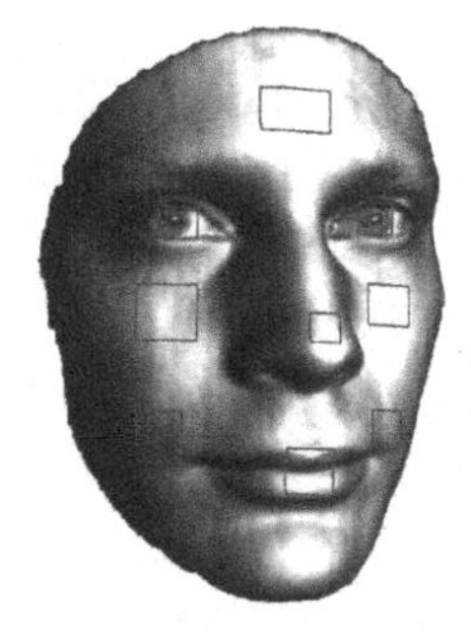
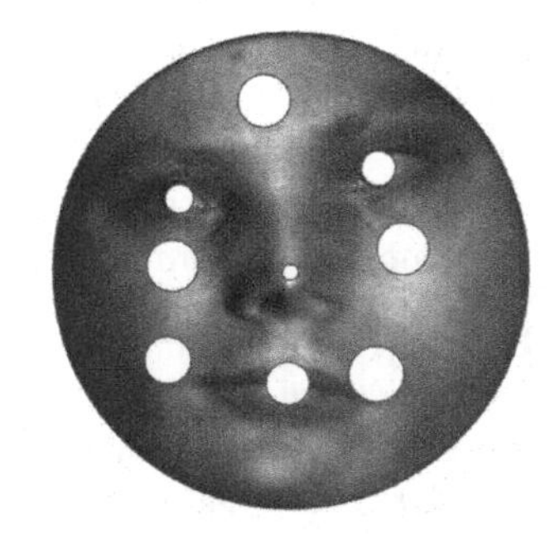

Fig. 7.12 A multiply connected domain is conformally mapped onto a planar disk with circular holes, the conformal module of the shape is represented as the centers and radii of the inner circular holes. Image from [324]

of a multiply connected domain. Figure 7.12 shows the computed conformal module for a multiply connected domain with 9 holes.

Require: A triangular mesh M, which is a multiply connected domain, the boundary of the mesh is $\partial M = c_0 - c_1 - c_2 \cdots - c_n$.
1: **repeat**
2: Randomly choose two circles $c_i, c_j, 1 \leq i < j \leq n$, fill all other holes, to get $\bar{M}$.
3: Use Algorithm 22 to map $\bar{M}$ to a canonical planar annulus $\tilde{M}$.
4: Remove all the filled disks from $\tilde{M}$
5: Update $M \leftarrow \tilde{M}$.
6: **until** The curvatures on the boundaries are close to constants
7: Suppose the center of c_1 is z_0, define a Möbius transformation

$$z \to \frac{z - z_0}{1 - \bar{z}_0 z}$$

to map the center of c_1 to the origin.
8: Rotate the planar image, such that the center of c_2 is on the real axis.

Algorithm 23: Conformal Module for Multiply Connected Domain by Euclidean Ricci Flow (Generalized Koebe's Method)

Theorem 7.12 (Multiply Connected Domain [90]) *Suppose* $(S, \mathbf{g})$ *is a multiply connected domain, the boundary is ordered as*

$$\partial S = c_0 - c_1 - c_2 \cdots - c_n, n > 2,$$

where c_0 *is the outer boundary. Then* S *can be conformally mapped to a unit disk with circular holes. A canonical mapping can be obtained, such that* c_0 *is mapped to the unit circle, the center of* c_1 *is the origin, the center of* c_2 *is a real number. Then the centers and radii of* $c_1, c_2, \ldots, c_n$ *are the conformal module. The Teichmüller space of triply connected domains is* $3n - 3$ *dimensional (because the center of* c_1 *and the x-component of* c_2 *are fixed).*

Figure 7.13 shows the computed conformal modules of a list of 3D faces scanned from different persons with similar expressions. The contours of lips and eyes have been identified and removed. Denote c_1 the boundary of the mouse hole, c_2 and c_3 the boundaries of left and right eyes holes, respectively. Denote r_k the mapped radius and (x_k, y_k) the mapped center of hole c_k. After normalization, c_1 and c_2 are centered at the origin and positive y-axis, respectively. The conformal module of a face is given as $(r_1, y_2, r_2, x_3, y_3, r_3)$. The distance between two faces is approximated by the Euclidean distance of their conformal modules. Figure 7.13 sort faces by their distances to the leftmost one.

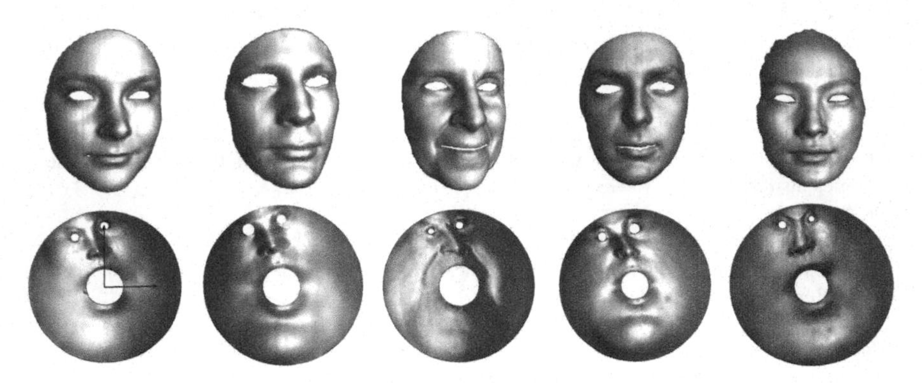

Fig. 7.13 Conformal modules for faces scanned from different persons with similar expressions. c_1 is the mouth boundary, c_2, c_3 are the left and right eyes. After normalization, c_1 and c_2 are centered at the origin and positive y-axis, respectively. The conformal module of a face is given as $(r_1, y_2, r_2, x_3, y_3, r_3)$. The distance between two surfaces is approximated by the Euclidean distance of their conformal modules. The surfaces are sorted by their distances to the leftmost face. Image from [324]

7.4 Shape Signature

Shape analysis of planar objects benefits applications in computer vision including image classification, recognition, and retrieval. Effective shape analysis of planar objects requires an efficient representation of the observed silhouettes and a robust metric measuring their dissimilarity.

Recently, many different representations for 2D shapes and various measures of dissimilarity between them have been proposed. For example, Zhu et al. [338] propose to represent shapes using their medial axis and compare their skeletal graphs through a branch and bound strategy. Liu et al. [174] use shape axis trees that are defined by the locus of midpoints of optimally corresponding boundary points to represent shapes. Belongie et al. [21] propose to represent and match 2D shapes for object recognition, based on the shape context and the Hungarian method. Mokhtarian [198] introduce a multi-scale, curvature-based shape representation technique for planar curves, which is especially suitable for recognition of a noisy curve. Besides, various statistical models for shape representation are also proposed by different research groups [68, 73, 235]. These approaches provide a simple way to represent shapes with finite dimensional spaces, although they cannot capture all the variability of shapes. Yang et al. [311] propose a signal representation approach called the Schwarz representation and apply it to shape matching problems. Lee et al. [162] propose to represent curves based on their complete silhouettes, through the use of harmonic embedding. Mumford et al. [243] propose a conformal approach to model simple closed curves that capture subtle variability of shapes up to scaling and translation. They also introduce a natural metric, called the Weil-Petersson metric, on the proposed representation space.

Most of the above methods work only on simple closed curves and generally cannot deal with multiply-connected objects. In real world applications, objects from their observed silhouette are usually multiply-connected domains (i.e. domains with holes in the interior). For example, the silhouette of a human face with the mouth and eyes is a multiply-connected shape. A lot of images from the real world consist of multiply-connected objects. In order to analyze shapes or images with arbitrary topologies effectively, it is necessary to develop an algorithm that can deal with multiply-connected domains.

We introduce a shape representation to model planar objects with arbitrary topologies. The key idea is to conformally map the exterior and interiors of a multiply-connected domain to unit and punctual disks, respectively using holomorphic 1-forms. A set of diffeomorphisms from a unit circle $\mathbb{S}^1$ to itself together with conformal modules define the shape signature of the multiply-connected domain. The signature uniquely represents shapes with arbitrary topologies up to scaling and translation.

7.4.1 Beltrami Equation

Consider a complex valued function $\phi : \mathbb{C} \to \mathbb{C}$ maps the z-plane to the w-plane, where $z = x + iy$, $w = u + iv$. The *complex partial derivative* is defined as:

$$\frac{\partial}{\partial z} := \frac{1}{2}\left(\frac{\partial}{\partial x} - i\frac{\partial}{\partial y}\right), \frac{\partial}{\partial \bar{z}} = \frac{1}{2}\left(\frac{\partial}{\partial x} + i\frac{\partial}{\partial y}\right) \tag{7.8}$$

The *Beltrami equation* for ϕ is defined by:

$$\frac{\partial \phi}{\partial \bar{z}} = \mu(z)\frac{\partial \phi}{\partial z} \tag{7.9}$$

where μ is called the *Beltrami coefficient.*

If μ is zero, then ϕ is called a *holomorphic or conformal mapping*. Otherwise, if $\|\mu\|_\infty < 1$, then ϕ is called a *quasiconformal mapping*. In terms of the metric tensor, consider the effect of the pullback under ϕ of the usual Euclidean metric ds_E^2; the resulting metric is given by:

$$\phi^*(ds_E^2) = |\frac{\partial \phi}{\partial z}|^2 |dz + \mu(z) d\bar{z}|^2. \tag{7.10}$$

The resulting metric, relative to the background Euclidean metric dz and $d\bar{z}$, has eigenvalues $(1 + |\mu|)^2 \frac{\partial f}{\partial z}$ and $(1 - |\mu|)^2 \frac{\partial f}{\partial z}$. μ is called the *Beltrami coefficient*, which is a measure of non-conformality. In particular, the map ϕ is conformal around a small neighborhood of p when $\mu(p) = 0$. Infinitesimally, around a point p, ϕ may be expressed with respect to its local parameter as follows:

$$\begin{aligned} f(z) &= f(p) + f_z(p)z + f_{\bar{z}}(p)\bar{z} \\ &= f(p) + f_z(p)(z + \mu(p)\bar{z}). \end{aligned} \tag{7.11}$$

Obviously, f is not conformal if and only if $\mu(p) \neq 0$. Inside the local parameter domain, f may be considered as a map composed of a translation to $f(p)$ together with a stretch map $S(z) = z + \mu(p)\bar{z}$, which is postcomposed by a multiplication of $f_z(p)$, which is conformal. All the conformal distortion of $S(z)$ is caused by $\mu(p)$. $S(z)$ is the map that causes f to map a small circle to a small ellipse. From $\mu(p)$, we can determine the angles of the directions of maximal magnification and shrinking and the amount of them as well. Specifically, the angle of maximal magnification is $\arg(\mu(p))/2$ with magnifying factor $1 + |\mu(p)|$; The angle of maximal shrinking is the orthogonal angle $(\arg(\mu(p)) - \pi)/2$ with shrinking factor $1 - |\mu(p)|$. The distortion or dilation is given by:

$$K = (1 + |\mu(p)|)/(1 - |\mu(p)|). \tag{7.12}$$

Thus, the Beltrami coefficient μ gives us important information about the properties of the map (See Fig. 7.14).

Given a compact simply-connected domain Ω in $\mathbb{C}$ and a Beltrami coefficient μ with $\|\mu\|_\infty < 1$, there is always a quasiconformal mapping from Ω to the unit disk $\mathbb{D}$, which satisfies the Beltrami equation in the distribution sense [90]. More precisely,

Theorem 7.13 (Measurable Riemann Mapping Theorem) *Suppose Ω is a simply connected domain in $\mathbb{C}$ that is not equal to $\mathbb{C}$, and suppose that $\mu : \Omega \to \mathbb{C}$ is Lebesgue measurable and satisfies $\|\mu\|_\infty < 1$, then there is a quasiconformal homomorphism ϕ from Ω to the unit disk, which is in the Sobolev space $W^{1,2}(\Omega)$ and satisfied the Beltrami equation 7.9 in the distribution sense.*

This theorem plays a fundamental role in this work. Suppose $f, g : \mathbb{C} \to \mathbb{C}$ are with Beltrami coefficients μ_f, μ_g respectively. Then the Beltrami ceofficient for the composition $\phi_2 \circ \phi_1$ is given by

$$\mu_{g \circ f} = \frac{\mu_f + (\mu_g \circ f)\tau}{1 + \bar{\mu_f}(\mu_g \circ f)\tau} \tag{7.13}$$

where $\tau = \frac{\bar{f_z}}{f_z}$.

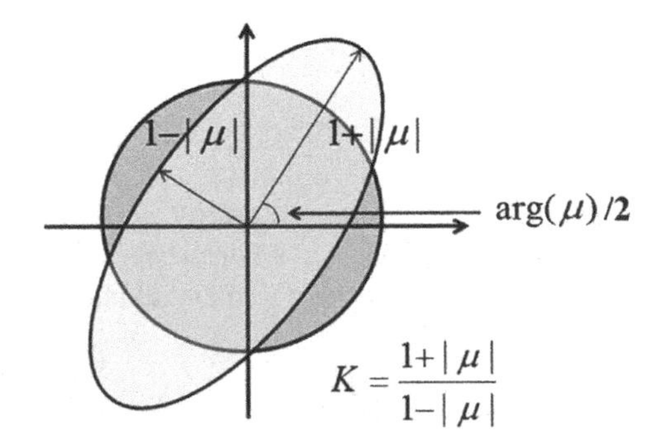

Fig. 7.14 Quasi-conformal maps infinitesimal circles to ellipses. The Beltrami coefficient measure the distortion or dilation of an ellipse under quasi-conformal map. Image from [184]

7.4.2 Conformal Module

Suppose Ω_1 and Ω_2 are planar domains. We say Ω_1 and Ω_2 are *conformally equivalent* if there is a biholomorphic diffeomorphism between them. All planar domains can be classified by the conformal equivalence relation. Each conformal equivalence class shares the same *conformal invariants*, the so-called *conformal module*. The conformal module is one of the key component for us to define the unique shape signature.

Suppose Ω is a compact domain on the complex plane $\mathbb{C}$. If Ω has a single boundary component, it is called a *simply connected domain*. Every simply connected domain can be mapped to the unit disk conformally and all such kind of mappings differ by a *Möbius transformation*: $z \to e^{i\theta}\frac{z-z_0}{1-\bar{z}_0 z}$.

Suppose Ω has multiple boundary components $\partial\Omega = \gamma_0 - \gamma_1 - \gamma_2 \cdots \gamma_n$, where γ_0 represents the exterior boundary component, then Ω is called a *multiply-connected planar domain*. A *circle domain* is a unit disk with circular holes. Two circle domains are conformally equivalent, if and only if they differ by a Möbius transformation. It turns out every multiply-connected domain can be conformally mapped to a circle domain, as described in the following theorem.

Theorem 7.14 (Riemann Mapping for Multiply Connected Domain) *If Ω is a multiply-connected domain, then there exists a conformal mapping $\phi : \Omega \to D$, where D is a circle domain. Such kind of mappings differ by Möbius transformations.*

Therefore, each multiply connected domain is conformally equivalent to a circle domain. The conformal module for a circle domain is represented as the centers and radii of inner boundary circles. All simply-connected domains are conformally equivalent. The topological annulus requires 1 parameter to represent the conformal module. In general case, because there are $n > 1$ inner circles, and the Mobius transformation group is 3 dimensional, therefore the conformal module requires $3n - 3$ parameters. We denote the conformal module of Ω as $Mod(\Omega)$.

Fix n, all conformal equivalence classes form a $3n - 3$ Riemannian manifold, the *Teichmüller space*. The conformal module can be treated as the Teichmüller coordinates. The Weil-Peterson metric [243] is a Riemannian metric for Teichmüller space, which induces negative sectional curvature. Therefore the geodesic between arbitrary two points is unique.

7.4.3 Holomorphic Differentials

In order to compute the conformal modules, one needs to find the holomorphic differential forms on the multiply connected domain. A *differential 1-form* on a planar domain ω is defined as

$$\tau = f(x, y)dx + g(x, y)dy,$$

where f, g are smooth functions. A *harmonic 1-form* is curl free and divergence free

$$\nabla \times \tau = 0, \nabla \cdot \tau = 0,$$

where the differential operator $\nabla = (\frac{\partial}{\partial x}, \frac{\partial}{\partial y})$. If τ is the gradient of another function defined on Ω, then it is called an *exact 1-form*.

The *Hodge star* operator acting on a differential 1-form gives the *conjugate differential 1-form*

$$^*\tau = -g(x, y)dx + f(x, y)dy,$$

intuitively, the conjugate 1-form $*\tau$ is obtained by rotating τ by a right angle everywhere. If τ is harmonic, so is its conjugate $*\tau$.

A *holomorphic 1-form* consists of a pair of conjugate harmonic 1-forms

$$\omega = \tau + i\, ^*\tau = \phi(z)dz,$$

where $\phi(z)$ is a holomorphic function. We further require that either τ or $^*\tau$ is orthogonal to all the boundaries. All the holomorphic 1-forms consist a group (with real coefficients), denoted as $\mathbb{H}(\Omega)$. A basis of $\mathbb{H}(\Omega)$ is given by

$$\{\omega_1, \omega_2, \ldots, \omega_n\},$$

such that $\int_{\gamma_j} \omega_i = \delta_i^j$, where δ_i^j is the Kronecker symbol. By using the holomorphic 1-forms, one can construct the following *circular slit map*.

Theorem 7.15 (Circular Slit Map) *Suppose a multiply connected domain Ω with more than one boundary components, then there exists a conformal mapping $\phi : \Omega \to \mathbb{C}$, such that γ_0, γ_1 are mapped to concentric circles, γ_k's are mapped to concentric circular slits. All such kind of mappings differs by a rotation.*

7.4.4 Conformal Welding

Suppose $\Gamma = \{\gamma_0, \gamma_1, \ldots, \gamma_k\}$ are non-intersecting smooth closed curves on the complex plane. Γ segments the plane to a set of connected components $\{\Omega_0, \Omega_1, \ldots, \Omega_s\}$. Each segment Ω_i is a multiply-connected domain. We assume Ω_0 contains the infinity point, $p \notin \Omega_0$. By using a Möbius transformation $\phi(z) = \frac{1}{z-p}$, p is mapped to ∞, Ω_0 is mapped to a compact domain. Replace Ω_0 by $\phi(\Omega_0)$. Construct $\phi_k : \Omega_k \to \mathbb{D}_k$ to map each segment Ω_k to a circle domain $\mathbb{D}_k, 0 \leq k \leq s$. Assume $\gamma_i \in \Gamma = \Omega_j \cap \Omega_k$, then $\phi_j(\gamma_i)$ is a circular boundary on the circle domain $\mathbb{D}_j$, $\phi_k(\gamma_i)$ is a another circle on $\mathbb{D}_k$. Let $f_i = \phi_j \circ \phi_k^{-1} : \mathbb{S}^1 \to \mathbb{S}^1$ be the diffeomorphism from the circle to itself, which is called the *signature of* γ_i.

Definition 7.16 (*Signature of a family of loops*) The signature of a family non-intersecting closed planar curves $\Gamma = \{\gamma_0, \gamma_1, \ldots, \gamma_k\}$ is defined as

$$S(\Gamma) := \{f_0, f_1, \ldots, f_k\} \cup \{Mod(\mathbb{D}_0), Mod(\mathbb{D}_1), \ldots, Mod(\mathbb{D}_s)\}.$$

The following main theorem plays the fundamental role for the current work.

Theorem 7.17 (Main Theorem) *The family of smooth planar closed curves Γ is determined by its signature $S(\Gamma)$, unique up to a Möbius transformation of the Riemann sphere $\mathbb{C} \cup \{\infty\}$.*

Note that if a circle domain $\mathbb{D}_k$ is disk, its conformal module can be omitted from the signature. The Möbius transformation of the Riemann sphere is given by $(az+b)/(cz+d)$, where $ad - bc = 1, a, b, c, d \in \mathbb{C}$. The proof of Theorem 7.17 can be found in the Appendix.

The theorem states that the proposed signature determines shapes up to a Möbius transformation. We can further do a normalization that fixes ∞ to ∞ and that the differential carries the real positive axis at ∞ to the real positive axis at ∞, as in Mumford's paper [243]. The signature can then determine the shapes uniquely up to translation and scaling.

7.4.5 Computing Shape Signatures of Planar Domains

Denote Ω a planar triangular mesh with n inner boundaries. Denote $\partial\Omega = \gamma_0 - \gamma_1 \cdots - \gamma_n$ the boundaries of Ω. They decompose Ω into a set of sub-domains Ω_k. We first compute conformal parameterizations of the planar object to circle domains using holomorphic 1-forms. Figure 7.15 shows a group basis of holomorphic 1-form for a 2-hole planar domain. We then apply a two-step algorithm to compute the shape signature of Ω.

Step 1: Conformal Maps from Ω_k to Circle Domains D_k

The conformal parameterization of Ω_k can be obtained easily by computing the circular slit map and performing the Koebe's iteration.

Circular slit:

The circular slit map can be obtained by finding a holomorphic 1-form ω, such that

$$Img(\int_{\gamma_0} \omega) = 2\pi, Img(\int_{\gamma_1} \omega) = -2\pi, img(\int_{\gamma_k} \omega) = 0, 2 \le k \le n-1. \tag{7.14}$$

To solve Eq. 7.14, we first compute the basis of the holomorphic 1-form group. ω is then a linear combination of the basis $\omega = \sum_{k=1}^{n} \lambda_k \omega_k$. The coefficients $\{\lambda_k\}$ can be obtained by solving the linear system Eq. 7.14. The circular slit map is given by $\phi(p) = exp(\int_q^p \omega), \forall p \in \Omega$, where q is a base point, and the integration path is arbitrarily chosen in Ω.

If Ω is a simply connected domain (topological disk), we compute the conformal mapping to map it to the unit disk in the following way. First, we punch a small hole

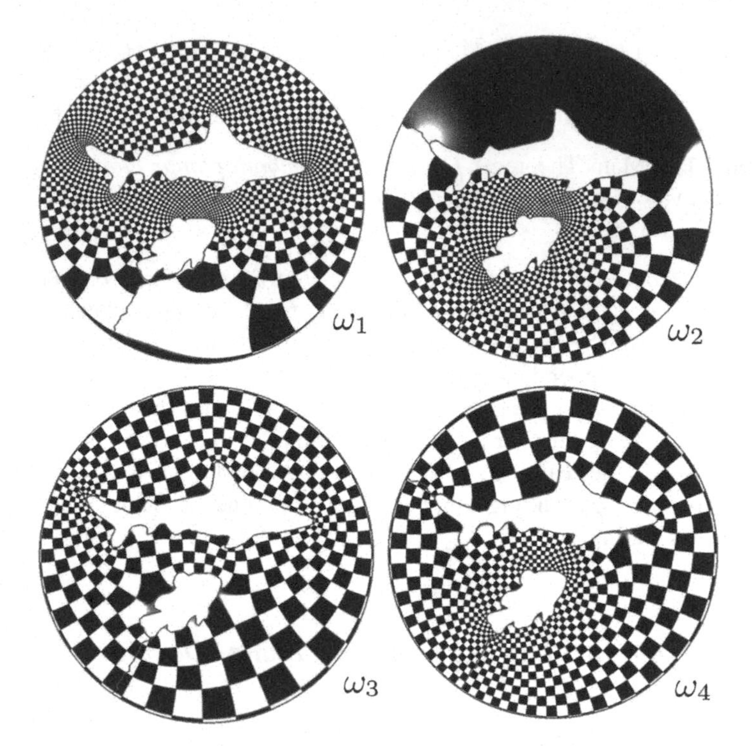

Fig. 7.15 Holomorphic 1-form basis. Image from [184]

in the domain, then treat it as topological annulus. Then we use circular slit map to map the punched annulus to the canonical annulus. By shrinking the size of the punched hole, the circular slit mappings converge to the conformal mapping.

Hole Filling:

After computing the circular slit map, the planar domain is mapped to the planar annulus with concentric circular slits. γ_0 is the unit circle, γ_1 is the inner circle, γ_k's are slits. We use Delaunay triangulation to generate a disk D_1 bounded by γ_1, $\partial D_1 = \gamma_1$, and glue Ω with D_1 along γ_1, $\Omega_1 := \Omega \cup_{\gamma_1} D_1$.

We then use circular slit map again to map Ω_1, such that γ_2 is opened to a circle. We compute a disk D_2 bounded by γ_2 and glue Ω_1 and D_2 to get Ω_2. By repeating circular slit map, at the step k, γ_k is opened to a circle. We compute a circular disk D_k bounded by γ_k, and glue Ω_{k-1} with D_k, $\Omega_k = \Omega_{k-1} \cup_{\gamma_k} D_k$.

Eventually, we can fill all the holes to get Ω_n. All the disks D_k in Ω_n are not exact circular.

Koebe's Iteration:

By Koebe's iteration, all the boundary components become rounder and rounder. Basically, each time we choose a disk D_k, the complement of D_k on Ω_n is a double connected domain. We map the complement to the canonical planar annulus, then

γ_k becomes a circle. We recompute the disk D_k bounded by the updated γ_k, and glue the annulus with the updated D_k. After this iteration, γ_k becomes a circle. Then we choose another disk D_j, and repeat this process to make γ_j a circle. This will destroy the perfectness of the circular shape of γ_k. But by repeating this process, all the γ_ks become rounder and rounder, and eventually converge to perfect circles. The convergence is exponentially fast. Detailed proof can be found in [118].

Step 2: Computing Conformal Modules and Signatures f_{ij} on Boundaries

With Ω_k conformally mapped to a circle domain, we then compute their conformal modules and the signature f_{ij} on each boundary. The conformal modules together with $\{f_{ij}\}$ give the complete signature $S(\Gamma)$. Figure 7.16 demonstrates the process of computing the shape signature. Given an image, we perform image segmentation and then calculate the contours of the objects in the image. Figure 7.16 shows the contour of each fish. For simplicity, we treat the outermost boundary of the image as a unit circle. Then all the contours segment the image into three sub-domains, namely, $\Omega_0, \Omega_1, \Omega_2$. We conformally map each connected sub-domain to a circle domain as shown in Fig. 7.16a. Ω_0 is mapped to a disk D_0 with two circular holes. The centers and radii (c_0, r_0) and (c_1, r_1) represent the conformal module of Ω_0. Ω_1 and Ω_2 are mapped to the unit disks D_1 and D_2 respectively. We denote the conformal maps of Ω_i by $\Phi_i : \Omega_i \to D_i$ $(i = 0, 1, 2)$. The contour of the small fish are mapped to the boundary of D_1 and one inner boundary of D_0. Its conformal welding is given by $f_{01} := \Phi_1 \circ \Phi_0^{-1}$, which is shown in Fig. 7.16b as the blue curve. Here, the diffeomorphisms from circle to circle is considered as a monotonic function from $[0, 2\pi]$ to itself. Similarly, the conformal welding f_{02} of the contour of the shark can also be computed, which is also shown in Fig. 7.16b as the red curve. The shape signature of the double fish image is given by $S(\Omega) = \{c_0, c_1, r_0, r_1, f_{01}, f_{02}\}$.

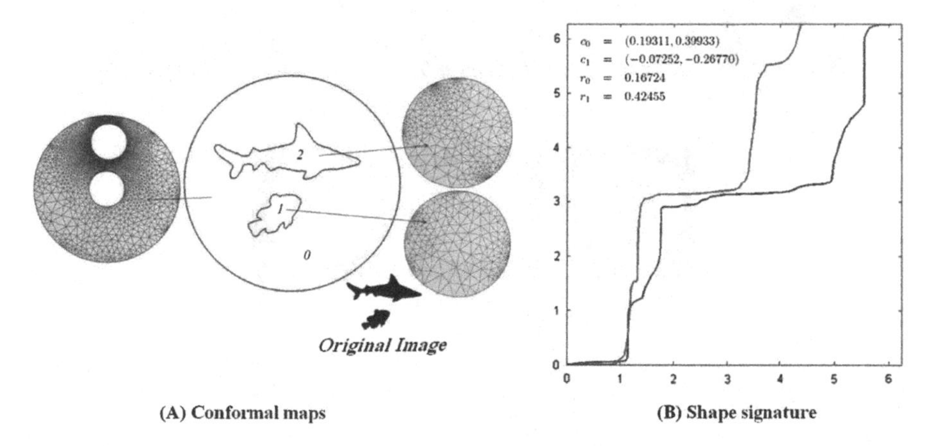

Fig. 7.16 **a** Each segment is mapped to a circle domains. **b** The conformal modules (centers and radii of inner circles) of the circle domains and conformal weldings define a shape signature. Image from [184]

7.4.6 Shape Clustering

The shape distance discussed provides a useful tool to quantitatively measure the geometric differences between different multiply-connected objects. Figure 7.17 shows a collection of multiply-connected shapes from three categories, fish, brain, and tool. Using the k-mean method, we cluster the randomly chosen collection of multiply-connected shapes into three categories based on the shape distance. Figure 7.18 shows the clustering result. Specifically, Fig. 7.18a gives the signatures corresponding to the outer boundaries of the collection of shapes. Figure 7.18b gives the signatures corresponding to the inner boundaries. They are colored according to the clustering result.

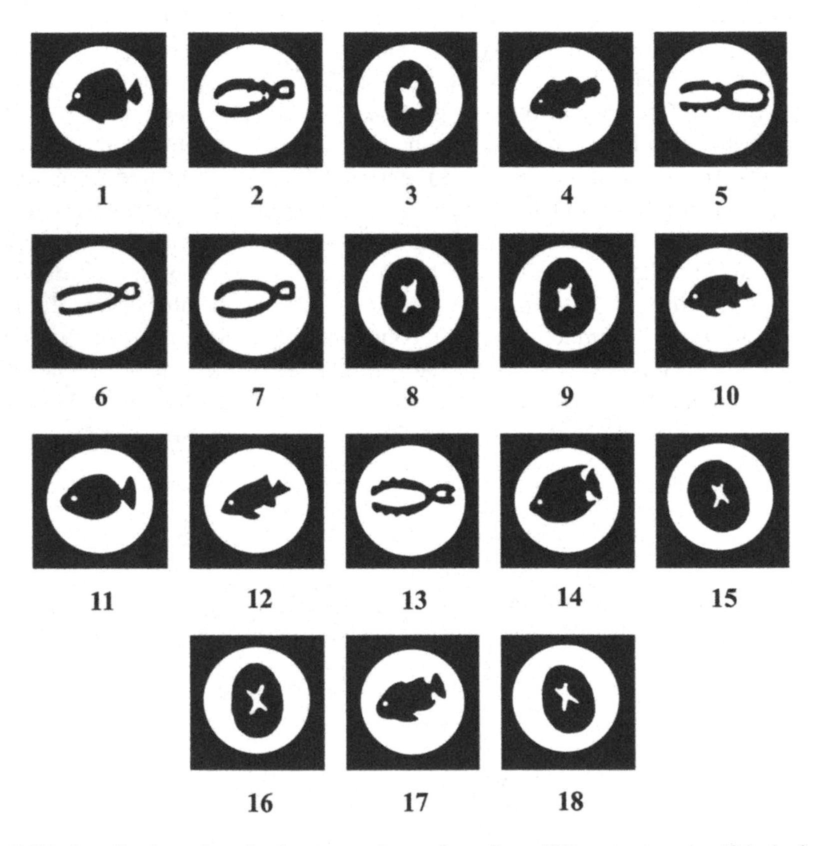

Fig. 7.17 A collection of randomly chosen shapes from three different categories (fish, brain and tool). The goal is to cluster them using the introduced shape distance. Image from [184]

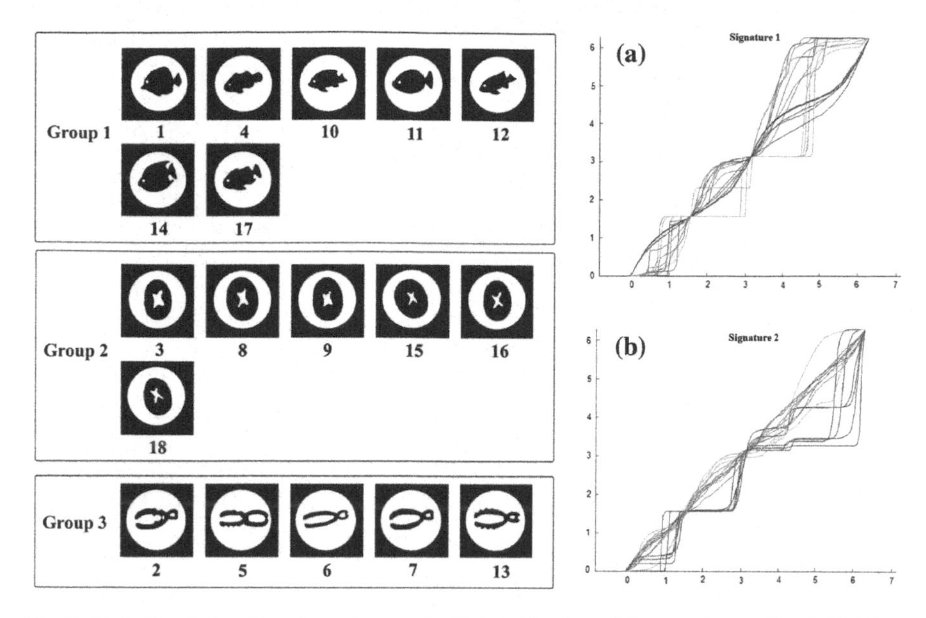

Fig. 7.18 Left column: The clustering result of the collection of shapes given in Fig. 7.17. Right column: The signatures of the collection of shapes given in Fig. 7.17, colored according to the clustering result. **a** shows the signature corresponding to the outer boundary. **b** shows the signature corresponding to the inner boundary. Image from [184]

7.5 Summary and Further Reading

Geometry plays a big role in computer vision, esp. 3D shade analysis, index, registration, and classification fields. We refer readers with interests in 3D facial shape index to [30].

Chapter 8
Geometric Modeling

Abstract In this chapter, we apply computational conformal geometry to construct manifold splines. We show that manifold splines afford a general theoretical and computational framework for modeling geometrically complicated surfaces of arbitrary topology. The technical challenge is how to extend polynomial-centric splines defined over open, planar domains to that over any manifold setting. Our solution is an affine structure for any manifold surface, serving as a parametric domain, so that piecewise spline functions can be naturally and elegantly blended to represent geometric shapes of arbitrary topology. Built upon our prior efforts, the primary foci are to broaden and strengthen the theoretical foundation as well as to devise practical algorithms for manifold splines. In particular, we advocate several novel mathematical tools to compute affine atlas for any domain manifold. At the theoretic level, we show that the lower bound of the number of singularities for surfaces with non-zero Euler number is one. We make use of discrete Ricci flow to actually reach this lower bound for manifold spline construction. At the practical level, we further relax this rather strict requirement by allowing users to control the number and positions of singular points. As a result, we construct polycube splines as a novel variant of manifold splines. Our computational tool is polycube maps, which can reduce both the area and angle distortion in the affine atlas. In order to demonstrate the general feasibility and efficiency of manifold splines, we design algorithms to extend various planar splines, such as triangular B-splines, Powell-Sabin splines, and T-splines to the manifold settings. We also highlight their modeling advantages and potentials in shape representation and analysis/synthesis through a wide array of experiments.

8.1 Introduction

Despite many theoretical and algorithmic advancements in shape computing in most recent years, one fundamental objective of the shape computing research community is to develop novel and powerful modeling, design, and simulation schemes that are capable of accurately representing complicated real-world physical objects

M. Jin et al., *Conformal Geometry*, https://doi.org/10.1007/978-3-319-75332-4_8

in a compact manner, and facilitating rapid computation of their desirable properties both globally and locally such as differential properties, smoothness requirements, and topological validity. From the standpoints of shape modeling, engineering design, finite element simulation, and scientific computation, elegant geometric properties such as high-order continuity and the ease of computing all the desirable properties rapidly are always desirable for the purpose of developing novel shape representations. Therefore, it is not surprising to see that spline-centric polar forms [238] are becoming the most popular computational tools in geometric modeling and shape design. Essentially, the methodology of polar forms naturally gives rise to parameterization-centered, piecewise polynomials defined over any open, planar parameter domain for the effective modeling and accurate computing of smooth spline surfaces.

However, examining all the real-world applications, we observe that the most natural shapes are 2-manifolds with complicated topologies and arbitrarily detailed geometric configurations, which cannot be completely covered by a single coordinate system (note that, it does not matter if the parametric surface is a polynomial or a non-polynomial, this fundamental principle remains the same). In contrast, a manifold might be covered by a family of *coordinate charts*, each coordinate chart covers only a portion of the manifold. Different charts may overlap with each other, a *coordinate transition function* transforms from one coordinate system to the other. If we simply adopt the algorithmic procedure of polar forms and other relevant computational techniques in a principled way, we can easily realize that conventional splines (defined over any open, planar domains) cannot be transferred over any manifold domains directly.

In order to model a manifold using piecewise polynomials, current approaches will typically segment the manifold to many patches, define a single coordinate system over each patch, such that each patch can be modelled by a spline patch. Finally, any generic approach will abut/glue all the spline patches together by adjusting the control points and the knots along their common boundaries. This whole process usually achieves only G^1, C^1 or C^2 continuity and is mainly performed manually, and it requires the users' skill and mathematical sophistication, and is tedious and error-prone.

It is highly desirable to design splines defined over any manifold domains directly, such that spline patches on different charts can be automatically glued together with high continuity, and the modeling process requires neither segmentation nor patching. Pioneering work has been done by Grimm and Hughes [97], which can model splines on arbitrary surfaces. Ying and Zorin [318] introduced a general method by constructing a conformal atlas. Other representative works of manifold-based construction include [58, 59, 64, 254, 283]. In those methods, smooth functions are defined on each chart and blended together to form a function coherently defined over the entire manifold domain. The methods are flexible for all manifolds with arbitrary topology. The functions can achieve any degree of desirable continuity without any singularity. The primary drawbacks of these existing methods are that surfaces constructed this way are not polynomials and their computation expenses are relatively high in comparison with conventional polynomial-based, spline surfaces.

Manifold splines are completely different from the above work in that: (1) The transition functions of manifold splines must be affine in order to produce polynomial surfaces defined over manifold domains. Therefore, this requirement appears to be stronger than that in previous cases. In principle, the property of topological obstruction prevents any meaningful manifold spline construction without introducing any extraordinary points; (2) Manifold splines aim to produce either polynomials or rational polynomials. On any chart, the basis functions are always polynomials or rational polynomials, which facilitates the evaluation and computation of various differential properties; (3) Manifold splines can accommodate high-order continuity and do not require any cutting and patching work to construct models of complicated topology. All the control points of manifold splines are free of any additional constraints, and thus, can participate in the interactive modeling and design procedure.

8.2 Triangular B-Splines

Essentially, splines have local support, so we shall define spline patches locally on the manifold and glue the locally-defined spline patches to cover the entire domain manifold. Furthermore, since splines are invariant under parametric affine transformations, we seek to glue the patches using affine transition functions. Therefore, if the domain surface admits an atlas on which all transition functions are affine, then we can glue the patches coherently. However, the existence of such an atlas is solely determined by the topology. In principle, we can glue the patches to cover the entire surface except a finite number of points, which are singular points and cannot be evaluated by the global splines on the manifold. These singular points represent the topological obstruction for the existence of the affine atlas.

8.2.1 Definition

Triangular B-spline surfaces can be defined on planar domains with arbitrary triangulations. The construction of triangular B-spline is as follows: let points $\mathbf{t}_i \in \mathbb{R}^2$, $i \in \mathbb{N}$, be given and define a triangulation

$$T = \{\Delta(I) = [\mathbf{t}_{i_0}, \mathbf{t}_{i_1}, \mathbf{t}_{i_2}] : I = (i_0, i_1, i_2) \in \mathcal{I} \subset \mathbb{N}^2\}$$

of a bounded region $D \subseteq \mathbb{R}^2$. Next, with every vertex $\mathbf{t}_i$ of T we associate a cloud of knots $\mathbf{t}_{i,0}, \ldots, \mathbf{t}_{i,n}$ such that $\mathbf{t}_{i,0} = \mathbf{t}_i$, and for every triangle $I = [\mathbf{t}_{i_0}, \mathbf{t}_{i_1}, \mathbf{t}_{i_2}] \in T$,

1. all the triangles $[\mathbf{t}_{i_0,\beta_0}, \mathbf{t}_{i_1,\beta_1}, \mathbf{t}_{i_2,\beta_2}]$ with $\beta = (\beta_0, \beta_1, \beta_2)$ and $|\beta| = \sum_{i=0}^{2} \beta_i \leq n$ are non-degenerate.
2. the set
$$interior(\cap_{|\beta|\leq n} X_\beta^I) \neq \emptyset,\ X_\beta^I = [\mathbf{t}_{i_0,\beta_0}, \mathbf{t}_{i_1,\beta_1}, \mathbf{t}_{i_2,\beta_2}] \quad (8.1)$$

3. If I has a boundary edge, say, $(\mathbf{t}_{i_0}, \mathbf{t}_{i_1})$, then the entire area $[\mathbf{t}_{i_0,0}, \ldots, \mathbf{t}_{i_0,n}, \mathbf{t}_{i_1,0}, \ldots, \mathbf{t}_{i_1,n})$ must lie outside of the domain.

Then the triangular B-spline basis function N^I_β, $|\beta| = n$, is defined by means of simplex splines $M(\mathbf{u}|V^I_\beta)$ as

$$N(\mathbf{u}|V^I_\beta) = |d^I_\beta| M(\mathbf{u}|V^I_\beta)$$

where $V^I_\beta = \{\mathbf{t}_{i_0,0}, \ldots, \mathbf{t}_{i_0,\beta_0}, \ldots, \mathbf{t}_{i_2,0}, \ldots, \mathbf{t}_{i_2,\beta_2}\}$ and

$$d^I_\beta = d(X^I_\beta) = det \begin{pmatrix} 1 & 1 & 1 \\ \mathbf{t}_{i_0,\beta_0} & \mathbf{t}_{i_1,\beta_1} & \mathbf{t}_{i_2,\beta_2} \end{pmatrix}$$

Assuming (8.1), these B-spline basis functions can be shown to be all non-negative and to form a partition of unity. Then, the DMS-spline is defined as

$$\mathbf{F}(\mathbf{u}) = \sum_{I \in \mathcal{I}} \sum_{|\beta|=n} \mathbf{c}_{I,\beta} N(\mathbf{u}|V^I_\beta) \tag{8.2}$$

where $\mathbf{c}_{I,\beta}$ is the control point. This spline is globally C^{n-1} if all the sets X^I_β, $|\beta| \leq n$ are affinely independent.

8.2.2 *Properties*

Triangular B-splines have the following valuable properties which are critical for geometric and solid modeling:

1. *Local support.* The spline surface has local support. In order to evaluate the image $F(\mathbf{u})$ of a point $\mathbf{u} \in \Delta^I$, we only need control points $\mathbf{c}^J_\beta$ (associated with knot set V^J_β on triangle J), where triangle J belongs to the 1-ring neighborhood of triangle I.
2. *Convex hull.* The polynomial surface is completely inside the convex hull of the control points.
3. *Completeness.* The B-spline basis is complete, namely, a set of degree n B-spline basis can represent any polynomial with degree no greater than n via a linear combination.
4. *Parametric affine invariance.* The choice of parameter is not unique: if one transforms the parameter affinely and the corresponding knots of control points are transformed accordingly, then the polynomial surface remains unchanged (see Fig. 8.1).
5. *Affine invariance.* If the control net is transformed affinely, the polynomial surface will be consistently transformed affinely.

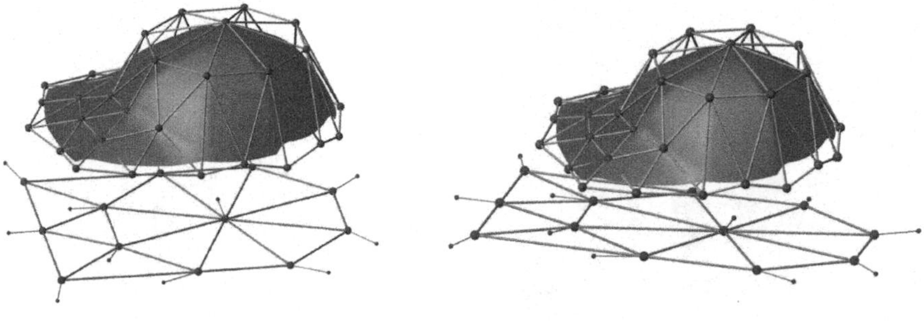

(a) Original triangular B-spline. (b) Transformed triangular B-spline.

Fig. 8.1 Parametric affine invariance: **a** and **b** are two triangular B-splines sharing the same control net, the two parametric domains differ only by an affine transformation. The same control nets result in the same polynomial surfaces shown in **a** and **b**. Image is taken from [103], included here by permission from Elsevier, Inc

Note that parametric affine invariance is different from affine invariance. The diagrams below illustrate the radical difference.

$$
\begin{array}{ccc}
\mathbf{u}, V^I_\beta & \xrightarrow{\phi} & \phi(\mathbf{u}), \phi(V^I_\beta) \\
\downarrow & & \downarrow \\
F & \xrightarrow[\phi]{} & F \circ \phi
\end{array}
\qquad\qquad
\begin{array}{ccc}
\mathbf{c}^I_\beta & \xrightarrow{\phi} & \phi(\mathbf{c}^I_\beta) \\
\downarrow & & \downarrow \\
F & \xrightarrow[\phi]{} & \phi \circ F
\end{array}
$$

(a) Parametric affine invariance (b) Affine invariance

The left one above represents parametric affine invariance, which refers to the property that, under a transformation between parameter domains, the shape of the polynomial surface remains the same; the right one above indicates affine invariance, which refers to the property that under a transformation of the control points, the polynomial surface will change accordingly.

The aforementioned properties are important for geometric and solid modeling applications. For example, the local support will allow designers to adjust the surface by moving nearby control points without affecting the global shape. Therefore, it is crucial to preserve these properties when we generalize the planar domain B-splines to manifold B-splines. We will prove that such a generalization does exist, and these desirable properties can be preserved (Fig. 8.2). The generalization completely depends on the so-called affine structure of the domain manifold. The local support and parametric affine invariance are crucial for constructing manifold splines.

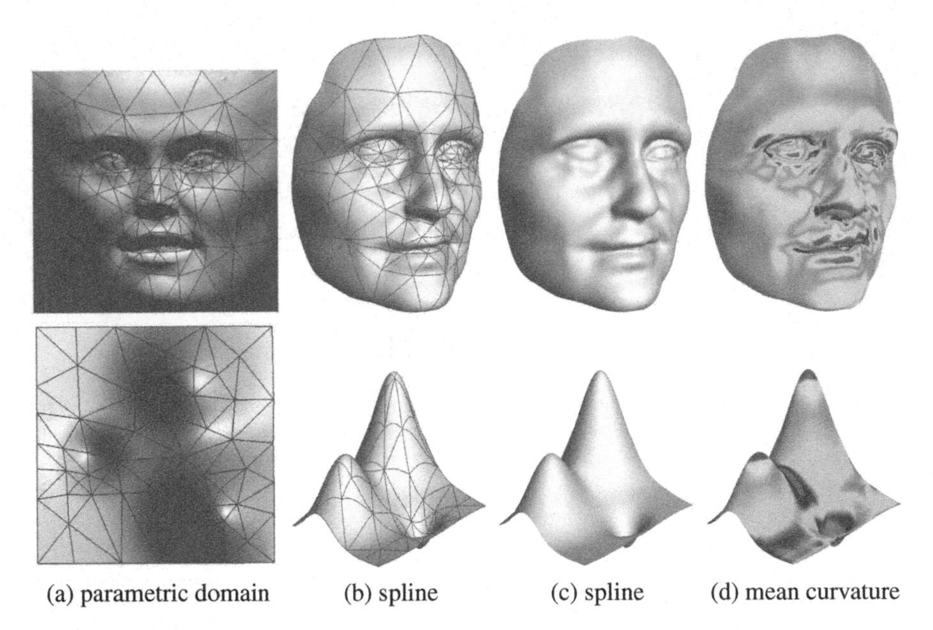

Fig. 8.2 Examples of triangular B-splines defined in $\mathbb{R}^2$. The red curves on the spline surfaces **b** correspond to the edges in the domain triangulation **a**. Note that there is no restriction on the triangulation of the parametric domain

8.2.3 Surface Reconstruction

The problem of reconstructing smooth surfaces from discrete scattered data arises in many fields of science and engineering. The problem can typically be stated as follows: given a set $P = \{\mathbf{p}_i\}_{i=1}^m$ of points $\mathbf{p}_i \in \mathbb{R}^3$, find a parametric surface $\mathbf{F} : \mathbb{R}^2 \to \mathbb{R}^3$ that approximates P. Triangular B-splines can be built over arbitrary triangulations and thus can be adapted to scattered data fitting. Then, we consider the following problem:

$$\min E(\mathbf{F}) = E_{dist}(\mathbf{F}) + \lambda \cdot E_{fair}(\mathbf{F}),$$

where

$$E_{dist}(\mathbf{F}) = \sum_{i=1}^{m} \|\mathbf{p}_i - \mathbf{F}(\mathbf{u}_i)\|^2,$$

$\mathbf{u}_i = (u_i, v_i)^T$ is the parameter of point $\mathbf{p}_i$ and $E_{fair}(\mathbf{F})$ is a fairness function with the smoothing factor $\lambda \geq 0$. The commonly-used fairness function is the thin-plate energy:

$$E_{fair}(\mathbf{F}) = \iint_{\Omega} \mathbf{F}_{uu}^2 + 2\mathbf{F}_{uv}^2 + \mathbf{F}_{vv}^2 du dv$$

where Ω is the parametric domain of $\mathbf{F}$.

The fitting result in a univariate free-knot spline is closely related to the knot placement. It is also true for triangular B-splines. To clarify our explanation, we call $\{\mathbf{t}_{i,0}|i \in \mathbb{N}\}$ the primary knots and $\{\mathbf{t}_{i,j}|i \in \mathbb{N}, 1 \leq j \leq n\}$ sub-knots. Since the sub-knots $\{\mathbf{t}_{i,n}|i \in \mathbb{N}\}$ do not contribute to the shape if the control net is continuous, we only consider $\{\mathbf{t}_{i,j}|i \in \mathbb{N}, 0 \leq j \leq n-1\}$ as free variables in the following. For a triangular B-spline, the primary knots are also the vertices of the domain triangulation.

Let us look at an example of the titanium heat dataset which is widely used to test univariate free-knot splines. This dataset is reconstructed by a degree 4 triangular B-spline with 32 domain triangles. If we construct the domain triangulation without any prior knowledge of the dataset, e.g., equal-distance primary knots (see Fig. 8.3c), and then, we fit the dataset with this domain triangulation and fixed knots, we obtain the reconstructed surface shown in Fig. 8.3d. Obviously, this approximation has many oscillations and poor approximation of the peak. However, if we place the primary knots along the "cross" ridges (see Fig. 8.3f) and use the same technique to fit the dataset, we achieve a much better result as shown in Fig. 8.3g and h. The oscillations disappear and the peak is correctly recovered. This example indicates that the fitting results are closely related to the knot configuration. More specifically, the primary knots should be placed along the features on the corresponding parametric plane.

We observe that the domain triangulation affects the fitting results significantly as shown in the titanium heat dataset (Fig. 8.3). Thus, the primary knots are more important than the sub-knots in determining the global shape. However, the sub-knots are useful for locally refining the shape, for example, the sharp features can only be obtained by degenerating some sub-knots (Fig. 8.4).

Gormaz [94] studied the intrinsic property of triangular B-spline. Although triangular B-spline has C^{n-1} continuity if there are no degenerate knots, the spline surfaces may not as smooth as one expects. The curvature of the images of the edges in the parametric domain is larger than vicinity. Figure 8.6 shows a degree 4 triangular B-spline, which is C^3-continuous everywhere. However, the surface is not smooth, because the high curvature concentrates along the edges of adjacent spline patches. This phenomenon is called "knot-line" of the triangular B-splines.

Given a degree n triangular B-spline surface $\mathbf{F}(\mathbf{u})$ defined on a planar triangulation T. Consider two triangles $\Delta(I) = [\mathbf{t}_0^I, \mathbf{t}_1^I, \mathbf{t}_2^I] \in T$ and $\Delta(J) = [\mathbf{t}_0^J, \mathbf{t}_1^J, \mathbf{t}_2^J] \in T$ such that $\Delta(I)$ and $\Delta(J)$ are adjacent. For example, suppose $\mathbf{t}_0^I = \mathbf{t}_0^J$ and $\mathbf{t}_1^I = \mathbf{t}_1^J$. Therefore, the sub-knots satisfy $\mathbf{t}_{0,i}^I = \mathbf{t}_{0,i}^J$ and $\mathbf{t}_{1,i}^I = \mathbf{t}_{1,i}^J$, $i = 1, \ldots, n$. Let $\mathbf{F}^I = \sum_{|\beta|=n} \mathbf{c}_{I,\beta} N(\mathbf{u}|V_\beta^I)$ be the polynomial on triangle I and similarly for $\mathbf{F}^J$. Let f^I and f^J be the polar forms of $\mathbf{F}^I$ and $\mathbf{F}^J$, respectively (see [237] for the details of polar form). Then, Gormaz proves the following result [94]:

Theorem 8.1 ([94]) *The triangular B-spline surface* $\mathbf{F}(\mathbf{u})$ *has no discontinuity of its nth derivative along the lines*

$$[\mathbf{t}_{0,\beta_0}^I, \mathbf{t}_{1,\beta_1}^I], \forall \beta, |\beta| = n, \beta_2 \leq r$$

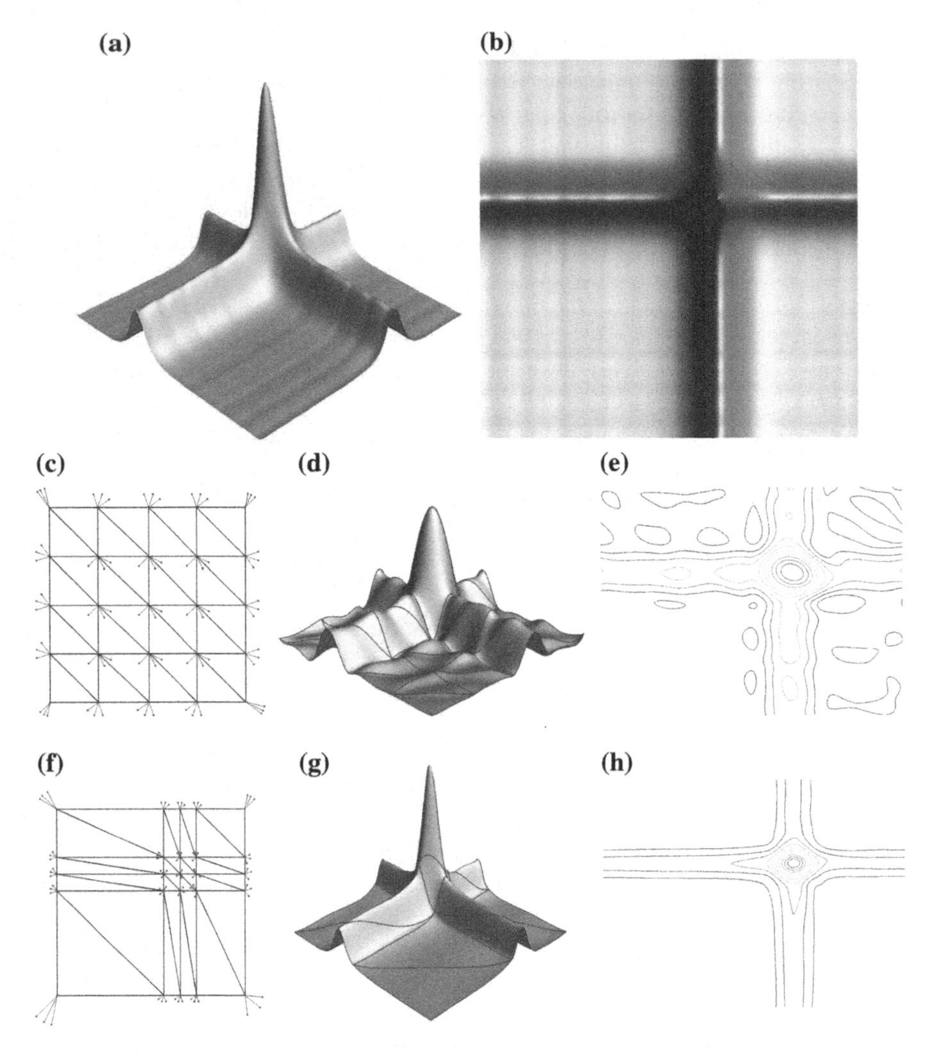

Fig. 8.3 The choice of primary knots has a large influence on the quality of the results of the surface fitting. **a**, **b** the 3D data and its parameterization; **c**, **f** domain triangulation and knot configuration; **d**, **g** reconstructed surfaces (the red lines indicate the domain triangulation); **e**, **h** contour plot. The root-mean-square error of **d** and **g** are 2.43×10^{-2} and 1.55×10^{-4}, respectively

if and only if

$$\mathbf{c}_{I,\beta} = f^J(\tilde{V}^I_\beta), \forall\beta, |\beta| = n, \beta_2 \leq r, \tag{8.3}$$

where $\tilde{V}^I_\beta = \{\mathbf{t}^I_{0,0}, \ldots, \mathbf{t}^I_{0,\beta_0-1}, \mathbf{t}^I_{0,0}, \ldots, \mathbf{t}^I_{0,\beta_0-1}, \mathbf{t}^I_{0,0}, \ldots, \mathbf{t}^I_{0,\beta_0-1}\}$.

Equation (8.3) defines the affine relations between the control points of $\mathbf{F}^I(\mathbf{u})$ and $\mathbf{F}^J(\mathbf{u})$. Given a $r \in [0, n)$, let the control points satisfying Eq. (8.3), then the discontinuity along certain knot lines disappear, and the curvature distribution along

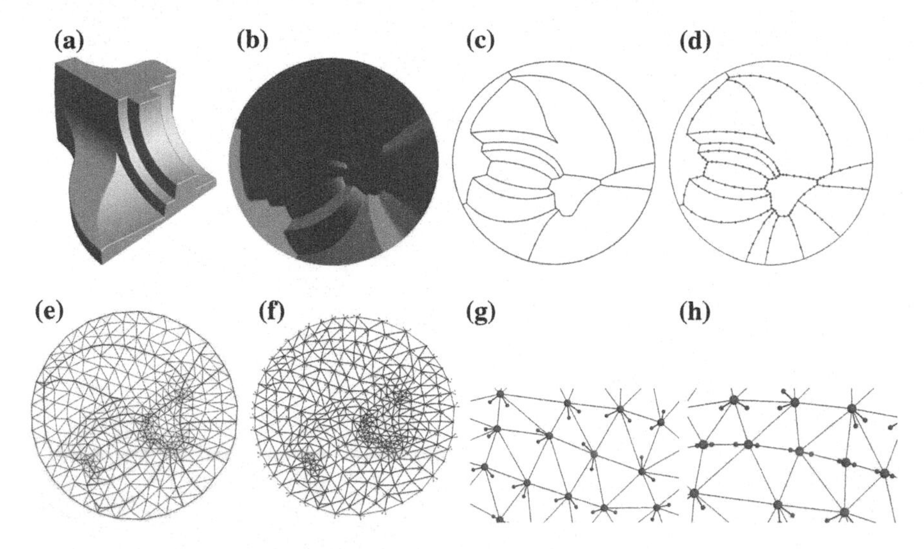

Fig. 8.4 Knot placement strategy. **a** Original model. **b** Conformal parameterization. **c** Sharp features on the parametric domain. **d** Place the primary knots along the sharp features. **e** Perform constrained Delaunay triangulation. **f** Assign the sub-knots. **g** Close-up view of the regular sub-knots. **h** Close-up view of the degenerated sub-knots. By placing the sub-knots along certain lines, we can model sharp features in the spline surface

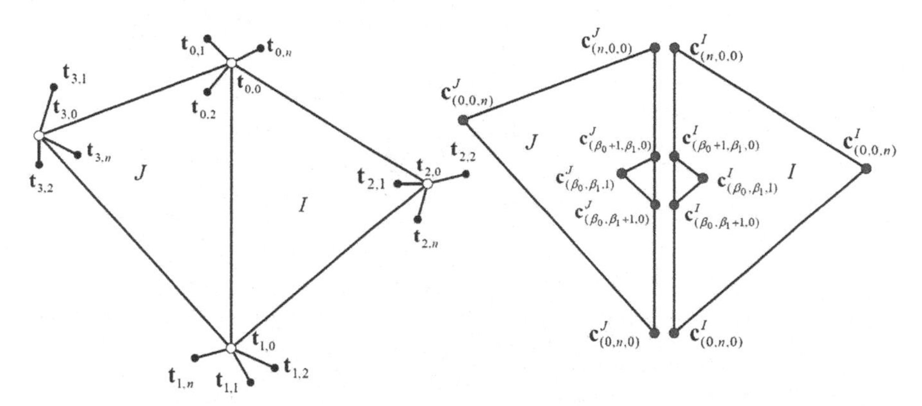

Fig. 8.5 Illustration of Eq. (8.3) for $r = 1$. *Left*, parametric domain; *Right*, control points

those lines improves. Figure 8.5 illustrates the case $r = 1$. For $\beta = (\beta_0, \beta_1, 1)$, Eq. (8.3) is written as

$$\begin{aligned} \mathbf{c}^I_{(\beta_0,\beta_1,1)} = & \frac{d(\mathbf{t}_{2,0}, \mathbf{t}_{1,\beta_1}, \mathbf{t}_{3,0})}{d(\mathbf{t}_{0,\beta_0}, \mathbf{t}_{1,\beta_1}, \mathbf{t}_{3,0})} \mathbf{c}^J_{(\beta_0+1,\beta_1,0)} \\ & + \frac{d(\mathbf{t}_{0,\beta_0}, \mathbf{t}_{2,0}, \mathbf{t}_{3,0})}{d(\mathbf{t}_{0,\beta_0}, \mathbf{t}_{1,\beta_1}, \mathbf{t}_{3,0})} \mathbf{c}^J_{(\beta_0,\beta_1+1,0)} + \frac{d(\mathbf{t}_{0,\beta_0}, \mathbf{t}_{1,\beta_1}, \mathbf{t}_{2,0})}{d(\mathbf{t}_{0,\beta_0}, \mathbf{t}_{1,\beta_1}, \mathbf{t}_{3,0})} \mathbf{c}^J_{(\beta_0,\beta_1,1)}, \end{aligned} \tag{8.4}$$

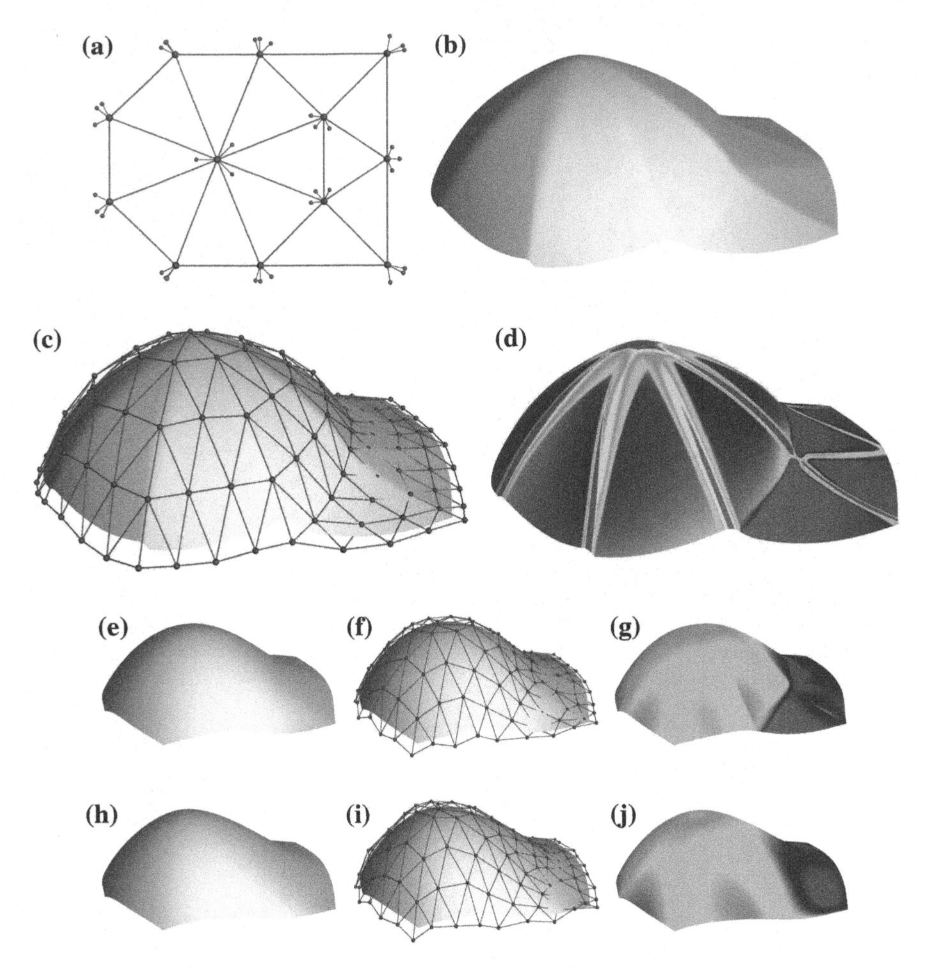

Fig. 8.6 Fairing a degree-4 triangular B-spline: **a** shows the parametric domain. (Due to the shared control points of the spline surface, only three sub-knots have contribution to the shape.) **b** and **c** show the spline surface and the control net respectively. **d** shows the mean curvature of the spline surface. Note that the curvature along the image of the edges on the domain triangulation is significantly larger than the vicinity. **e–g** show fairing the spline surface with $r = 1$. **h–j** show fairing the spline surface with $r = 2$. Although the control points are not changed too much, the surface quality improves significantly

where $d(\cdot, \cdot, \cdot)$ is the determinant function. It is easy to verify that Eq. (8.3) is just a linear combination of the control points for $0 \leq r \leq n-1$.

In the following, we consider the global fairing problem for triangular B-splines. Given a (planar, spherical or manifold) triangular B-spline surface $\mathbf{F}(\mathbf{u}) = \sum_I \sum_{|\beta|=n} \mathbf{c}_{I,\beta} N_{I,\beta}(\mathbf{u})$. We want to find a smooth surface $\tilde{\mathbf{F}}(\mathbf{u}) = \sum_I \sum_{|\beta|=n} \tilde{\mathbf{c}}_{I,\beta} N_{I,\beta}(\mathbf{u})$ such that $\tilde{\mathbf{F}}$ approximates the original surface $\mathbf{F}$ as much as possible. This leads to the following least square problem:

$$\min_{\tilde{\mathbf{c}}} \sum_{I} \sum_{|\beta|=n} \|\tilde{\mathbf{c}}_{I,\beta} - \mathbf{c}_{I,\beta}\|^2 \tag{8.5}$$
$$\text{subject to } \tilde{\mathbf{c}}_{I,\beta} = f^J(\tilde{V}^I_\beta), \forall I, \forall \beta, |\beta| = n, \beta_2 \le r$$

In the objective function, we minimize the squared distance between the control points of the original and the new spline surface, which implies that the minimal change of the shape. In the constraints, we use an integer r, $0 \le r \le n-1$, to control the smoothness of the spline surface. $r = 0$ implies that the control points along common edges of two adjacent triangles in the parametric triangulation are identical. The bigger value r, the smoother surface we will obtain. In our experiments, we can get visually pleasing surfaces with $r = 1$ or 2.

Since Eq. (8.3) corresponds to affine relations between the control points of $\mathbf{F}^I$ and $\mathbf{F}^J$, the constraints in Eq. (8.5) are just linear equations of the control points. Therefore, Eq. (8.5) is a linear constrained quadratic programming problem which has the following format:

$$\min_x \frac{1}{2} x^T Q x + c^T x + f \tag{8.6}$$
$$\text{subject to } Ax = b$$

Our problem is very special in that Q is an identity matrix. Therefore, it is very efficient to solve Eq. (8.5) using Lagrange multiplier approach.

Figure 8.6 illustrates the fairing algorithm to planar triangular B-spline. Compared to the shapes before and after fairing, the curvature concentration phenomena disappear, i.e., the knot-lines are totally eliminated.

8.3 Theoretical Foundation of Manifold Splines

In this section, we will systematically define manifold splines using our theoretical results on affine structure and triangular B-splines and show their existence is equivalent to that of affine structure. We first discuss the existence of affine structure for general manifolds, and then we compute the affine structure through the use of conformal structure for any manifold. For the consistency of our manifold spline theory, we shall utilize the parametric affine invariance and polynomial reproduction properties of general spline schemes (triangular B-splines in particular for this chapter).

8.3.1 Definitions and Concepts

A manifold spline is geometrically constructed by gluing spline patches in a coherent way, such that the patches cover the entire manifold. The knots and control points are also defined consistently across the patches and the surface evaluation is independent of the choice of chart. First of all, we define the local spline patch. After that, we define a global manifold spline which can be decomposed into a collection of local spline patches.

Definition 8.2 (*Spline Surface Patch*) A degree k spline surface patch is a triple $S = (U, C, F)$, where $U \subset \mathbb{R}^2$ is a planar simply-connected parametric domain. $F : U \to \mathbb{R}^3$ is a piecewise polynomial surface and C is the set of control points, $C := \{\mathbf{c}_\beta^I, X_\beta^I \in (\mathbb{R}^2)^{|\beta|}, |\beta| = k\}$. F can be evaluated from C by polar form.

Definition 8.3 (*Manifold Spline*) A manifold spline of degree k is a triple (M, C, F), where M is the domain manifold with an atlas $\mathcal{A} = \{(U_\alpha, \phi_\alpha)\}$. F is a map $F : M \to \mathbb{R}^3$ representing the entire spline surface. C is the control points set, each control point $\mathbf{c}_\beta^I$ is associated with a set of knots X_β^I which are defined on the domain manifold M directly,

$$C := \{\mathbf{c}_\beta^I, X_\beta^I \in M^{|\beta|}, |\beta| = k\}$$

such that

1. For each chart (U_α, ϕ_α), the restriction of F on U_α is denoted as $F_\alpha = F \circ \phi_\alpha^{-1}$, a subset of control points C_α can be selected from C, such that $(\phi_\alpha(U_\alpha), C_\alpha, F_\alpha)$ form a spline patch of degree k, where $C_\alpha := \{\mathbf{c}_\beta^I, \phi_\alpha(X_\beta^I) \in (\mathbb{R}^2)^{|\beta|}, |\beta| = k\}$.
2. The evaluation of F is independent of the choice of the local chart, namely, if U_α intersects U_β, then $F_\alpha = F_\beta \circ \phi_{\alpha\beta}$, where $\phi_{\alpha\beta}$ is the chart transition function.

The technical essence of the above definition is to replace a planar domain by the atlas of the domain manifold, and the surface evaluation of the spline patches is independent of the choice of charts (see Fig. 8.7). After the formal definition, we use one simple example to further illustrate the concept of our manifold splines (see Fig. 8.8).

One Dimensional Example. Here the domain manifold is a unit circle S^1. There are n distinct points $t_0, t_1, \cdots, t_{n-1}$ distributed on the circle in a counterclockwise way. All the summation and subtraction on indices are modular n. The intervals between points are arbitrary. The control net is a planar n-gon, the control points are denoted as $\mathbf{c}_0, \mathbf{c}_1, \cdots, \mathbf{c}_{n-1}$ also in a counterclockwise way, and the knots for c_i are $t_{i-2}, t_{i-1}, t_i, t_{i+1}, t_{i+2}$.

The affine atlas of S^1 is constructed in the following way: the arc segment $U_i = (t_{i-2} - \varepsilon, t_{i-1}, t_i, t_{i+1}, t_{i+2} + \varepsilon)$, $\varepsilon \in \mathbb{R}^+$ is mapped to an interval in $\mathbb{R}^1$ by $\phi_i : S^1 \to \mathbb{R}^1$, such that

$$\phi_i(t_i) = a, \phi_i(t) = a + b \int_{t_i}^{t} ds, a \in \mathcal{R}, b \in \mathbb{R}^+. \tag{8.7}$$

where a, b are arbitrarily chosen. The union of all local charts (U_i, ϕ_i) form an affine atlas $\mathcal{A} = \{(U_i, \phi_i)\}$. Note that by choosing different a, b, there might be infinite local charts in $\mathcal{A}$.

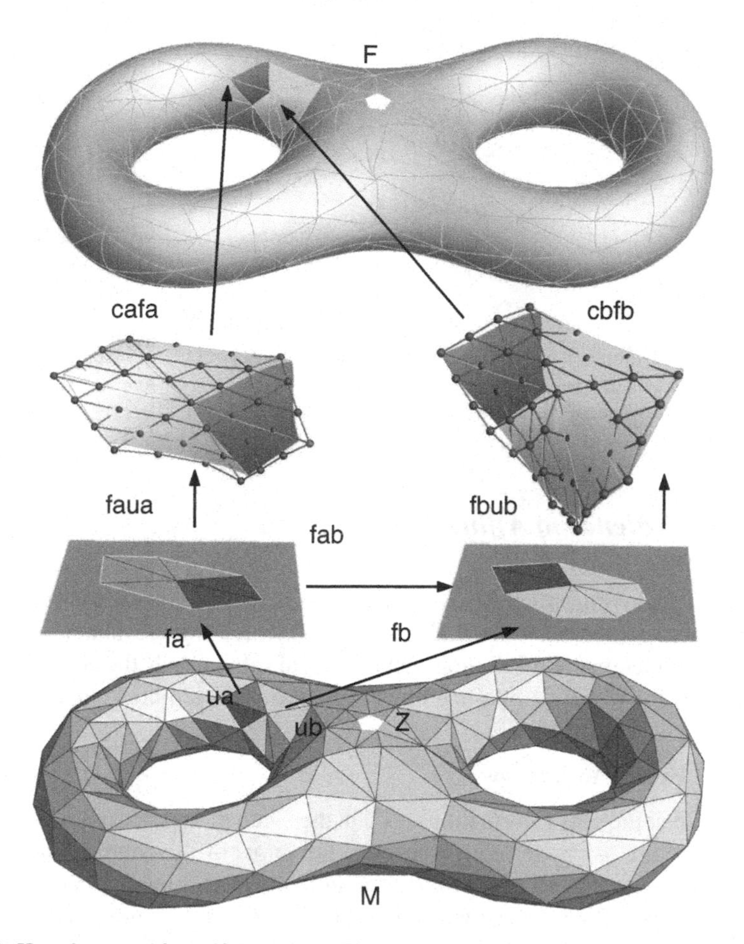

Fig. 8.7 Key elements of manifold splines: The parametric domain M is a triangular mesh with arbitrary topology as shown at the bottom. The polynomial spline surface F is shown at the top. Two overlapping spline patches $(\phi_\alpha(U_\alpha), C_\alpha, F_\alpha)$ and $(\phi_\beta(U_\beta), C_\beta, F_\beta)$ are magnified and highlighted in the middle. On each parameter chart (U_α, ϕ_α),(U_β, ϕ_β), the surface is a triangular B-spline surface. For the overlapping part, its two planar domains differ only by an affine transformation $\phi_{\alpha\beta}$. The zero point neighbor is Z. Image is taken from [103], included here by permission from Elsevier, Inc

The control net corresponding to local chart (U_i, ϕ_i) is the line segments $C_i = \{\mathbf{c}_{i-2}, \mathbf{c}_{i-1}, \mathbf{c}_i, \mathbf{c}_{i+1}, \mathbf{c}_{i+2}\}$. The piecewise polynomial curve is formed by n pieces of polynomials, the i-th piece $F_i : [t_i, t_{i+1}] \to \mathbb{R}^2$ is evaluated on (U_i, ϕ_i) with control polygon C_i using cubic B-spline.

Then we define the cubic B-spline curve on the unit circle consistently. It is C^2 continuous everywhere. The B-spline patches are $\{\phi_i(U_i), C_i, F_i\}$.

The above example can be trivially extended to construct a two-dimensional surface in a similar way. The key step is to find an affine atlas for the domain manifold. The next section will discuss the existence of such an atlas for general 2-manifolds in detail.

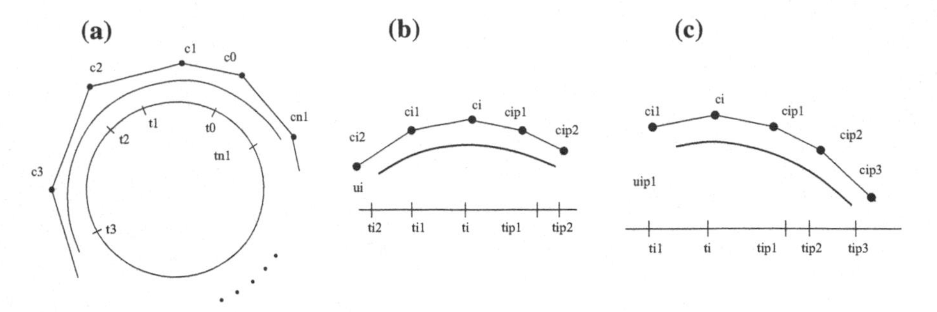

Fig. 8.8 Manifold spline on S^1: **a** The domain manifold is a unit circle S^1 with n distinct knots $t_0, \cdots, t_{n-1}$; **b** The i-th spline patch $U_i = (t_{i-2} - \varepsilon, \cdots, t_{i+2} + \varepsilon)$; **c** The $i+1$-th spline patch $U_{i+1} = (t_{i-1} - \varepsilon, \cdots, t_{i+3} + \varepsilon)$. Image is taken from [103], included here by permission from Elsevier, Inc

8.3.2 *Equivalence to Affine Atlas*

The central issue of constructing manifold splines is that the atlas must satisfy some special properties in order to meet all the requirements for the evaluation independence of chart selection. We will show that for a local spline patch, the only admissible parameterizations differ by an affine transformation. This requires that all the chart transition functions are affine.

Admissible Parameterizations

It is obvious that the only informations used in the evaluation process are barycentric coordinates of the parameter with respect to the knots of the control points. If we change the parameter by an affine transformation, the evaluation is invariant and the final shape of the spline surface will not be modified. On the other hand, an affine transformation is the only parametric transformation that will keep the consistency between the spline surface and its parameters. In other words, affine transformations are the only admissible parametric transformations for a spline patch.

Lemma 8.4 *Assume there are two spline surface patches of C^k continuity, $k > 0$,*

$$S = (U, C, F) \text{ and } \tilde{S} = (\tilde{U}, \tilde{C}, \tilde{F}).$$

The parametric transformation

$$\phi : U \to \tilde{U}$$

is invertible. Suppose $S, \tilde{S}$ share the same knot configuration, namely, the triangulation $\tilde{\mathcal{T}}$ is induced from $\mathcal{T}$ by ϕ, and the knots $\tilde{t}^I_{i,j}$ are induced from $t^I_{i,j}$ by ϕ

$$\tilde{t}^I_{i,j} = \phi(t^I_{i,j}), \tag{8.8}$$

the control points with corresponding knots coincide $\mathbf{c}^I_\beta = \tilde{\mathbf{c}}^I_\beta$, then

1. *if ϕ is affine, then $F = \tilde{F} \circ \phi$ holds for arbitrary control nets.*
2. *if $F = \tilde{F} \circ \phi$ holds for arbitrary control nets, then ϕ is affine.*

In other words, the following diagram commutes for arbitrary control nets

$$\begin{array}{ccc} U \subset \mathbb{R}^2 & \xrightarrow{\phi} & \tilde{U} \subset \mathbb{R}^2 \\ \downarrow F & & \downarrow \tilde{F} \\ F(U) \subset \mathbb{R}^3 & \xrightarrow[id]{} & \tilde{F}(\tilde{U}) \subset \mathbb{R}^3 \end{array} \tag{8.9}$$

if and only if ϕ is affine.

Proof The sufficient condition part is obvious, because the evaluation of the splines only involves barycentric coordinates. Affine transformations preserve the barycentric coordinates; therefore the diagram is commutative.

The proof for the necessary condition requires the completeness of the spline scheme. We set all control points of C to be zero except the one corresponding to knots X^I_β. Correspondingly, we set all control points of $\tilde{C}$ to be zero except one corresponding to knots $\tilde{X}^I_\beta$. Then we get the basis functions $F(\mathbf{u}) = N^I_\beta(\mathbf{u})$, $\tilde{F} = \tilde{N}^I_\beta(\tilde{\mathbf{u}})$, by $F = \tilde{F} \circ \phi$, we get

$$N^I_\beta(\mathbf{u}) = \tilde{N}^I_\beta(\tilde{\mathbf{u}}).$$

Therefore, all basis functions of S equal the corresponding basis functions of $\tilde{S}$. Suppose $\mathbf{u} = (u_1, u_2)$, then u_1 is a polynomial of (u_1, u_2). By completeness of the spline scheme, u_1 can be represented as the linear combination of $N^I_\beta(\mathbf{u})$, therefore it can be represented as the linear combination of $\tilde{N}^I_\beta(\tilde{\mathbf{u}})$. As a result, u_1 and u_2 can be represented as piecewise polynomials of $\tilde{\mathbf{u}}$ of C^k continuity. Because S and $\tilde{S}$ are symmetric, $\tilde{\mathbf{u}}$ are also piecewise polynomials of $\mathbf{u}$ of C^k continuity. Therefore, $\mathbf{u}$ and $\tilde{\mathbf{u}}$ can linearly represent each other piecewisely with C^k continuity. So, because the parameter transition ϕ is piecewise linear and C^k continuous, ϕ must be a global linear map over all pieces. In other words, ϕ is affine.

Theorem 8.5 *The sufficient and necessary condition for a manifold M to admit manifold spline is that M is an affine manifold.*

Proof Consider two intersecting local charts (U_α, ϕ_α) and (U_β, ϕ_β), where the manifold spline F restricted on them are F_α and F_β, respectively. We select a subset of control points C whose knots are contained in $U_\alpha \bigcap U_\beta$. The spline patches $(\phi_\alpha(U_\alpha \bigcap U_\beta), C, F_\alpha)$ and $(\phi_\beta(U_\alpha \bigcap U_\beta), C, F_\beta)$ satisfy the condition in Lemma 1, therefore, the chart transition function $\phi_{\alpha\beta}$ must be affine.

This theorem indicates that the existence of manifold splines depends on the existence of affine atlas. If the domain manifold M is an affine manifold, we can easily generalize the planar triangular B-spline surfaces to be defined on M directly. The major differences are as follows:

1. The knots associated with each vertex $\mathbf{t}_i^I$ are defined on the manifold directly.
2. The knots associated with each pole X_β^I are defined on M directly.
3. The barycentric coordinates $\lambda_{\beta,i}^I$ used in the evaluation process are defined on any chart of $\mathcal{A}$. Because $\mathcal{A}$ is affine, the value of the barycentric coordinates is independent of the choice of the chart.

8.3.3 *Existence*

From the previous discussion, it is clear that in order to define a manifold spline, an affine atlas of the domain manifold must be found first. According to characteristic class theory [195], general closed 2-manifolds do not have an affine atlas. On the other hand, all open surfaces admit an affine atlas. In order to define manifold splines, the domain manifold has to be modified to admit an atlas by removing a finite number of points. This offers a theoretical evidence to the existence of singular points due to the topological obstruction.

A classical result from characteristic class theory claims that the only closed surface admitting affine atlas is of genus one.

Theorem 8.6 (Benzécri) *Let S be a closed two dimensional affine manifold, then* $\chi(S) = 0$.

This result is first proven by Benzécri [23]. Shortly after his proof, J. Milnor presented a much more broader result using vector bundle theories [196]. In this framework, the topological obstruction of a global affine atlas is the Euler class. In fact, by removing one point from the closed domain manifold, we can convert it to an affine manifold.

Theorem 8.7 (Open Surfaces are Affine Manifold) *Let M be an orientable open 2-manifold, then M is affine manifold.*

Proof Figure 8.9 illustrates the proof by constructing an affine atlas for the open surface M in (a). One boundary may be a closed curve or a single point as shown in (a) by a dark spot. We deform (a) continuously to generate (b) by gradually enlarging the hole. (b) is homeomorphic to the ribbon figure in (c), which is immersed in $\mathbb{R}^2$. Then we cut each annulus of (c) to get a fundamental domain as shown in (d).

The colored disks U_α are open sets of M, another open set U can be defined to cover $M \setminus \bigcup U_\alpha$. (d) shows the way U and U_α's are mapped to $\mathbb{R}^2$. It is obvious that all chart transition functions are combinations of translations and rotations.

For surfaces with multiple boundaries, we can fill all of the boundaries with disks except one, and the proof is similar.

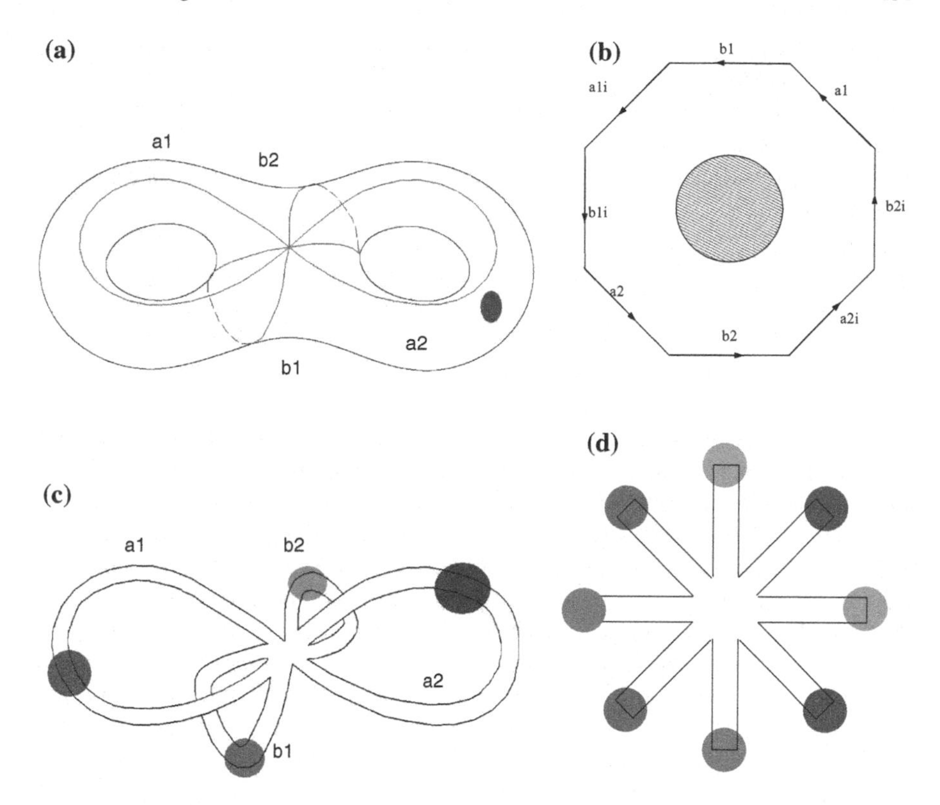

Fig. 8.9 Open surfaces are affine manifolds: An genus 2 surface M with one boundary in **a** is isomorphic to an octagon with a hole in **b**, then the octagon is immersed in $\mathbb{R}^2$ as the ribbon figure in **c**. The colored disks indicate the open sets U_α of M. There is another open set U, which covers $M \setminus \bigcup U_\alpha$. U and U_α are mapped to $\mathbb{R}^2$ as shown in **d**. All transitions are rotation and translation. This illustrates the affine atlas for the open surface of M. Image is taken from [103], included here by permission from Elsevier, Inc

8.4 Constructing Affine Atlas

As mentioned above, the key element of manifold splines is the affine structure of domain manifold. This section presents three algorithms, namely holomorphic 1-form, discrete Ricci flow, and polycube map, to compute the affine structure of arbitrary oriented 2-manifold. From the chart-relation's point of view, these methods differ in three aspects, the number and the locations of singularities, the angle/area distortion, and the type of transition functions. Each method has its own merits and users may choose one or another depending on their specific application needs. The holomorphic 1-form method induces the affine structure with the fixed number of extraordinary points, i.e., $|2g-2+b|$, where g is the genus of the domain manifold, b is the number of boundaries. For genus-zero surfaces, we usually intentionally cut two boundaries on the model. Note that, although we do not modify the geometry of the original model, the number of extraordinary points drop to zero. The affine atlas

Table 8.1 Comparison of the methods to compute affine structures. g, genus of the domain manifold M; b, number of boundaries of M

	Holomorphic 1-form	Discrete Ricci flow	Polycube map
# of singularities	$\|2g - 2 + b\|$	User-specified	Many
Singularity location	Difficult to control	User-specified	Easy to control
Area distortion	Large	Large	Low
Angle distortion	No	No	Low
Transition Function	Translation	Translation for $\chi = 0$; Translation and rotation for $\chi \neq 0$	Translation And $k90°$-rotation, $k = 1, 2, 3$

of the bird model shown in Fig. 8.19 is constructed using this technique. Although conformal structure preserves the angles very well, they inevitably introduce large area distortion if the model has some long, extruding parts, e.g., the tail and feet of the bird model. These large area distortions usually make the spline construction very difficult, since we need to introduce more control points in such areas. Polycube maps are ideal to reduce both the area and angle distortion in the affine atlas, as shown in the Buddha and 3-hole torus models in Fig. 8.19. Thus, it facilitates the spline construction procedures. However, the side-effect to reduce the area distortion is to introduce more extraordinary points simultaneously. Usually, the lesser the area distortion, the more number of extraordinary points. Ricci flow method can be used to balance the aforementioned two requirements. The user can specify the number and the locations of extraordinary points. Both conformal structure and polycube map induces affine atlas with transition functions of simple translations. Ricci flow method generates affine atlas whose transition functions are combination of translations and rotations (Table 8.1). The following table summaries the salient differences among these methods.

8.4.1 Computing Affine Structure Using Holomorphic 1-form

Conformal structure is an intrinsic structure of the surface. A conformal atlas is an atlas such that all transition functions are holomorphic (analytic). All oriented surfaces have conformal structure and are called Riemann surfaces [142]. Conformal structure is closely related to affine structure. In particular, an affine atlas can be easily computed by using special differential complex forms defined on the conformal atlas [105].

Theorem 8.8 *Given a closed genus $g(\geq 1)$ surface M, and a holomorphic 1-form ω, the zero set of ω is Z, then the size of Z is no more than $2g - 2$ and there exists an affine atlas on $M \setminus Z$ deduced by ω.*

Given a holomorphic 1-form ω on a surface M, assume its zero point set is Z. Then, an affine atlas $\mathcal{A}$ for $M \setminus Z$ can be constructed straightforwardly by integrating the holomorphic 1-form. The algorithm is as follows:

1. Compute the holomorphic 1-form ω using Gu-Yau's method [105].
2. Locate the zero points of ω by checking the winding number. Denote the zero points by Z.
3. Remove zero points Z and the faces attaching to them.
4. Construct an open covering for $M \setminus Z$. For each vertex p, take the union of all faces within its 1-ring neighbor to form the chart U_p.
5. For each chart U_p, set $\phi_p(p) = 0$ and $\phi_p(q_i) = \omega[p, q_i]$ for any vertex $q_i \in U_p$, where $\omega[p, q_i]$ is the holomorphic 1-form on edge connecting p and q_i. The transition function between two overlapping charts is just a translation.

8.4.2 Computing Affine Structure Using Discrete Ricci Flow

Ricci flow was introduced to differential geometry field by Hamilton in [110]. In this research, the key motivation for using Ricci flow is its computational power to compute the affine atlas of a mesh with any desired number of singularities.

Ricci Flow On Surfaces. The basic idea of Ricci flow is rather simple. We can deform the surface driven by its curvature to the desired shape. Suppose S is a closed surface with Riemannian metric g, and $u : S \to \mathbb{R}$ is a real-valued function defined on S. Then, $\bar{g} = e^{2u} g$ is another metric which is conformal to g. The Gaussian curvature and geodesic curvature under $\bar{g}$ are

$$\bar{K} = e^{-2u}(-\Delta u + K) \quad (8.10)$$

$$\bar{k}_g = e^{-u}(\partial_n u + k_g), \quad (8.11)$$

where K, k_g are the Gaussian and geodesic curvature under metric g, n is the tangent vector orthogonal to the boundary.

Riemann uniformization theorem [142] states that for any surface S, there exists a unique conformal metric, such that it induces constant Gaussian curvature K and zero geodesic curvature,

$$\bar{K} = \begin{cases} +1, & \chi(S) > 0 \\ 0, & \chi(S) = 0 \\ -1, & \chi(S) < 0 \end{cases}$$

and geodesic curvature $\bar{k}_g$ equals to zero. Such kind of metric is called the Riemannian uniformization metric of S.

Given the target curvatures $\bar{K}$ and $\bar{k}_g$, in order to find the conformal factor e^{2u}, we need to solve the above partial differential equations. However, these equations are highly non-linear and a conventional finite element method cannot be applied directly. Ricci flow is a powerful tool to find the desired conformal metric for prescribed curvatures. The Ricci flow is defined as

$$\frac{du(t)}{dt} = \bar{K} - K(t), \quad (8.12)$$

where the area preserving constraint is explicitly formulated as

$$\int_S dA(t) = \int_S e^{2u} dA, \tag{8.13}$$

where $K(t)$ is the Gaussian curvature induced by the metric $e^{2u(t)}g$.

Discrete Ricci Flow. In practice, a surface is represented as a triangle mesh, which is a simplicial complex embedded in $\mathbb{R}^3$. The vertex, edge, and face sets are denoted by V, E and F, respectively. We denote a vertex as v_i, an edge connecting v_i and v_j as e_{ij}, a face with vertices v_i, v_j and v_k as f_{ijk}.

The Riemannian metric is approximated by the edge lengths, $l : E \to \mathbb{R}$. On each triangle f_{ijk}, the edge lengths l_{ij}, l_{jk} and l_{ki} satisfy the triangle inequality.

Suppose v_i is an interior vertex with surrounding faces $F = \{f_{ijk}\}$, where the corner angle of f_{ijk} at v_i is θ_i^{jk}. Then the discrete Gaussian curvature, $K : V \to \mathbb{R}$, is defined as

$$K_i = \begin{cases} 2\pi - \sum_{f_{ijk} \in F} \theta_i^{jk} & v_i \in M \setminus \partial M, \\ \pi - \sum_{f_{ijk} \in F} \theta_i^{jk} & v_i \in \partial M, \end{cases}$$

Similar to the smooth case, discrete Gaussian curvatures satisfy the Gauss–Bonnet formula,

$$\sum_{v_i \in M \setminus \partial M} K_i + \sum_{v_j \in \partial M} K_j = 2\pi\chi(M), \tag{8.14}$$

where $\chi(M)$ is the Euler number of mesh M.

Ricci flow conformally deforms the surface. Conformal mapping has an important property: it transforms infinitesimal circles on the surface to infinitesimal circles, while preserving the intersection angles among the circles. Based on this property, Thurston introduced the circle packing metric in [272]: a circle with the radius γ_i is associated with each vertex v_i. For an each edge e_{ij}, two circles intersect at the angle Φ_{ij}, called edge weight. The edge length of e_{ij} is determined by γ_i, γ_j and Φ_{ij},

$$l_{ij} = \sqrt{r_i^2 + r_j^2 + 2r_i r_j \cos \Phi_{ij}}. \tag{8.15}$$

It can be shown that for any face f_{ijk} with vertex radii $\{\gamma_i, \gamma_j, \gamma_k\}$ and edge weights $\{\Phi_{ij}, \Phi_{jk}, \Phi_{ki}\}$, if edge weights are acute angles, then the edge lengths $\{l_{ij}, l_{jk}, l_{ki}\}$ satisfy the triangle inequality, i.e., $l_{ij} + l_{jk} > l_{ki}$.

We use $\Gamma : V \to \mathbb{R}^+$ to denote the vertex radii, $\Phi : E \to [0, \frac{\pi}{2}]$ the edge weights, then a circle packing metric is represented as (M, Γ, Φ). Two circle packing metrics (M, Γ_1, Φ_1) and (M, Γ_2, Φ_2) are conformal to each other, if $\Phi_1 \equiv \Phi_2$. Namely, a discrete conformal mapping will change the vertex radii only and preserve the intersection angles.

For a mesh with circle packing metric (Γ, Φ), given the target curvature $\bar{K}$ for each vertex, we want to compute a conformal circle packing metric $(\bar{\Gamma}, \Phi)$ which induces the prescribed curvature $\bar{K}$. Discrete Ricci flow is able to solve the vertex radii Γ. We use $e^u\Gamma$ to denote the conformal metric with vertex radius $e^{u_i}\gamma_i$ at vertex i. Let

$$u_i = \begin{cases} \ln \tan \frac{\gamma_i}{2}, & \text{Spherical geometry} \\ \ln \gamma_i, & \text{Euclidean geometry} \\ \ln \tanh \frac{\gamma_i}{2}, & \text{Hyperbolic geometry} \end{cases} \tag{8.16}$$

Then discrete Ricci flow is formulated as

$$\frac{du_i(t)}{dt} = \bar{K}_i - K_i(t),$$

where u_i satisfies the constraint $\sum_i u_i = 0$ (equivalent to the area preserving constraint). Note that by choosing $\bar{K} = +1, 0, -1$ and u_i above, we can compute the Riemannian uniformization metric using discrete Ricci flow. It has been proven in [49] that the convergence rate of Ricci flow is exponential, for any point p on the surface,

$$|K(t, p) - K(\infty, p)| < c_1 e^{-c_2 t},$$

where c_1, c_2 are two constants determined by the surface itself.

After presenting the necessary theory, next we shall explain our algorithmic details to construct affine atlas for two different cases: (1) surfaces with zero Euler number and (2) surfaces with non-zero Euler number.

Affine Structures for Surfaces with Zero Euler Number

According to the Gauss–Bonnet theorem, the total Gaussian curvature of a surface with zero Euler number is zero. Thus, we can find a special metric such that the curvatures for the interior and boundary vertices are zero.

For a genus one closed surface, we can periodically flatten the mesh onto the plane using such a metric. The algorithm is as follows:

1. Set the target curvature $\bar{K}$ to be zero everywhere.
2. Compute the conformal metric using discrete Ricci flow.
3. Compute the cut graph of the domain manifold. The fundamental domain is denoted by $\bar{S}$.
4. Suppose γ is an edge on the cut graph. Then it corresponds to two boundary segments $\bar{\gamma}^+$ and $\bar{\gamma}^-$ on the fundamental domain. Compute the unique translation $g : \mathbb{R}^2 \to \mathbb{R}^2$ to align $\bar{\gamma}^+$ with $\bar{\gamma}^-$, i.e., $g(\bar{\gamma}^+) = \bar{\gamma}^-$. Find all such translations for each arc on the cut graph and denote them as $G = \{g_1, g_2, \dots, g_n\}$.
5. Shift the copies of the embedded fundamental domains using the translations generated by G and glue them together along the aligned boundary segments. This process induces a tessellation of the plane.

For a genus-one open surface, we can set the target curvature as zero for interior vertices as well as boundary vertices. Again, we use the Ricci flow to compute the desired metric and embed the fundamental domain to the Euclidean plane. As a result, the surface can be embedded periodically on a stripe with parallel straight boundaries.

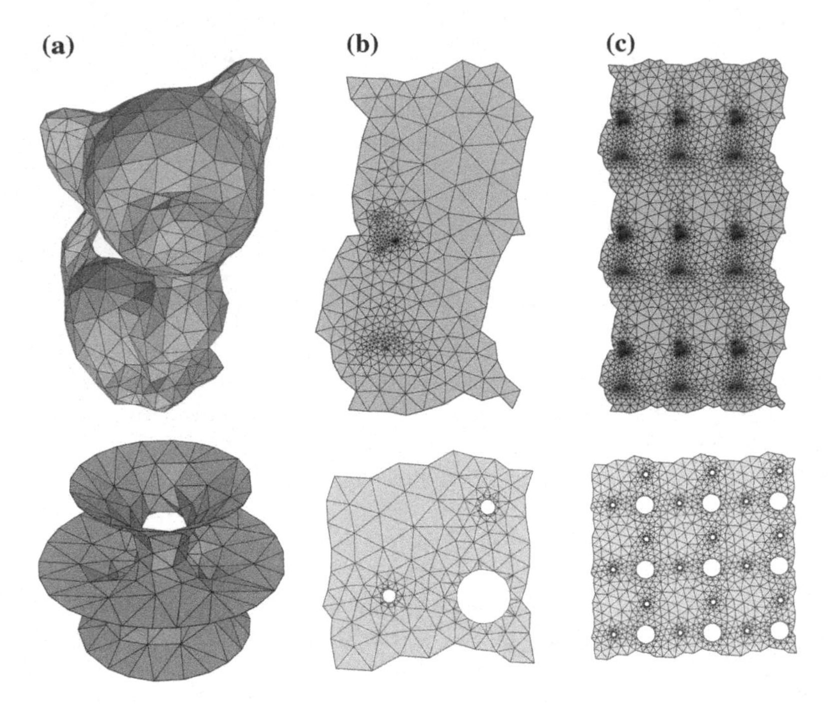

Fig. 8.10 Computing the affine structures for genus-one surfaces using discrete Ricci flow. The cat model is a closed surface. The hypersheet model has three boundaries. The homology basis is denoted by $\{a_1, b_1\}$. Slicing the model along a_1 and b_1, we get a topological disk $\bar{M}$. Then we embed $\bar{M}$ to the Euclidean plane (shown in **b**). The affine structures of both models have one central chart, which corresponds to $\bar{M}$, and two edge charts, which correspond to the homology basis $\{a_1, b_1\}$. The universal covering space is shown in **c**. The transition function between the central chart and edge charts is just a translation. Images of the second row from [102]

Figure 8.10 shows the affine structures of genus-one surfaces. The affine structure contains one central chart, and two edge charts which cover the homology basis $\{a_1, b_1\}$. The transition functions are just translations.

Affine Structures for Surfaces with Non-zero Euler Number

If a surface S has non-zero Euler number, according to the Gauss–Bonnet Theorem, its parameterization must contain some singularities at either interior or boundary vertices, where the curvatures are not zero. The total Gaussian curvature at singularities equals to $2\pi\chi(S)$.

For a high genus mesh, ideally, we can choose only one singular point and concentrate all curvatures on it. If the mesh is open, we can assign the target curvatures for all the interior vertices to be zero and assign the target curvatures for boundary vertices such that the total boundary curvature equals to $2\pi\chi$. By this way, all the curvature will be pushed to the boundary. Ricci flow only changes the vertex radii, therefore, the result metric is conformal to the original one, no angle distortion will be introduced. But the area distortion is unavoidable. The uniformity of the parameterization varies drastically depending on the location of singularities.

The algorithm to compute affine atlas for surfaces with non-zero Euler number is as follows:

1. We can select the extraordinary points $Z = \{v_1, v_2, \ldots, v_k\}$, $k \geq 0$ anywhere on the domain manifold.
2. Assign the target curvature of extraordinary points such that
$$\sum_{i=1}^{k} \bar{K}(v_i) = 2\pi\chi,$$
where χ is the Euler number of M. Set the target curvature of other vertices are zero.
3. Solve discrete Ricci flow on M to get the flat metric.
4. Compute the canonical homology basis $\{a_i, b_i\}_{i=1}^{g}$ of M.
5. Slice M along the homology basis to form a topological disk $\bar{M}$. The boundary of $\bar{M}$ has canonical form $\partial\bar{M} = a_1 b_1 a_1^{-1} b_1^{-1} a_2 b_2 a_2^{-1} b_2^{-1} \ldots a_g b_g a_g^{-1} b_g^{-1}$.
6. Compute the planar embedding of $\bar{M} \setminus Z$ using the flat metric in step 3.
7. Construct the central chart which covers $\bar{M}$ and $2g$ edge charts which cover $\{a_1, b_1, \ldots, a_g, b_g\}$. The transition functions between the central chart and edge charts are translations and rotations.

Figure 8.11 shows an example of genus-two surface. The affine structure contains a central chart and 4 edge charts. The transition functions are translations and rotations.

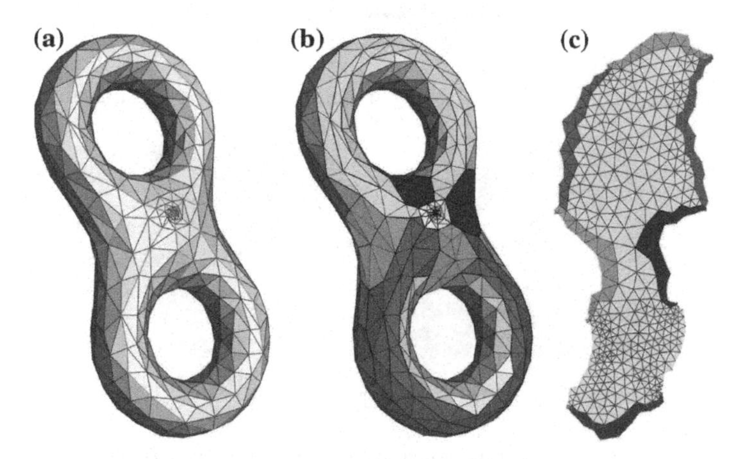

Fig. 8.11 Affine atlas of the two-hole torus model acquired by using Ricci flow. **a** shows the extraordinary point and the cut graph, which is a set of canonical homology basis curves passing through the extraordinary point. **b** shows the edge charts covering the cut graph. We compute the circle packing metric using discrete Ricci flow, which further induces the flat metric and the planar embedding of the central chart as shown in **c**. The overlapping relation between the central chart and edge charts are also colored in **c**. Image from [102]

8.4.3 Computing Affine Structure Using Polycube Maps

The concept of polycube maps is pioneered by Tarini et al. in [265]. The goal is to minimize both the angular distortion and area distortion of the map between polycube and the original model.

Polycube, although geometrically simple, induces the affine structure. Because of its regularity, the polycube is now only covered by charts which are uniquely associated with faces and edges belonging to one of the cubes. As a result of the polycube map, all the corner points are now becoming singular. As shown in Fig. 8.12, each face and edge on the polycube are associated with its own local chart. Each face

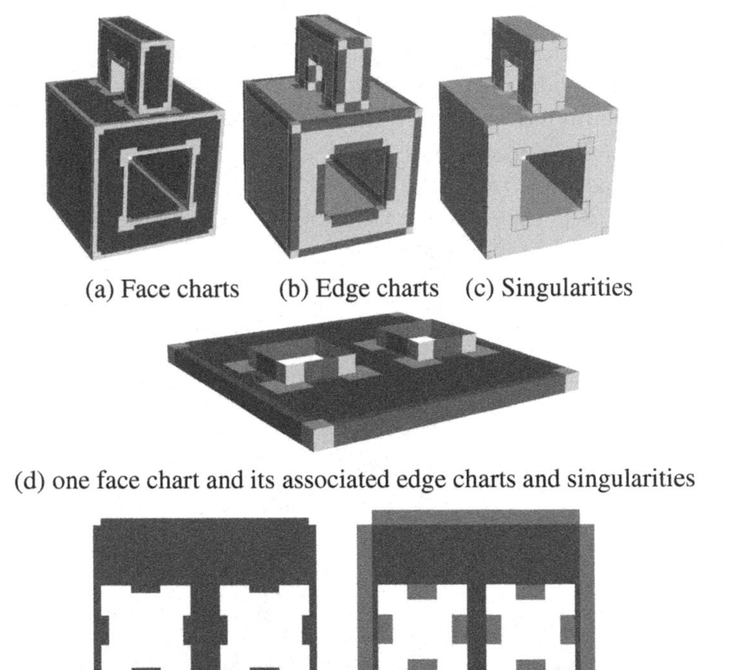

Fig. 8.12 Polycube map induces affine structure. The polycube is only covered by face and edge charts. Each face chart (drawn in blue) covers only interior points of the corresponding face and leaves off all the edges of the face. Each edge chart (drawn in red) covers interior points of the edges but leaves off corner vertices. The corners (drawn in yellow) are singularities which are NOT covered by any charts. We highlight one face chart and its associated edge charts and singularities in **d** and **e**. By flattening the edge charts, we get the planar domain shown in **f**. Note that the transition functions between overlapped edge and face charts are just translations. Therefore, by removing all the corners, the open polycube $P\backslash C$ has the affine structure. Image from [285]

chart covers only interior points of corresponding face and leaves off edges of the face. Each edge chart covers interior points of the edge but leaves off corner vertices. Note that there is no vertex chart for the corner vertex, i.e., the corners are singular points. Note that the transition functions between overlapped edge and face charts are simply translations. Therefore, polycube map naturally induces the affine structure.

Construction of the polycube map is equivalent to a bijective map between the 3D model and the polycube. The key difference between the techniques employed in [265] and ours in this paper is that Tarini et al.'s technique is trying to find the one-to-one mapping of the 3D shape and polycube extrinsically, which typically requires the projection of points from one shape to the other. As a result, their method is usually quite difficult to handle cases where the two shapes differ too much and the point projection does not establish the one-to-one correspondence. In contrast, our method aims to compute such a mapping in an intrinsic way. We first conformally map the 3D shape and the polycube to the same canonical domains (e.g., sphere, Euclidean plane, or hyperbolic disk), then we construct a map between these two domains, which induces an one-to-one map between the 3D shape and the polycube. Since our method avoids the direct projection of the 3D shape to the polycube, the polycube can be constructed independent of the actual geometry of 3D shape, allowing different complexity and resolution for the polycube.

The overall flow of our algorithm for establishing the one-to-one mapping between models and their parametric domains can be summarized as follows:

1. Given a 3D model M from data acquisition, construct a polycube P which roughly resembles the geometry of M and is of the same topology of M.
2. Make use of the above discrete Ricci flow to compute the Riemannian uniformization metric of M and embed M in the canonical domain D_M, which is $\mathbb{S}^2$, $\mathbb{E}^2$ or $\mathbb{H}^2$, i.e., $\phi_M : M \to D_M$.
3. Compute the uniformization metric of P and embed P in the canonical domain D_P, i.e., $\phi_P : P \to D_P$.
4. Construct the map $\phi_{D_M \to D_P} : D_M \to D_P$.
5. Finally, the composition $\phi_P^{-1} \circ \phi_{D_M \to D_P} \circ \phi_M$ gives the desired polycube map from M to P.

Note that, our construction method varies depending on the topology of surfaces. Genus-zero surfaces are mapped to the unit sphere $\mathbb{S}^2$ with positive curvature $\bar{K} = 1$. Genus-one surfaces are mapped to Euclidean plane $\mathbb{E}^2$ with zero curvature $\bar{K} = 0$. Surfaces of high genus are mapped to hyperbolic disk $\mathbb{H}^2$ with negative curvature $\bar{K} = -1$.

Once the target curvature $\bar{K}$ is given, we compute the conformal map $\phi_M : M \to D_M$ and $\phi_P : P \to D_P$ by solving the uniformization metric $e^{2u}\mathbf{g}$ using discrete Ricci flow. Then by specifying three feature points on M and P, we can uniquely find the map $\phi : D_M \to D_P$. Finally, the polycube map from M to P is expressed as $\phi_P^{-1} \circ \phi_{D_M \to D_P} \circ \phi_M$. Figure 8.13 demonstrates the procedure to compute polycube maps of various topology.

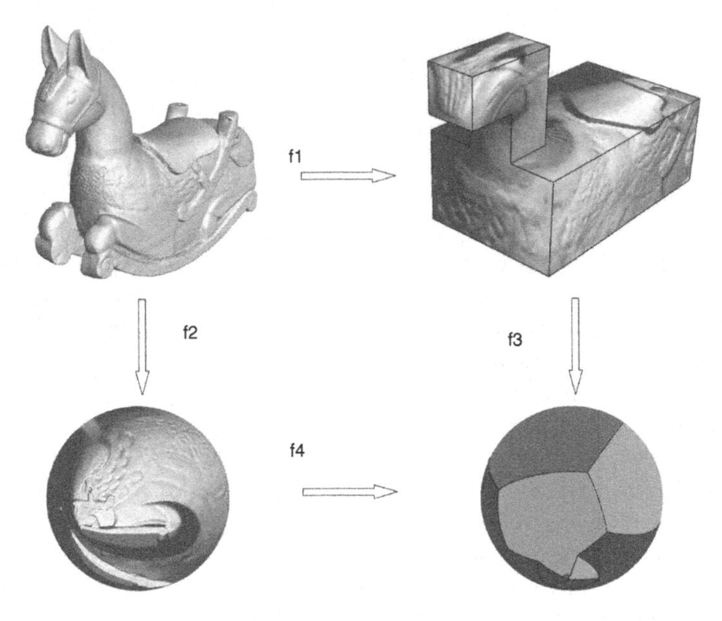

(a) Genus zero conformal polycube map

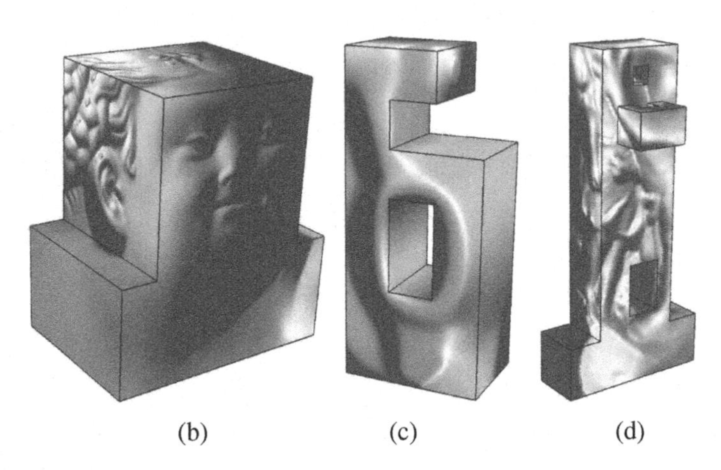

(b) (c) (d)

Fig. 8.13 Constructing polycube maps of various topology using geometric structures. **a** shows the construction pipeline of genus-zero case. Both the original mesh M and the polycube P are conformally mapped to the canonical domains, i.e., $\mathbb{S}^2$, $\mathbb{E}^2$ or $\mathbb{H}^2$. Denote these maps by $\phi_M : M \to D_M$ and $\phi_P : P \to D_P$. By finding the optimal map between D_M and D_P, we get the polycube map $\phi_{M\to P} = \phi_P^{-1} \circ \phi_{D_M \to D_P} \circ \phi_M$. This construction method varies depending on different types of surfaces. Genus-zero surfaces are mapped to the unit sphere $\mathbb{S}^2$ with positive curvature $\bar{K} = 1$. Genus-one surfaces are mapped to Euclidean plane $\mathbb{E}^2$ with zero curvature $\bar{K} = 0$. Surface of high genus are mapped to hyperbolic disk $\mathbb{H}^2$ with negative curvature $\bar{K} = -1$. **b**, **c** and **d** show the construction results of polycube maps of various topology. Image is taken from [285], included here by permission from Elsevier, Inc

8.5 Extending Planar Splines to Manifold Domains

As pointed out in Sect. 8.3, if a particular planar spline scheme is invariant under the parametric affine transformation, it can be easily generalized to manifold domain of arbitrary topology with no more than Euler number of singular points. Fortunately, most of the known planar spline schemes satisfy the parametric affine invariant property. This section provides the details on how to construct triangular B-splines, Powell–Sabin splines, and T-splines over manifold of arbitrary topology.

The general flow chart to construct manifold splines is as follows:

1. Construct the domain manifold.
2. Compute the affine structure of the domain manifold.
3. Construct a planar spline patch on each chart. These spline patches are automatically glued together by the affine transition functions across the overlapped charts.
4. Modify the control points to obtain a desired shape.

Given an affine manifold M covered by a set of charts (U_α, ϕ_α), a manifold spline $\mathbf{F}$ is expressed as follows:

$$\mathbf{F}(\mathbf{u}) = \sum_i \mathbf{c}_i B_i(\phi_i(\mathbf{u})), \quad \mathbf{u} \in M,$$

where $\mathbf{c}_i \in \mathbb{R}^3$ are the control points (Table 8.2).

8.5.1 Manifold Triangular *B*-spline

Using triangular B-splines, users can represent shapes over triangulated planar domains with lower-degree piecewise polynomials (rather than frequently-used tensor-product surface construction over regular domains) that nonetheless maintain higher-order continuity across the boundary of their piecewise patchwork.

Given a domain manifold M, a manifold triangular B-spline surface is defined as follows:

$$\mathbf{F}(\mathbf{u}) = \sum_I \sum_{|\beta|=n} \mathbf{c}_{I,\beta} N_{I,\beta}(\tau_I(\mathbf{u})), \quad \mathbf{u} \in M,$$

Table 8.2 Building block of manifold splines

Spline scheme	Basis function	Continuity	Domain
Triangular B-spline	Simplex spline	C^k	Triangular mesh
Powell–Sabin spline	Bézier spline	C^1	Triangular mesh
T-spline	B-spline	C^k	T-mesh

where I is the triangle index and $N_{I,\beta}$ is the normalized simplex spline. The algorithm for construction manifold triangular B-spline is as follows:

1. The initial control points $\mathbf{c}_{I,\beta}$ are chosen by uniformly subdivided the domain manifold M according to the user-specified degree n. Each domain triangle is associated with $(n+1)(n+2)/2$ control points.
2. We compute the new position of control points $\tilde{\mathbf{c}}_{I,\beta}$ by solving the following least square problem:

$$\min_{\tilde{\mathbf{c}}} \sum_{I} \sum_{|\beta|=n} \|\tilde{\mathbf{c}}_{I,\beta} - \mathbf{c}_{I,\beta}\|^2 \tag{8.17}$$

$$\text{subject to } \tilde{\mathbf{c}}_{I,\beta} = f^J(V_\beta^I), \forall I, \forall \beta, |\beta| = n, \beta_2 \leq r$$

where $V_\beta^I = \{\mathbf{t}_{0,0}^I, \ldots, \mathbf{t}_{0,\beta_0-1}^I, \ldots, \mathbf{t}_{2,0}^I, \ldots, \mathbf{t}_{2,\beta_2-1}^I\}$ and $\mathbf{t}_{i,j}^I$ are the knots for triangle I.

Note that, the initial manifold triangular B-spline surfaces acquired in Step-1 usually have very bad curvature distribution, especially along the edges of the domain triangles. The purpose of Step-2 is to fair the spline surface by modifying the control points. In the objective function Eq. (8.17), we minimize the squared distance between the control points of the original and the new spline surface, which implies that the minimal change of the shape. In the constraints, we use an integer $r, 0 \leq r \leq n-1$, to control the fairness of the spline surface. The bigger the value r, the better quality surface we can obtain. In our experiments, we can get visually pleasing surfaces with $r = 1$ for cubic splines or $r = 2$ for splines of degree 5 or above. Figures 8.19 and 8.18 show the examples of manifold triangular B-splines.

8.5.2 *Manifold Powell–Sabin Spline*

Powell–Sabin splines are functions in the space $S_2^1(\Delta_{ps})$ of C^1 continuous piecewise quadratic functions on a Powell–Sabin refinement [218]. Such a refinement Δ_{ps} can be obtained from an arbitrary triangulation Δ by splitting each triangle into six sub-triangles with a common interior point. Dierckx presents the normalized B-spline basis for Powell–Sabin splines [67]. This representation has a very nice geometric interpretation involving the tangent control triangles for manipulating the Powell–Sabin surfaces. In contrast to triangular Bézier splines, where imposing smoothness requirements across the patches requires that a large number of nontrivial equations involving many control points must be enforced, the C^1 continuity of a Powell–Sabin spline is guaranteed for any choice of the control points.

Manifold Powell–Sabin spline inherits the promising properties of its planar counterpart such that it is globally defined and interpolates both the positions and normals of the vertices of the domain manifold [115]. The algorithm for constructing manifold Powell–Sabin spline is summarized as follows:

1. Compute the affine atlas of the domain manifold M. For each vertex $\mathbf{V}_i \in M$, denote by (U_i, ϕ_i) its coordinate chart which contains the 1-ring neighbors of $\mathbf{V}_i$).
2. For each vertex $\mathbf{V}_i \in M$, compute the three linear independent triplets, $(\alpha_{ij}, \beta_{ij}, \gamma_{ij})$, $j = 1, 2, 3$. Build the basis functions using Dierckx's algorithm [67].
3. For each vertex $\mathbf{V}_i$ with normal $\mathbf{n}_i$, assign the control points $(\mathbf{C}_{i1}, \mathbf{C}_{i2}, \mathbf{C}_{i3})$ which satisfy

$$\mathbf{V}_i = \sum_{j=1}^{3} \alpha_{ij} \mathbf{C}_{ij} \tag{8.18}$$

and

$$\frac{(\mathbf{C}_{i1} - \mathbf{C}_{i2}) \times (\mathbf{C}_{i1} - \mathbf{C}_{i3})}{\| (\mathbf{C}_{i1} - \mathbf{C}_{i2}) \times (\mathbf{C}_{i1} - \mathbf{C}_{i3}) \|} = \mathbf{n}_i. \tag{8.19}$$

It can be proven that the control triangle $(\mathbf{C}_{i1}, \mathbf{C}_{i2}, \mathbf{C}_{i3})$ is tangent to the spline surface $\mathbf{s}$ at $\mathbf{V}_i$, i.e.,

$$\mathbf{s}(\phi_i(\mathbf{V}_i)) = \mathbf{V}_i \tag{8.20}$$

$$\frac{\mathbf{s}_u(\phi_i(\mathbf{V}_i)) \times \mathbf{s}_v(\phi_i(\mathbf{V}_i))}{\| \mathbf{s}_u(\phi_i(\mathbf{V}_i)) \times \mathbf{s}_v(\phi_i(\mathbf{V}_i)) \|} = \mathbf{n}_i \tag{8.21}$$

The detailed proof is in [115]. Figure 8.14 demonstrates the manifold Powell–Sabin spline for the genus-two frog model.

8.5.3 *Manifold T-spline*

T-splines, proposed by Sederberg et al. in [236], are a generalization of NURBS surfaces that are capable of significantly reducing the number of superfluous con-

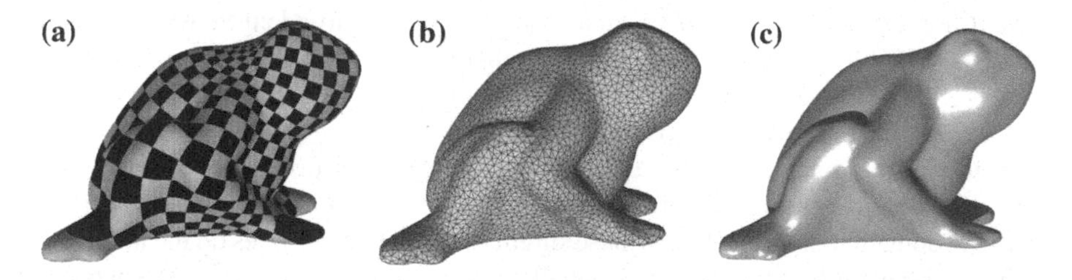

Fig. 8.14 Example of manifold Powell–Sabin spline. **a** shows the global conformal parameterization. **b** shows the domain manifold which is an arbitrary triangular mesh. **c** shows the manifold Powell–Sabin spline interpolating both the positions and normals of the vertices of the domain manifold. The spline is C^1 continuous everywhere

trol points by using the T-junction mechanism. Consequently, T-splines enable much better local refinement capabilities than NURBS. Furthermore, using the techniques presented in [236], it is possible to merge adjoining T-spline surfaces into a single T-spline by adding a few more control points. However, this patching process requires that the knot intervals of the to-be-merged edges must establish an one-to-one correspondence between the two surfaces. Manifold T-spline is a natural and necessary integration of T-splines into our manifold spline framework, with a goal to retain all the desirable properties while overcoming the aforementioned modeling drawbacks at the same time.

Unlike the manifold triangular B-spline which does not have any restriction on the domain manifold, manifold T-splines, however, require each face of the domain manifold to be a rectangle. We have proposed three methods to construct affine atlas of arbitrarily oriented 2-manifold, and among them, two techniques are ideal for computing the affine structure for T-mesh. First, the conformal structure induces the natural tensor-product structures on the domain manifold with Euler number of zero points. There are regular vertices (of valence 4) everywhere except the zero points (valence 8) and T-junctions (valence 3) along the cut graph. The detailed discussion can be found in [116]. Second, polycube map is another natural way to construct the T-spline. Since each face of polycube is just a planar region with poly-rectangular shape, we can easily construct a planar T-spline for each face and then glue the adjacent faces using the edge charts. Note that, this gluing procedure is totally automatic. Figure 8.19 shows the examples of manifold T-splines.

8.6 Handling Extraordinary Points

As mentioned above, manifold splines *MUST* have singularities if the Euler number of domain manifold is not zero. The existence of extraordinary points is due to the topological obstruction. Unfortunately, they cannot be totally avoided in the current manifold spline framework.

The extraordinary points of affine structure induced from conformal structure behaves similar to the map $z \rightarrow z^2, z \in \mathbb{C}$. Thus, if we draw a closed curve c around the extraordinary point Z on M, its image $\phi(c) \in \mathbb{R}^2$ is a closed curve which passes around $\phi(Z)$ at least twice (The winding number of the image curve around $\phi(Z)$ is no less 2). Therefore, the map $\phi : Z \rightarrow \mathbb{R}^2$ is not a one-to-one map and we cannot easily construct spline patches covering Z.

In theory, these singularities are just points on the spline surfaces. In construction algorithms, however, we must remove the singularities and their connected faces from the domain manifold. Hence, these singularities appear as holes on the manifold splines. These holes can be arbitrarily small. Given an arbitrary value ε, we can make the 1-ring neighbors of the extraordinary points small enough so that the diameter of the hole on the manifold spline is less than ε. Although making senses in theory, this approach is far from adequate in practice for real-world applications, since it requires an extremely large number of charts in the vicinity of all the extraordinary points.

To streamline the implementation, we use a simple technique to handle the extraordinary points in this paper. We use a spline surface to fill the hole such that it satisfies the user-specified boundary conditions. We shall minimize the thin-plate energy of the hole-filling surfaces to improve the overall smoothness of the surface. Thus, our goal is to solve the following optimization problem:

$$E(s) = \iint_{\Omega} \left(\frac{\partial^2 s}{\partial u^2}\right)^2 + 2\left(\frac{\partial^2 s}{\partial u \partial v}\right)^2 + \left(\frac{\partial^2 s}{\partial v^2}\right)^2 dudv. \tag{8.22}$$

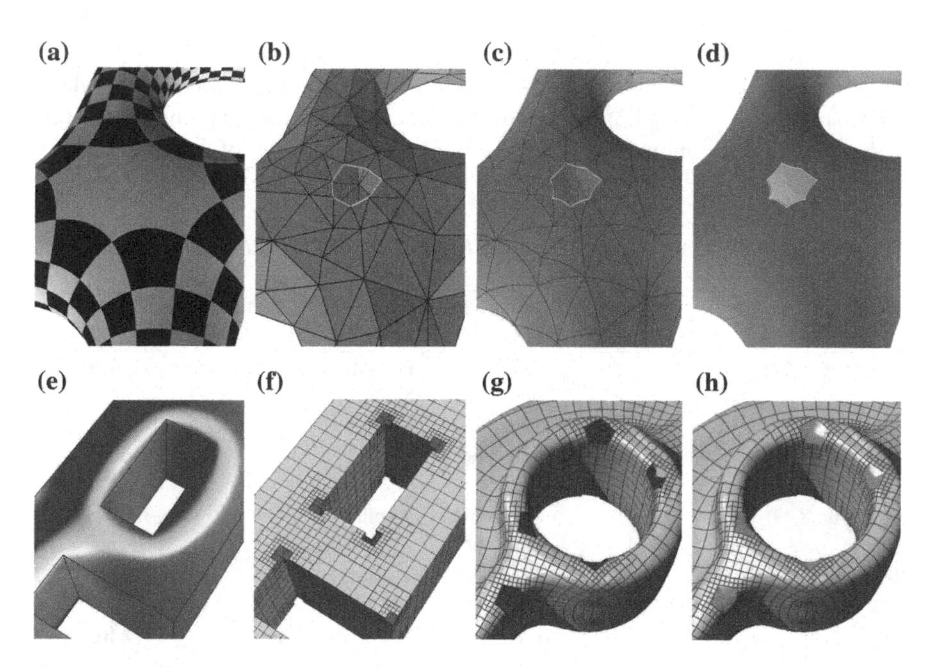

Fig. 8.15 Handing the extraordinary points of the manifold splines. The first row shows the example of a manifold triangular B-spline, whose affine atlas is constructed using conformal structure. The singularity is shown in **a**. We remove the extraordinary point Z and its connected faces on the domain manifold M (shown in **b**). As a result, the manifold triangular B-spline is an open surface shown in **c**. We then construct a cubic planar triangular B-spline surface which minimizes the thin-plate energy function while satisfying the boundary condition. **d** shows the result after the hole-filling is completed. Hole area is colored in yellow. The second row shows an example of manifold T-spline, whose affine atlas is constructed using polycube maps, where all the corners are extraordinary points (shown in **e**). **f** shows the domain manifold after removing all the corners. **h** shows the open manifold T-spline surface with many holes. **h** shows the final result after hole-filling (hole areas are all colored in green). Images of the second row from [285]

Our strategy to fill the hole is to find s^* by solving the following minimization problem:

$$min\{E(s) : s|_{\partial\Omega} = f, \frac{\partial s}{\partial u} \times \frac{\partial s}{\partial v}|_{\partial\Omega} = n\}. \tag{8.23}$$

where f and n are the boundary positions and normals.

In our current implementation, we use a cubic triangular B-spline to handle the hole-filling problem because of its flexibility in the domain construction and its potential to match with any number of sides of holes. The boundary conditions are represented by several sampling points on the boundary of the spline surface. The boundary position constraints naturally lead to a system of linear equations on the control points. The normal constraints are expressed as

$$\frac{\partial s}{\partial u} \times n = 0, \quad \frac{\partial s}{\partial v} \times n = 0.$$

Therefore, Eq. (8.23) is a linear least-square problem with linear constraints, which can be solved easily using Lagrange Multiplier method. We should point out that this method outputs G^1 continuous along the boundaries of the holes of the extraordinary

Table 8.3 Spline Configurations. g: genus; N_b: # of boundaries; N_s: # of singularities; N_v: # of vertices in the domain manifold; N_c: # of control points

Model	g	N_b	N_s	Affine Atlas	Spline Kernel	Continuity	N_c
Frog (Fig. 8.14)	0	2	0	Conformal structure	Powell–Sabin spline	C^1	17, 901
Buddha (Fig. 8.19)	0	0	16	Polycube	T-spline	C^2	11, 067
Bird (Fig. 8.19)	0	2	0	Conformal structure	T-spline	C^2	3, 940
Rockerarm (Fig. 8.15)	1	0	24	Polycube map	T-spline	C^2	4, 132
Hypersheet (Fig. 8.19)	1	3	0	Ricci flow	Triangular B-spline	C^2	1, 446
Two-hole Torus (Fig. 8.15)	2	0	2	Conformal structure	Triangular B-spline	C^2	2, 321
Genus-two Bottle (Fig. 8.19)	2	0	2	Conformal structure	Triangular B-spline	C^2	6, 743
Three-hole Torus (Fig. 8.19)	3	0	32	Polycube map	T-spline	C^2	5, 180
Sculpture (Fig. 8.16)	3	0	1	Ricci Flow	Triangular B-spline	C^2	8, 962
Four-kids (Fig. 8.18)	8	0	14	Conformal Structure	Triangular B-spline	C^2	21, 992

Fig. 8.16 Modeling the genus-three sculpture model using manifold triangular B-splines. We choose the singular point on the side of the model and then set its target curvature to be $-8\pi/S$ (where S is the area of the surface) and the curvatures for the other points to be zero. By performing the Ricci flow, we get the affine atlas with only one extraordinary point. This spline surface is C^2 continuous everywhere except at the extraordinary point (highlighted in green)

points. We could achieve higher-order continuity by adding more constraints, such as curvatures, into the above minimization problem at the expense of extra computational costs. Figure 8.15 shows the procedure to handle extraordinary points for manifold T-splines and manifold triangular B-splines (Table 8.3).

8.7 Discussions

In this section, we address some key aspects for manifold splines.

8.7.1 Domain Manifold

In the construction of manifold splines, the topology of the domain manifold is much more critical than its geometry. Although for most of the examples we have shown in this paper, the domain manifolds resemble the geometry of the spline surfaces, which facilitates the spline construction. In theory, the only requirement is that the domain manifold has the same topology as the spline surface. Thus, we can model totally different surfaces using the same domain manifold. Figure 8.17 shows such a case. The domain manifold is a genus-one surface with three boundaries shown in Figure 8.17c. We model two different spline surfaces (see Fig. 8.17a and e) with the same topology. The differences between these two spline surfaces are just the position of the control points (shown in Fig. 8.17b and d). Thus, by linearly interpolating the control points, we get the morphing series between two spline surfaces shown in Fig. 8.17 f–j.

We should also point out that the domain manifold is not necessary to be embedded in the Euclidean space. We can even construct manifold splines using abstract domain manifolds. For example, we can construct manifold spline on a manifold of genus-one, where the Gaussian curvature of each vertex is zero. Not that this is an abstract manifold and cannot be embedded in $\mathbb{R}^3$.

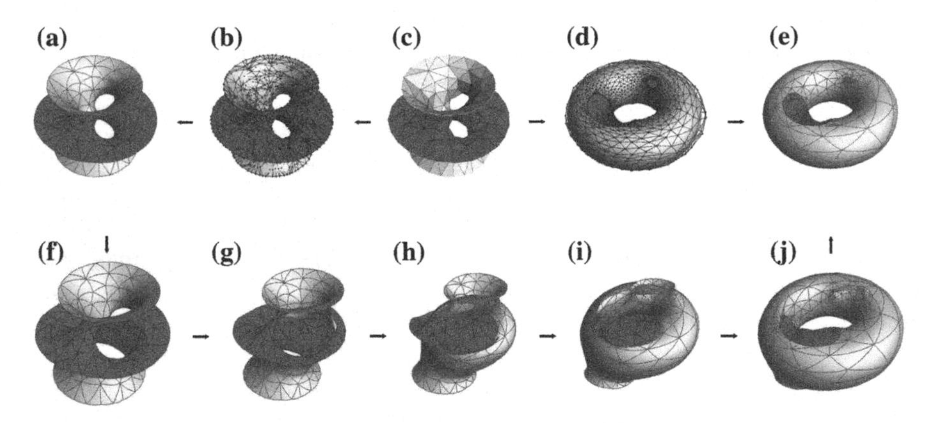

Fig. 8.17 The domain manifold is not necessarily resembling the spline surface to be modeled. We construct two manifold triangular B-spline surfaces, **a** and **e**, which have different control points shown in **b** and **d**, but share the same domain manifold in **c**. The second row shows the morphing sequence between these two spline surfaces by linearly interpolating the control points of **b** and **d**

8.7.2 Manifold Splines Versus Planar Splines

The planar splines which satisfy the parametric affine invariant property can be viewed as special cases of manifold splines in the following aspects:

1. Planar parametric domain is a special case of oriented 2-manifold such that it can be covered by a single coordinate system, while a general manifold is covered by several coordinate charts.
2. Planar parametric domain is an open manifold, thus it has affine structure. Hence, planar splines do not have extraordinary points.

8.7.3 Manifold Splines Versus Spherical Triangular B-Splines

Spherical triangular B-splines [114] are ideal tools to model genus-zero surfaces without singularities because of their topological equivalence to a sphere. Manifold splines are different from spherical triangular B-splines in that

1. Spherical triangular B-splines are *trivariate* functions defined on the sphere, i.e., $\mathbf{F}(\mathbf{u})$, $\mathbf{u} = (u, v, w) \in \mathbb{S}^2$ satisfying $u^2 + v^2 + w^2 = 1$. Manifold splines are *bivariate* functions defined on the charts of the domain manifold, i.e., $\mathbf{u} \in M$, $\phi(\mathbf{u}) \in \mathbb{R}^2$. Thus the traditional planar splines can be viewed as the special cases of manifold splines.
2. Spherical triangular B-splines have no singularities but can only model genus-zero (closed or open) shapes. For genus-zero closed shapes, manifold splines, however, have at least one singular point. In practice, we often change the topology of the shapes. For example, we cut two boundaries into the genus zero closed shape, and by double covering, the modified shape is of genus one. Then we can compute the affine structure without any singularity. Finally, we need to fill the holes caused by the two boundaries in the domain manifold.
3. Spherical triangular B-splines are extrinsic splines, whose constructions require the embedding of the domain manifold into $\mathbb{R}^3$. Manifold splines are intrinsic splines, which are defined on the charts of the domain manifold and do not require the embedding of the domain manifold.
4. The basis function of spherical triangular B-spline is invariant under rotation on $\mathbb{S}^2$. The basis function of manifold spline is invariant under affine transformation on $\mathbb{R}^2$.

8.7.4 Manifold Splines Versus Subdivision Surfaces

We now analyze classical Catmull–Clark subdivision surfaces from the point of view of manifold splines.

Suppose S_0 is the base mesh, $S_1, S_2, \cdots, S_n$ are subdivision surfaces after n steps. We use V_n to denote the vertices set of S_n, E_n the extraordinary points on S_n. We call a neighborhood compose of 3 by 3 faces a *normal neighborhood* if all the 16 vertices are not extraordinary. Given a point p in S_0, it will be mapped to S_n in a natural way. We define the atlas of S_0 as the followings: for any point $p \in S_0, p \notin E_0$, there must be a finite k, such that p is contained in a face $f_p \in S_k$, such that f_p is the center face of a normal neighborhood $N_p \subset S_k$. Then we define a local chart $(U_p, \phi_p), f_p \subset U_p \subset N_p$, ϕ_p maps N_p to a regular integer 3 by 3 grids on $\mathbb{R}^2$, whose grid points are integers. The atlas is $\mathcal{A} = \bigcup_p (U_p, \phi_p), p \notin E_0$. The control net is $C = \bigcup_{k=0}^{\infty} V_k / E_k$. Then on each chart (U_p, ϕ_p), the limit Catmull–Clark subdivision surface is a uniform bicubic B-spline patch; there are infinite local charts; all the transition functions are translation, rotation, and scaling on $\mathbb{R}^2$. Therefore, subdivision surfaces also depend on the affine structure.

The extraordinary points of subdivision surfaces depend on the connectivity of the control mesh. The extraordinary points of manifold splines solely depend on the topology of the domain manifold. There is no singularity for surfaces with zero Euler number. The theoretical lower bound of the number of singularities for surfaces with non-zero Euler number is one. We can achieve this lower bound by using discrete Ricci flow to compute the affine atlas.

8.7.5 General Manifold Spline Program

Klein's Erlangen program is an influential research programme which guided the research in geometry in nineteenth century. According to Klein, geometry is characterized by its group of symmetries (by symmetry it means a transformation of space which leaves all the objective properties of space invariant) and investigates the mathematical objects which are invariant under this given group.

Till now, we have studied the planar splines and proposed their manifold generalizations via affine atlas. The key rationale that we only focus on affine structure is that all the existing planar splines are parametric affine invariant, i.e., the function value remains unchanged if we perform an affine transformation on the knots.

Stronly inspired by Klein's Erlangen program, it is natural for us to propose the following general manifold spline program in which the spline basis functions are characterized by the geometric structures.

Definition 8.9 (*Geometric Structure*) Suppose M is a manifold, X is a topological space, G is a transformation group on X, a (G, X) *atlas* is an atlas $\{(U_\alpha, \phi_\alpha)\}$, such that

1. Local coordinates are in X,

$$\phi_\alpha : U_\alpha \to X.$$

2. Transition functions are in group G,

Fig. 8.18 Modeling the genus-8 sculpture model using manifold triangular B-splines. We compute the affine atlas using conformal structure, thus it has 14 extraordinary points. This spline surface is C^2 continuous everywhere except at the extraordinary points (colored in green)

$$\phi_{\alpha\beta} \in G.$$

Two (G, X) atlas are equivalent, if their union is still a (G, X) atlas. Each equivalent class of (G, X) atlas is a (G, X) structure.

General Manifold Spline Program *Manifold splines study the functional basis defined on different (G, X) structures which are invariant under the transformations in G.*

The manifold splines developed in this paper and our previous work fall into the general category of **affine manifold spline**, i.e., the spline is invariant under parametric affine transformation. Due to the topological obstruction, the only closed affine 2-manifold is torus. Therefore, extraordinary points are inevitable to closed manifolds other than tori.

Unlike affine atlas, spherical atlas exists for genus zero surfaces, hyperbolic atlas exists for surfaces of high genus and projective atlas exists for surfaces of all kind of topology. Thus, studying the spline functions in spherical, hyperbolic, and projective

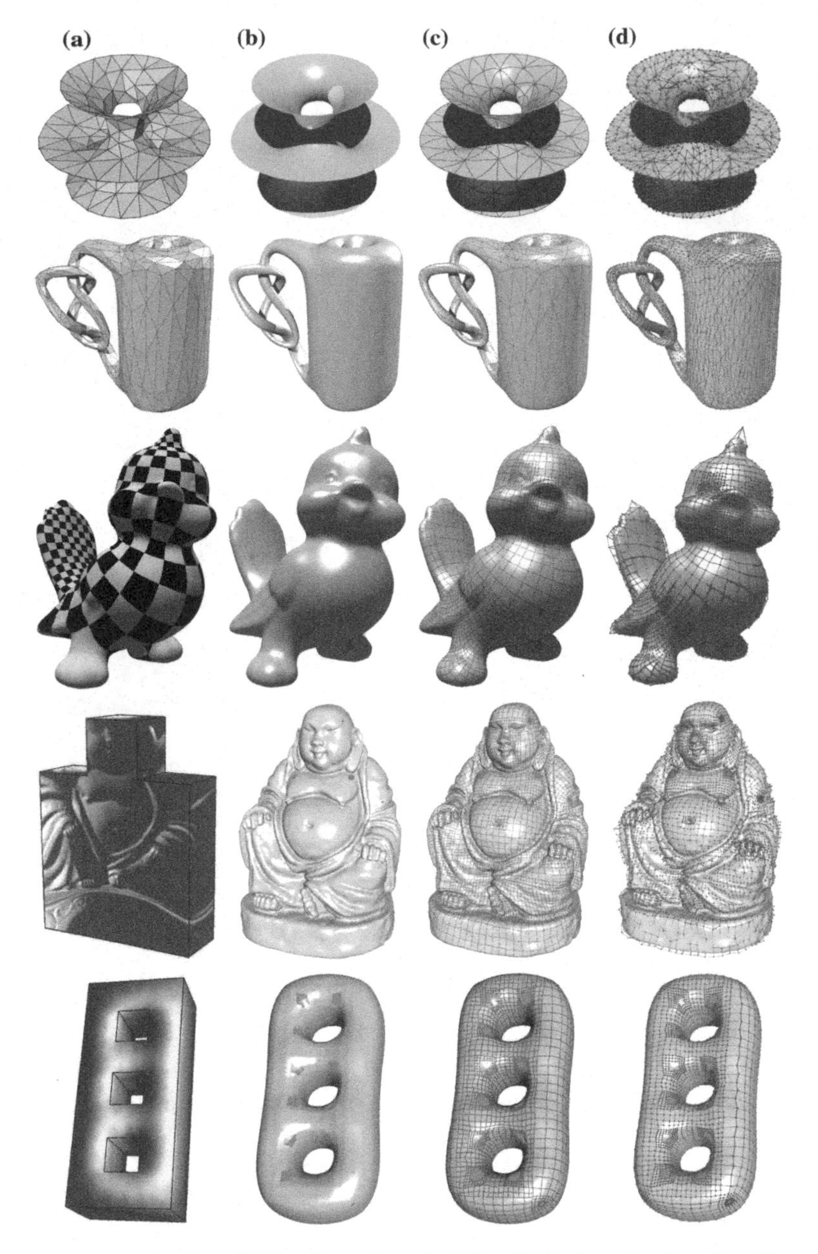

Fig. 8.19 Examples of manifold splines. *Row 1∽2 Manifold triangular B-spline*: Hypersheet model is of genus-1 with three boundaries. The affine atlas is constructed using discrete Ricci flow. Genus-2 bottle (the second handle is inside the bottle) is a closed surface. The affine atlas is constructed using conformal structure. *Row 3∽5 Manifold T-spline*: The affine atlas of the bird model is constructed using conformal structure. The affine atlas of the buddha and 3-hole torus models are constructed using polycube maps where the corners are extraordinary points. **b** shows the manifold splines. Extraordinary points are drawn in green. The red curves in **c** illustrate spline patchwork. **d** shows the splines overlaid with control points. All splines are C^2 continuous everywhere except at the extraordinary points (colored in green)

space could lead to **spherical, hyperbolic and projective manifold splines**. These splines are ideal to model shapes of various kinds of topology without singularities. These observations shall lead to some future research directions.

8.8 Summary

We systematically broaden and strengthen the theoretical foundation for manifold splines in this paper. By employing novel mathematical tools such as discrete Ricci flow and polycube maps, we have developed a general computational framework for modeling surfaces of arbitrary topology. The key component in the building pipeline for manifold splines is affine atlas. We have developed several techniques to compute the affine structure for any domain manifold. As a result, users can easily control the number and locations of extraordinary points when designing manifold spline surfaces. We have presented several algorithms to generalize various planar splines, such as triangular B-spline, Powell–Sabin spline, T-spline, and point-based spline to domain manifolds of arbitrary topology. Each member in our manifold spline family has its own modeling advantages for interactive design and shape representation, while accommodating specific needs from different users. In order to have a rich and deep understanding of manifold splines, we have made a thorough comparison between manifold splines and other commonly used surface modeling schemes to highlight the major advantages of manifold splines. Ultimately, we are able to articulate the general manifold spline program which is much broader than the currently-available spline schemes, pinpointing possible future research directions. Through our extensive modeling experiments, we demonstrated that manifold splines are both theoretically fundamental and practically powerful for the accurate and efficient modeling of arbitrarily complicated surfaces with any topological type.

Chapter 9
Medical Imaging

Abstract This chapter introduces the applications of computational conformal geometry on medical image analysis research. Specifically, we focus on magnetic resonance imaging (MRI) analysis of brain images where surface-based approaches attract continued interest due to their sub-voxel accuracy and the capability of detecting subtle local changes in brain anatomical shapes. Several research topics, including brain surface conformal parameterization, brain surface registration, global transformation–invariant brain shape descriptors, and point-to-point local brain surface deformation measurements, together with their experimental results, are detailed in this chapter.

9.1 Introduction

Over the past decade, exciting opportunities have emerged in studying 3D medical imaging data thanks to the rapid progress made in medical imaging techniques. Among various imaging modalities, magnetic resonance imaging (MRI) technique provides detailed anatomy and function information without exposing the subject to ionizing radiation so MRI is widely used in hospital and clinics for medical diagnosis, staging of disease and follow up examinations.

In brain imaging research, when registering structural MR images, the volume-based methods, e.g. [51], have much difficulty with the highly convoluted cortical surfaces due to the complexity and variability of the sulci and gyri. Early research [78, 266, 278] has demonstrated that surface-based brain mapping may offer advantages over volume-based brain mapping as a method to study the structural features of the brain, such as surface deformation, as well as the complexity and change patterns in the brain due to disease or developmental processes. Surface-based human brain mapping methods and statistical morphometric analyses are attracting more interests, due to their sub-voxel accuracy and the capability of detecting subtle local changes in brain anatomical shapes.

One way to analyze and compare brain surface data is to map them into a canonical space while retaining geometric information on the original structures as far as possible [52, 78, 266]. Brain conformal mapping [8, 100, 125, 295, 296] has been

M. Jin et al., *Conformal Geometry*, https://doi.org/10.1007/978-3-319-75332-4_9

widely studied because of its wide adaptability and convenient computation algorithms. Computational conformal geometry provides a complete and solid theoretical and algorithmic solution to conformally map brain surfaces to different canonical domains. The obtained conformal parameterizations have been used for a variety of research, e.g. visualization, brain surface registration, surface-based morphology statistics, etc.

Since image intensities vary among scans, surface-based morphology may provide robust and biologically sound shape statistics to characterize variations of brain shapes. Generally speaking, surface-based morphology statistics can be classified into two classes: one is the class of transformation-invariant global shape descriptors that requires no surface registration; the second class of features is some local measurements defined on particular locations after the brain surface registration among the population. The first class of features is usually concise and intrinsic to surface structure and the second class of features may lend themselves to immediate visualization. The choice between different types of features usually depends on specific applications.

Computational conformal geometry research has extensive applications in the human brain mapping field, including brain structure and function visualization, comparison of 3D anatomical models across subjects, construction of population-based brain atlases, and detection of group patterns in structural magnetic resonance imaging (MRI) data, etc. Here we attempt to give an overview of brain conformal mapping algorithms and software, highlighted by some recently developed techniques, such as brain surface conformal parameterization, brain surface registration, various global shape descriptors, and multivariate local surface deformation measurements.

In a typical surface-based morphometry pipeline, after MRI intensity is corrected with nonparametric nonuniform intensity normalization method, the images are usually spatially normalized into the stereotaxic space using a global affine transformation. Afterwards, an automatic tissue-segmentation algorithm is used to classify each voxel as cerebrospinal fluid (CSF), grey matter (GM), white matter (WM), or different subcortical structures such as hippocampus, lateral ventricle, etc. Usually, marching cube algorithm [178] is used to generate the cortical or subcortical surface meshes. Because the human cerebral cortex has a 3D highly convoluted topology structure, additional algorithms, such as Laplace–Beltrami operator based method [252], are applied to remove the geometric and topological outliers and generate robust and accurate meshes for the following surface-based morphometry analyses.

In brain imaging research, some common landmark curves defined on cortical surfaces are frequently needed to help brain cortical surface registration. A variety of brain landmark protocols are studied (e.g. [205, 268, 278]). There is a set of 'Core 6' landmark curves including the Central Sulcus (CeS), Anterior Half of the Superior Temporal Gyrus (aSTG), Sylvian Fissure (SF), Calcarine Sulcus (CaS), Medial Wall Ventral Segment, and Medial Wall Dorsal Segment, are automatically traced on each cortical surface using the Caret package [278]. In Caret software, the PALS-B12 atlas is used to delineate the "core 6" landmarks, which are well-defined and geographically consistent, when compared with other gyral and sulcal features

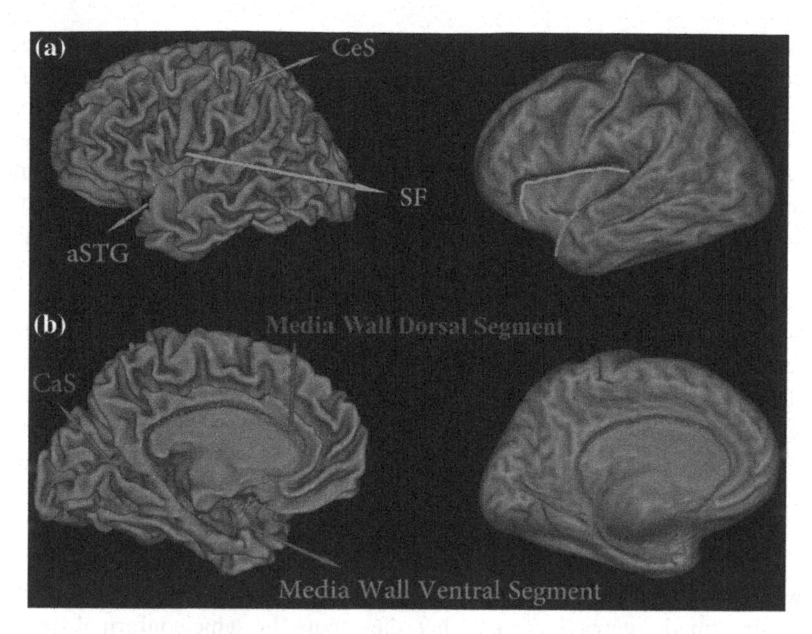

Fig. 9.1 A left cortical surface with six landmark curves, which are automatically labeled with Caret [278], showing in two different views on both the original and inflated surfaces. The inflated surface is obtained with FreeSurfer [78]. Image from [249]

on human cortex. The stability and consistency of the six landmarks was validated in [277]. An illustration of the landmark curves on a left cortical surface is shown in Fig. 9.1 with two views. We show the landmarks with both the original and inflated cortical surfaces for clarity.

In brain imaging, sometimes we cut open cortical or subcortical surfaces along some specified landmark curves. It is called *topology optimization*. For subcortical surfaces, these cutting curves are usually located on some anatomically extreme points. For cortical surfaces, we usually cut along these specified and automatically identified landmark curves. After the cutting, surfaces become open boundary surfaces. With conformal geometry, these boundaries are either used as boundary condition to enforce surface registration [295, 297, 299], or compute intrinsic surface feature descriptors [249, 325].

9.2 Brain Surface Conformal Parameterization

Parameterization of brain cortical and subcortical surfaces is a fundamental problem for surface-based morphometry. Sometimes it is also called brain surface flattening. The goal of surface parameterization is to find some mappings between brain surfaces and some common flattening surfaces, i.e. some surfaces with constant

Gaussian curvature. After that, these common spaces serve as canonical spaces for surface registration and morphometry analysis. Brain surface parameterization has been studied extensively. A good surface parameterization preserves the geometric features and facilitates the following surface signal processing. Some research proposed near-isometric mappings [18] or area preserving mappings [33]. Another branch of research used concepts from conformal geometry to compute brain surface conformal parameterization [8, 125]. In addition to angle preserving property, conformal parameterization provides a rigorous framework for representing, splitting, matching and measuring brain surface deformations.

According to Poincarè uniformization theory, a general surface can be conformally mapped to one of three canonical spaces, the unit sphere, the Euclidean plane, and the hyperbolic space. In the following, we will introduce a variety of parameterization approaches that have been applied in human brain mapping research.

9.2.1 Spherical Brain Conformal Parameterization

It is well known that all orientable surfaces are Riemann surfaces. If two surfaces can be conformally mapped to each other, they share the same conformal structure. Therefore, computing conformal mappings is equivalent to computing conformal structures for surfaces. For genus zero closed surfaces, harmonic maps are equivalent to conformal maps [232].

We use K to represent the simplicial complex, u, v to denote the vertices, and $\{u, v\}$ to denote the edge spanned by u, v. We use f, g to represent the piecewise linear functions defined on K, use $\mathbf{f}$ to represent vector value functions. We use Δ_{PL} to represent the discrete Laplacian operator. All piece linear functions defined on K form a linear space, denoted by $C^{PL}(K)$.

Definition 9.1 Suppose a set of string constants $k_{u,v}$ are assigned for each edge $\{u, v\}$, the inner product on C^{PL} is defined as the quadratic form

$$< f, g >= \frac{1}{2} \sum_{\{u,v\} \in K} k_{u,v}(f(u) - f(v))(g(u) - g(v)) \tag{9.1}$$

The energy is defined as the norm on C^{PL}.

Definition 9.2 Suppose $f \in C^{PL}$, the string energy is defined as:

$$E(f) =< f, f >= \sum_{\{u,v\} \in K} k_{u,v} ||f(u) - f(v)||^2 \tag{9.2}$$

By changing the string constants $k_{u,v}$ in the energy formula, we can define different string energies.

Definition 9.3 Suppose edge $\{u, v\}$ has two adjacent faces T_α, T_β, with $T_\alpha = \{v_1, v_2, v_3\}$, define the parameters

$$a^{\alpha}_{v_1,v_2} = \frac{1}{2}\frac{(v_1 - v_3)\cdot(v_2 - v_3)}{|(v_1 - v_3)\times(v_2 - v_3)|} \tag{9.3}$$

$$a^{\alpha}_{v_2,v_3} = \frac{1}{2}\frac{(v_2 - v_1)\cdot(v_3 - v_1)}{|(v_2 - v_1)\times(v_3 - v_1)|} \tag{9.4}$$

$$a^{\alpha}_{v_3,v_1} = \frac{1}{2}\frac{(v_3 - v_2)\cdot(v_1 - v_2)}{|(v_3 - v_2)\times(v_1 - v_2)|} \tag{9.5}$$

T_β is defined similarly. If $k_{u,v} = a^{\alpha}_{u,v} + a^{\beta}_{u,v}$, the string energy obtained is called the *harmonic energy*.

Our goal is try to minimize the harmonic energy directly on the sphere domain. First, we use either star map or Gauss map to generate an initial spherical map [100]. Second, we compute the harmonic energy with discrete Laplace–Beltrami operator [215]. We minimize the harmonic energy with the steepest descent algorithm. To ensure the unique spherical conformal mapping, we use absolute derivative in the steeping descent step and also use Möbius transformation to normalize the results map in each iteration. Figure 9.2a illustrates a cortical surface and its spherical conformal mapping result. The normal information from the original surface is preserved on the sphere so that the correspondence is illustrated by the shading effects.

9.2.2 Planar Brain Conformal Parameterization with Holomorphic Functions

Mathematically, the differential form is a concept used in multivariable calculus that is independent of any chosen set of coordinates - it has many applications in geometry,

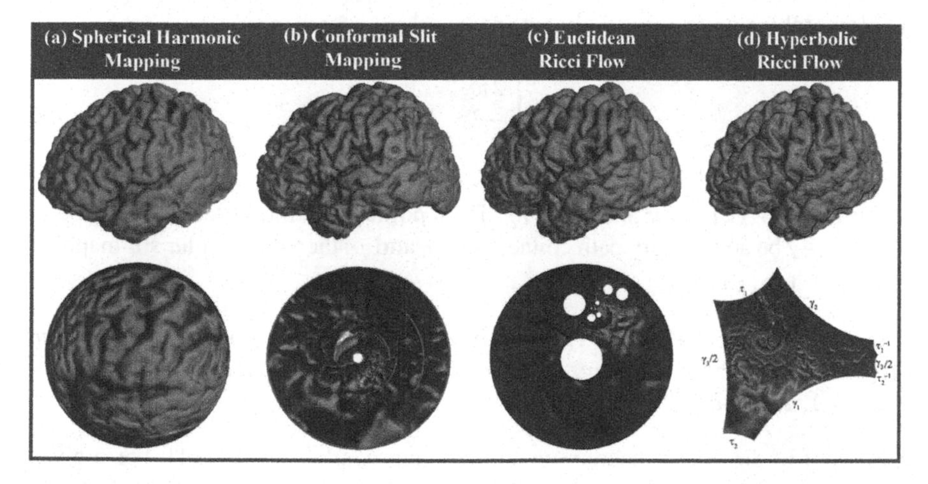

Fig. 9.2 Brain surface conformal parameterization results. Image from [100, 290, 296, 298]

topology and physics. The general setting for the study of differential forms is on a differential manifold. A differential 1-form in the local parameters (x, y) may be defined as

$$\omega = f(x, y)dx + g(x, y)dy$$

where f, g are smooth functions. Differential 1-forms are naturally dual to vector fields on a manifold. The algebra of differential forms along with the exterior derivative defined on it is preserved under smooth functions between two manifolds. Specifically, is a *closed 1-form*, if in each local parameter (x, y), $\frac{\partial f}{\partial y} - \frac{\partial g}{\partial x} = 0$. If ω is the gradient of another function defined on S, it can be called an *exact 1-form*. An exact 1-form is also a *closed 1-form*. If a closed 1-form ω satisfies $\frac{\partial}{\partial x} + \frac{\partial g}{\partial y} = 0$, then it is a *harmonic 1-form*. The gradient of a harmonic function is an *exact harmonic 1-form*. The Hodge star operator acting on a differential 1-form gives the conjugate differential 1-form

$$^*\omega = -g(x, y)dx + f(x, y)dy$$

Intuitively, the conjugate 1-form $^*\omega$ is obtained by rotating ω by a right angle everywhere. If ω is harmonic, so is its conjugate $^*\omega$.

Built upon the harmonic 1-form concept, the holomorphic 1-form is a differential form on a manifold which can have complex coefficients. It consists of a pair of conjugate harmonic 1-forms

$$\tau = \omega\sqrt{-1}^*\omega.$$

If two harmonic fields are orthogonal everywhere, they form a holomorphic 1-form. An intrinsic way to compute conformal parameterization is to search for a holomorphic 1-form that satisfies certain properties. In conformal slit mapping, we shall find certain holomorphic 1-forms with special behavior on the boundaries of the surface.

Suppose S is an open surface with n boundaries $\gamma_1, \ldots, \gamma_n$. We can uniquely find a holomorphic one-form ω, such that

$$\int_{\gamma_k} \omega = \begin{cases} 2\pi & k = 1 \\ -2\pi & k = 2 \\ 0 & otherwise \end{cases} \tag{9.6}$$

Definition 9.4 (*Circular Slit Mapping*) Fix a point p_0 on the surface, for any point $p \in S$, let γ be an arbitrary path connection p_0 and p, then the circular slit mapping is defined as $\phi(p) = e^{\int_\gamma \omega}$.

Theorem 9.5 *The function ϕ effects a one-to-one conformal mapping of M onto the annulus $1 < |z| < e^{\lambda_0}$ minus $n - 2$ concentric arcs situated on the circles $|z| = e^{\lambda_i}, i = 1, 2, \ldots, n - 2$.*

The proof of the above theorem on slit mapping can be found in [1]. For a given choice of the inner and outer circle, the circular slit mapping is uniquely determined up to a rotation around the center. The parallel slit mapping can be defined in a similar way.

Definition 9.6 (*Parallel Slit Mapping*) Let $\bar{S}$ be the universal covering space of the surface S, $\pi : \bar{S} \to S$ be the projection and $\bar{\omega} = \pi^*\omega$ be the pull back of ω. Fix a point $\bar{p}_0$ on $\bar{S}$, for any point $p \in \bar{S}$, let $\bar{\gamma}$ be an arbitrary path connection $\bar{p}_0$ and $\bar{p}$, then the parallel slit mapping is defined as $\bar{\phi}(\bar{p}) = \int_{\bar{\gamma}} \bar{\omega}$.

To parameterize brain cortical surfaces, we employ the concepts of holomorphic 1-form and conformal slit mapping to compute a conformal parameterization of cortical surfaces. Next, we briefly describe the algorithm implementation to compute the 1-forms and slit mapping.

Suppose M is a triangle mesh with $n+1$ boundaries, denoted as $\partial M = \gamma_0 - \gamma_1 - \cdots - \gamma_n$. We use v_i to denote a vertex, which is also a 3D coordinate vector, $[v_i, v_j]$ denote an edge connecting vertices v_i and v_j, $[v_i, v_j, v_k]$ denote a triangle formed by vertices v_i, v_j, and v_k. The angle at vertex v_i in triangle $[v_i, v_j, v_k]$ is denoted as ${}^{i}_{jk}$. The discrete functions defined on vertices, edges, and faces are called discrete 0-forms, 1-forms, and 2-forms, respectively.

The algorithm pipeline is as follows:

1. Compute the basis for all exact harmonic 1-forms;
2. Compute the basis for all harmonic 1-forms;
3. Compute the basis for all holomorphic 1-forms;
4. Construct the slit mapping.

Here we give a brief description of the pipeline implementation. Without loss of generality, we map the boundary γ_0 to the outer circle of the circular slit domain, γ_1 to the inner circle and all other boundaries to the concentric slits. Briefly, first, for each inner boundary $\gamma_1, , \gamma_n$, a harmonic function is computed by solving a Dirichlet problem. By definition, the exact harmonic 1-form ω_k can be computed as the gradient of the harmonic function on boundary γ_k, $(k = 1, 2, , n)$. We denote the exact harmonic 1-forms basis as $\{\omega_1, \omega_2, , \omega_n\}$. Second, we compute the closed but non-exact harmonic 1-form τ_k along the path connecting the inner boundary γ_k, $(k = 1, 2, , n)$ to outer boundary γ_0. The closed harmonic 1-form basis is the union of the exact harmonic 1-forms basis $\{\omega_1, \omega_2, , \omega_n\}$ and the closed but non-exact harmonic 1-forms basis $\{\tau_1, \tau_2, , \tau_n\}$. The computed closed harmonic 1-forms basis is used to improve the computational accuracy of the conjugate 1-form ${}^*\omega_k$ of the exact harmonic 1-form ω_k because of the inaccuracy of directly applying the brute-force Hodge star operator. We denote the conjugate 1-forms basis as $\{{}^*\omega_1, {}^*\omega_2, \ldots {}^*\omega_n\}$. Then, the holomorphic 1-forms basis is $\{\omega_1 + \sqrt{-1}^*\omega_1, \ldots, \omega_n + \sqrt{-1}^*\omega_n\}$. Next, for surface conformal slit mapping, we need to find a special holomorphic 1-form $\omega = \sum_{i=1}^{n} \lambda_i(\eta_i + \sqrt{-1}^*\eta_i)$, such that the imaginary part of its integral satisfies

$$Im\left(\int_{\gamma_k} \omega\right) = \begin{cases} -2\pi & k = 1 \\ 0 & k > 1 \end{cases}$$

To get the coefficients λ_i, we solve the following linear system for $\lambda_i, i = 1, \ldots, n$:

$$\begin{pmatrix} \alpha_{11} & \alpha_{12} & \cdots & \alpha_{1n} \\ \alpha_{21} & \alpha_{22} & \cdots & \alpha_{2n} \\ \vdots & \vdots & \ddots & \vdots \\ \alpha_{n1} & \alpha_{n2} & \cdots & \alpha_{nn} \end{pmatrix} \begin{pmatrix} \lambda_1 \\ \lambda_2 \\ \vdots \\ \lambda_n \end{pmatrix} = \begin{pmatrix} -2\pi \\ 0 \\ \vdots \\ 0 \end{pmatrix}$$

where $\alpha_{kj} = \int_{\gamma_j} {}^*\eta_k$.

It can be proven that this linear system has a unique solution, which reflects the fact that γ_1 is mapped to the inner circle of the circular slit domain. Further, the system implies the following equation $\lambda_1\alpha_{01} + \lambda_2\alpha_{02} + \cdots + \lambda_n\alpha_{0n} = 2\pi$, which means that γ_0 is mapped to the outer circle in the circular slit domain. The circular slit mapping is a complex-valued function $\phi : M \to C$. Choosing a base vertex v_0 arbitrarily, and for each vertex $v \in M$ choosing the shortest path γ from v_0 to v, we can compute the map as the following: $\phi(v) = e^{\int_\gamma \omega}$.

Figure 9.2b shows an example of conformal slit mapping [298], where there are a cortical surface with several landmark curves labeled as blue lines and its circular slit mapping result. The boundary curves on the parameter domain annotate important anatomical features.

9.2.3 *Hyperbolic Brain Conformal Parameterization with Ricci Flow Method*

The Ricci flow is an intrinsic geometric flow that deforms the metric of a Riemannian manifold, such as a 3D surface. The Ricci flow was introduced by Richard Hamilton for general Riemannian manifolds in his seminal work [109] in 1982, and it has gained increasing attention and interest in the engineering field [131, 256, 296, 324, 339]. It plays an important role in the proof of the Poincaré conjecture for three-dimensional manifolds [209–211]. Compared to other conformal parameterization methods used in brain imaging [7, 100, 126, 295, 298], the Ricci flow method can handle surfaces with complicated topologies (boundaries and landmarks) without producing singularities. It also provides a universal and flexible way to compute conformal Riemannian metrics with prescribed Gaussian curvatures. In the discrete case, it is equivalent to optimizing a convex energy. The global optimum exists and is unique, so the computation is stable and efficient.

In conformal geometry, when the Riemannian metric is conformally deformed, curvatures are also changed accordingly. Suppose $\mathbf{g}$ is changed to $\tilde{\mathbf{g}} = -e^{2u}\mathbf{h}$, where $\mathbf{h} = \mathbf{g}(0)$. Then, the Gaussian curvature will become

$$\tilde{K} = e^{-2u}(-\Delta_{\mathbf{h}}u + K_{\mathbf{h}}), \tag{9.7}$$

where $\Delta_{\mathbf{h}}$ is the Laplacian–Beltrami operator under the original metric $\mathbf{g}(0)$. The geodesic curvature will become

$$\tilde{k}_g = e^{-u}(\partial_{\mathbf{r}} u + k_h), \tag{9.8}$$

where $\mathbf{r}$ is the tangent vector orthogonal to the boundary. According to the Gauss-Bonnet theorem, the total curvature is still $2\pi\chi(S)$, where $\chi(S)$ is the Euler characteristic of S.

Suppose S is a smooth surface with a Riemannian metric $\mathbf{g}$. The Ricci flow deforms the metric $\mathbf{g}(t)$ according to the Gaussian curvature $K(t)$ (induced by itself) [48],

$$\frac{d\mathbf{g}_{ij}(t)}{dt} = (c - K(t))\mathbf{g}_{ij}(t). \tag{9.9}$$

where t is the time parameter and $c = \frac{2\pi\chi(S)}{\text{Area}(S)}$, $\chi(S)$ is the Euler characteristic of S. There is an analogy between the Ricci flow and the heat diffusion process. Suppose $T(t)$ is a temperature field on the surface. The heat diffusion equation is $\frac{dT(t)}{dt} = -\Delta_{\mathbf{g}} T(t)$, where $\Delta_{\mathbf{g}}$ is the Laplace–Beltrami operator induced by the surface metric. The temperature field becomes more and more uniform with the increase of t, and it will become constant eventually.

In a physical sense, the curvature evolution induced by the Ricci flow is exactly the same as heat diffusion on the surface, as follows:

$$\frac{dK(t)}{dt} = -\Delta_{\mathbf{g}(t)} K(t), \tag{9.10}$$

where $\Delta_{\mathbf{g}(t)}$ is the Laplace–Beltrami operator induced by the metric $\mathbf{g}(t)$. If we replace the metric in Eq. (9.9) with $\mathbf{g}(t) = e^{2u(t)}\mathbf{g}(0)$, then the Ricci flow can be simplified as

$$\frac{du(t)}{dt} = -2K(t), \tag{9.11}$$

which states that the metric should change according to the curvature. Combining Eqs. (9.7), (9.9) and (9.11), we can have the non-linear evolution of $K(t)$.

$$\frac{dK(t)}{dt} = 2K(t)^2 - e^{-2u}\Delta_{\mathbf{g}} K(t) \tag{9.12}$$

Prior work have prove that the surface Ricci flow defined in is convergent and leads to a conformal uniformization metric on surfaces with nonpositive Euler numbers [110] and on surfaces with positive Euler numbers (9.9).

The Ricci flow can be easily modified to compute a metric with a *user-defined* curvature $\bar{K}$ as,

$$\frac{du(t)}{dt} = 2(\bar{K} - K). \tag{9.13}$$

With this modification, the solution metric $\mathbf{g}(\infty)$ can be computed, which induces the curvature $\bar{K}$. It leads to an unified discrete Ricci flow method to compute surface conformal parameterization on general surfaces.

In triangular meshes, the discrete Gaussian curvature K_i on a vertex $v_i \in \Sigma$ may be computed from the angle deficit,

$$K_i = \begin{cases} 2\pi - \sum_{f_{ijk} \in F} \theta_i^{jk}, & v_i \notin \partial\Sigma \\ \pi - \sum_{f_{ijk} \in F} \theta_i^{jk}, & v_i \in \partial\Sigma \end{cases} \tag{9.14}$$

where θ_i^{jk} represents the corner angle attached to vertex v_i and $\partial\Sigma$ represents the boundary of the mesh. The discrete Gaussian curvatures are determined by the discrete metrics. Meanwhile, one can use cosine laws to compute θ_i^{jk} with edge lengths [131].

Further we can use the concept of circle packing metric [273] to approximate conformal metric deformation. Conformal metric deformations, in the smooth case, preserve infinitesimal circles and the intersection angles among them. The discrete conformal deformation of metrics uses circles with finite radii to approximate infinitesimal circles. Let $\mathbf{r}$ be a function defined on the vertices, $\mathbf{r} : V \to \mathbb{R}^+$, which assigns a radius r_i to the vertex v_i. Similarly, let Φ be a function defined on the edges, $\Phi : E \to [0, \frac{\pi}{2}]$, which assigns an acute angle ϕ_{ij} to each edge e_{ij} and is called a weight function on the edges. The set $\{\mathbf{r}, \Phi\}$ is called the *circle packing metric* of the mesh Σ. Two circle packing metrics $(\mathbf{r}_1, \Phi_1)$ and $(\mathbf{r}_2, \Phi_2)$ on the same mesh are *conformally equivalent* if $\Phi_1 \equiv \Phi_2$. A conformal deformation of a circle packing metric only modifies the vertex radii and preserves the intersection angles on the edges.

Suppose $\mathbf{u} : V \to \mathbb{R}$ denotes the discrete conformal factor vector, $\mathbf{u} = \{u_1, u_2, \ldots, u_n\}$, where n is the number of vertices on Σ, and define [246, 324]

$$u_i = \log\left(\tanh\frac{r_i}{2}\right) \tag{9.15}$$

The discrete Ricci flow is defined as follows:

$$\frac{du_i(t)}{dt} = -2K_i(t), \tag{9.16}$$

The discrete Ricci flow is exactly the same form as the Ricci flow equation in the continuous setting. Let the conformal factor vector be $\mathbf{u}_0 = \{0, 0, \ldots, 0\}$ at time 0, the discrete Ricci energy is defined as in [324]

$$f(\mathbf{u}) = \int_{\mathbf{u}_0}^{\mathbf{u}} \sum_{i=1}^{n} K_i du_i, \tag{9.17}$$

The discrete Ricci flow (Eq. (9.16)) is the negative gradient flow of the Ricci flow. In brain imaging research, discrete Ricci flow based on the gradient descent method was introduced in [291] and later Newton's optimization method was adopted in [246, 292, 296]. Figure 9.2c shows a brain surface conformal parameterization result with Euclidean Ricci flow method where brain surfaces are cut open along some given landmark curves (blue curves). Figure 9.2d shows a brain surface conformal parameterization result with hyperbolic Ricci flow method.

In summary, we introduced several different brain surface conformal parameterization methods. Such a set of global brain surface conformal parameterization methods are technically sound and numerically stable. They may increase computational accuracy and efficiency when solving partial differential equations using grid-based or metric-based computations. They also provide solid foundation for other brain imaging research including brain surface registration and brain surface morphometry study.

9.3 Brain Surface Registration

Brain surface registration or warping can be achieved by first mapping each of the 3D surfaces to a canonical parameter space such as a sphere [17, 78, 259] or a planar domain [205, 270]. A flow, computed in the parameter space of the two surfaces, induces a correspondence field in 3D. The flow can be computed by aligning curvature, sulcal depth or other geometric maps of the surfaces, as applied in FreeSurfer [63, 78], or by aligning the surface parameterizations, as applied in spherical harmonics (SPHARM) [259], or by aligning meaningful landmark curves, as applied in the cortical pattern matching algorithm [270]. Another set of brain surface warping methods, e.g. [276], are based on the large deformation diffeomorphic metric mapping (LDDMM) framework [194]. They compute diffeomorphic registrations between individual and template surfaces by generating time-dependent diffeomorphisms in a metric space.

Conformal geometry provides theoretically rigorous and computationally efficient solutions to compute canonical space for brain surface registration. Conformal geometry preserves local geometry structure which is useful to construct point-to-point correspondence defined by geometry similarities. Besides, conformal geometry may enforce landmark curves by converting landmark matching as boundary matching conditions.

An accurate cortical surface registration is critical for the brain imaging research. However, an automatic computation of such a biologically meaningful correspondence is by no means trivial. Methods for inter individual registration of cortical surfaces can be divided into two categories: (1) Landmark-based methods incorporating with precise knowledge of sulcal anatomy [139, 140, 183, 207, 268, 294, 296, 298]; (2) Techniques that align curvature, sulcal depth or other geometric maps [11, 79, 186, 247, 275, 316]. Conformal geometry research may benefit both approaches. Next, we will briefly review four different brain surface registration methods based on conformal geometry. Among them, the first there methods use brain landmark curves as matching constraints and the last method only uses geometry features for matching and reduces the need to locate brain landmark curves. Specifically, the brain conformal mapping optimization with landmark approach uses the landmark matching as soft conditions while the constrained harmonic map and hyperbolic harmonic map methods take a hard landmark matching approach by converting the it as a boundary matching condition via conformal geometry.

9.3.1 *Optimization of Brain Conformal Mapping with Landmarks*

Surface-based approaches often map cortical surface data to a parameter domain such as a sphere, providing a common coordinate system for data integration [78, 316]. There is increasing evidence for a good relationship between primary sulci and the boundaries of cytoarchitectonic fields in primary cortices [77] and it suggests that landmark-based method may provide better results. Brain spherical conformal mapping [100] offers a convenient method to parameterize cortical surfaces without angular distortion, generating an orthogonal grid on the cortex that locally preserves the metric. Here we introduce a new method, based on a new energy functional, to optimize the conformal parameterization of cortical surfaces by using landmarks [294].

Suppose C_1 and C_2 are two cortical surfaces we want to register. For a diffeomorphism between two genus zero surfaces, a map is conformal if it minimizes the harmonic energy $E_{harmonic}$ (Definition 9.3). We let $f_1 : C_1 \rightarrow S^2$ be the conformal parameterization of C_1 mapping it onto S^2. We manually label the landmarks on the two cortical surfaces as discrete point sets. We denote them as $\{p_i \in C_1\}$, $\{q_i \in C_2\}$, with p_i matching q_i. We proceed to compute a map $f_2 : C_2 \rightarrow S^2$ from C_2 to S^2, which minimizes the harmonic energy as well as minimizing the so-called landmark mismatch energy. The landmark mismatch energy measures the Euclidean distance between the corresponding landmarks. Alternatively, landmark errors could be computed as geodesic distances with respect to the original surfaces, rather than on the sphere; here we chose to perform distance computations on the sphere. Using our algorithm, the computed map should effectively preserve the conformal property and match the geometric features on the original structures as far as possible.

Let $h : C_2 \rightarrow S^2$ be any homeomorphism from C_2 onto S^2. We define the landmark mismatch energy of h as, $E_{landmark}(h) = 1/2 \sum_{i=1}^{n} ||h(q_i)) - f_1(p_i)||^2$. where the norm represents distance on the sphere. By minimizing this energy functional, the Euclidean distance between the corresponding landmarks on the sphere is minimized.

To optimize the conformal parameterization, we propose to find $f_2 : C_2 \rightarrow S^2$ which minimizes the following new energy functional (instead of the harmonic energy functional), $E_{new}(f_2) = E_{harmonic}(f_2) + \lambda E_{landmark}(f_2)$, where λ is a weighting factor (Lagrange multiplier) that balances the two penalty functionals. It controls how much landmark mismatch we want to tolerate. When $\lambda = 0$, the new energy functional is just the harmonic energy. When λ is large, the landmark mismatch energy can be greatly reduced. But more conformality will be lost (here we regard deviations from conformality to be quantified by the harmonic energy).

Now, let K represent the simplicial realization (triangulation) of the brain surface C_2, let u, v denote the vertices, and $[u, v]$ denote the edge spanned by u, v. Our new energy functional can be written as:

$$\begin{aligned} E_{new}(f_2) &= 1/2 \sum_{[u,v]\in K} k_{u,v}||f_2(u) - f_2(v)||^2 + \lambda/2 \sum_{i=1}^{n} ||f_2(q_i) - f_1(p_i)||^2 \\ &= 1/2 \sum_{[u,v]\in K} k_{u,v}||f_2(u) - f_2(v)||^2 + \lambda/2 \sum_{u\in K} ||f_2(u) - L(u))||^2 \chi_M(u) \end{aligned}$$

where $M = \{q_1, \ldots, q_n\}$; $L(q_i) = p_i$ if $u = q_i \in M$ and $L(u) = (1, 0, 0)$ otherwise. The first part of the energy functional is defined as in [100]. Note that by minimizing this energy, we may give up some conformality but the landmark mismatch energy is progressively reduced.

Suppose we would like to compute a mapping f_2 that minimizes the energy $E_{new}(f_2)$. This can be solved easily by steepest descent.

Definition 3.1: Suppose $f \in C^{PL}$, where C^{PL} represent a vector space consists of all piecewise linear functions defined on K. We define the Laplacian as follows: $\Delta f(u) = \sum_{[u,v]\in K} k_{u,v}(f(u) - f(v)) + \lambda \sum_{u\in K}(f_2(u) - L(u))\chi_M(u)$.

Definition 3.2: Suppose $\vec{f} \in C^{PL}$, $\vec{f} = (f_0, f_1, f_2)$, where the f_i are piecewise linear. Define the Laplacian of $\vec{f}$ as $\Delta \vec{f} = (\Delta f_0(u), \Delta f_1(u), \Delta f_2(u))$.

Now, we know that $f_2 = (f_{20}, f_{21}, f_{22})$ minimizes $E_{new}(f_2)$ if and only if the tangential component of $\Delta f_2(u)$= $(\Delta f_{20}(u), \Delta f_{21}(u), \Delta f_{22}(u))$ vanishes. That is $\Delta(f_2) = \Delta(f_2)^{\perp}$.

In other words, we should have $P_{\vec{n}} \Delta f_2(u) = \Delta f_2(u) - (\Delta f_2(u) \cdot \vec{n})\vec{n} = 0$. We use a steepest descent algorithm to compute $f_2 : C_2 \to S^2$: $\frac{df_2}{dt} = -P_{\vec{n}} \Delta f_2(t)$.

Algorithm 1 *Algorithm to Optimize the Combined Energy E_{new}*

Input: *(mesh K, step length δt, energy difference threshold δE).*
Output: *($f_2 : C_2 \to S^2$), which minimizes E. The computer algorithm proceeds as follows:*
1. Given a Gauss map $I : C_2 \to S^2$. Let $f_2 = I$, compute $E_0 = E_{new}(I)$.
2. For each vertex $v \in K$, compute $P_{\vec{n}} \Delta f_2(v)$.
3. Update $f_2(v)$ by $\delta f_2(v) = -P_{\vec{n}} \Delta f_2(v)\delta t$.
4. Compute energy E_{new}.
5. If $E_{new} - E_0 < \delta E$, return f_2. Otherwise, assign E to E_0. Repeat steps 2 to 5.

In our experiment, we tested our algorithm on a set of left hemisphere cortical surfaces generated from brain MRI scans of 40 healthy adult subjects, aged 27.5 $\pm$ 7.4SD years (16 males, 24 females), scanned at 1.5 T (on a GE Signa scanner). Data and cortical surface landmarks were those generated in a prior paper [268], where the extraction and sulcal landmarking procedures are fully detailed. Using this set of 40 hemispheric surfaces, we mapped all surfaces conformally to the sphere and minimized the compound energy matching all subjects to a randomly selected

Table 9.1 Numerical data from our experiment. The landmark mismatch energy is greatly reduced while the harmonic energy is only slightly increased. The table also illustrates how the results differ with different values of λ. The landmark mismatch error can be reduced by increasing λ, but conformality will increasingly be lost

	$\lambda = 3$	$\lambda = 6$	$\lambda = 10$
$E_{harmonic}$ of the initial (conformal) parameterization:	100.6	100.6	100.6
$\lambda E_{landmark}$ of the initial (conformal) parameterization:	81.2	162.4	270.7
Initial compound energy ($E_{harmonic} + \lambda E_{landmark}$):	181.8	263.0	371.3
Final $E_{harmonic}$	109.1 (↗ 8.45%)	111.9 (↗ 11.2%)	123.0 (↗ 22.2%)
Final $\lambda E_{landmark}$	11.2 (↘ 86.2%)	13.7 (↘ 91.6%)	15.6(↘ 95.8%)
Final compound energy ($E_{harmonic} + \lambda E_{landmark}$)	120.3 (↘ 33.8%)	125.6(↘ 52.2%)	138.6 (↘ 62.7%)

individual subject (alternatively, the surfaces could have been aligned to an average template of curves on the sphere). An important advantage of this approach is that the local adjustments of the mapping to match landmarks do not greatly affect the conformality of the mapping. In Fig. 9.3a, the cortical surface C_1 (a control subject) is mapped conformally ($\lambda = 0$) to the sphere. In (b), another cortical surface C_2 is mapped conformally to the sphere. Note that the sulcal landmarks appear very different from those in (a) (see landmarks in the green square). This means that the geometric features are not well aligned on the sphere unless a further feature-based deformation is applied. In Fig. 9.3c, we map the cortical surface C_2 to the sphere with our algorithm, while minimizing the compound energy. This time, the landmarks closely resemble those in (a) (see landmarks in the green square).

In Fig. 9.4, statistics of the angle difference are illustrated. Note that under a conformal mapping, angles between edges on the initial cortical surface should be preserved when these edges are mapped to the sphere. Any differences in angles can be evaluated to determine departures from conformality. Figure 9.4a shows the histogram of the angle difference using the conformal mapping, i.e. after running the algorithm using the conformal energy term only. Figure 9.3b shows the histogram of the angle difference using the compound functional that also penalizes landmark mismatch. Despite the fact that inclusion of landmarks requires more complex mappings, the angular relationships between edges on the source surface and their images on the sphere are clearly well preserved even after landmark constraints are enforced.

We also tested with other parameter λ with different values. Table 9.1 shows numerical data from the experiment. From the Table, we observe that the landmark mismatch energy is greatly reduced while the harmonic energy is only slightly increased. The table also illustrates how the results differ with different values of λ. We observe that the landmark mismatch error can be reduced by increasing λ, but conformality is increasingly lost.

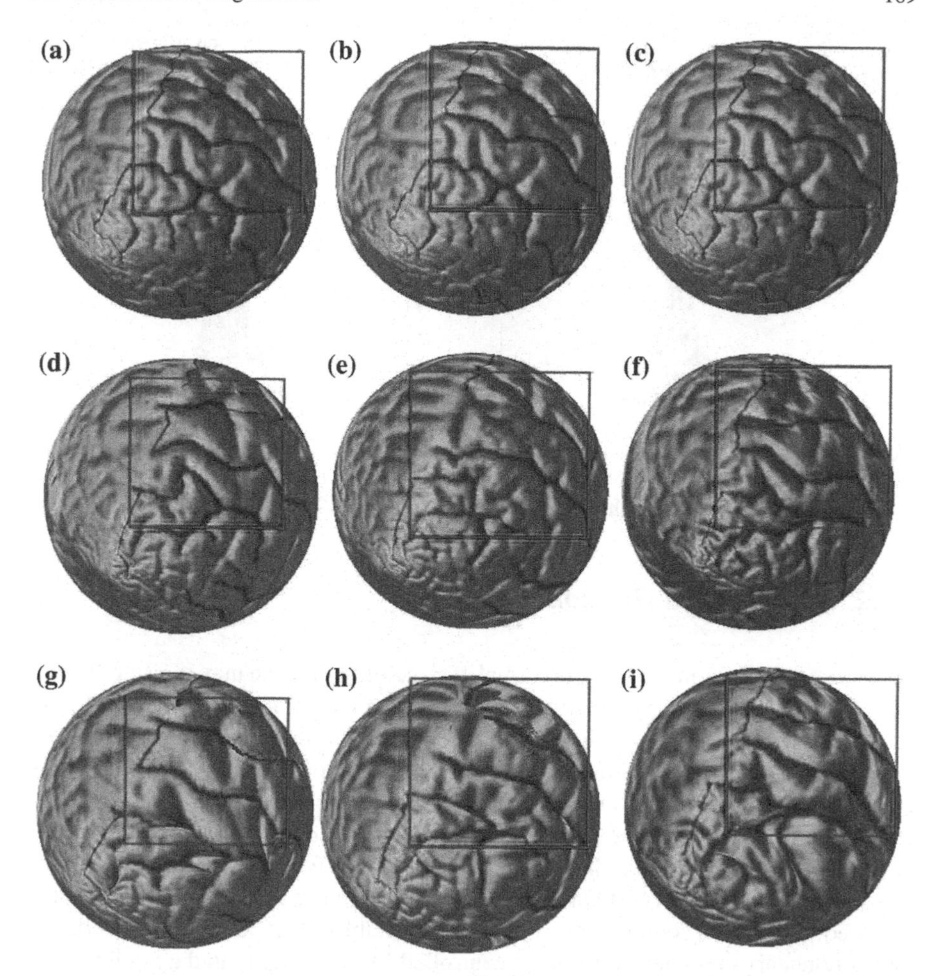

Fig. 9.3 In **a**, the cortical surface C_1 (the control) is mapped conformally ($\lambda = 0$) to the sphere. In **d**, another cortical surface C_2 is mapped conformally to the sphere. Note that the sulcal landmarks appear very different from those in **a** (see landmarks in the green square). In **g**, the cortical surface C_2 is mapped to the sphere using our algorithm (with $\lambda = 3$). Note that the landmarks now closely resemble those in **a** (see landmarks in the green square). **b** and **c** shows the same cortical surface (the control) as in **a**. In **e** and **f**, two other cortical surfaces are mapped to the spheres. The landmarks again appears very differently. In h and i, the cortical surfaces are mapped to the spheres using our algorithm. The landmarks now closely resemble those of the control. Image from [294]

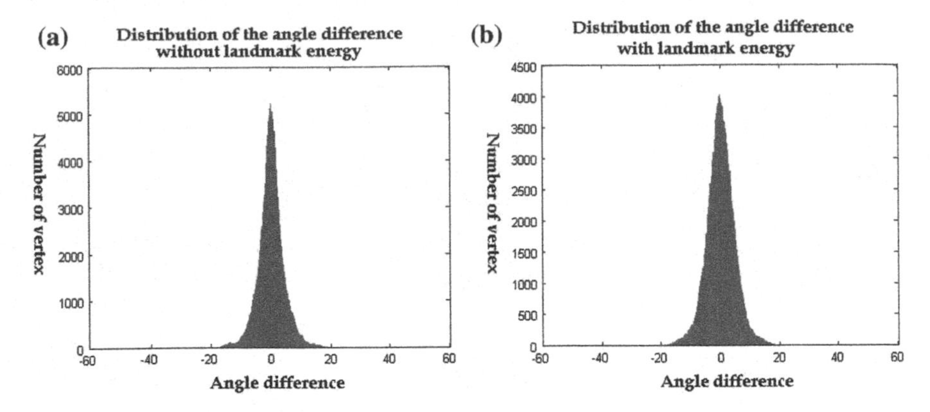

Fig. 9.4 Histogram **a** shows the statistics of the angle difference using the conformal mapping. Histogram **b** shows the statistics of the angle difference using our algorithm ($\lambda = 3$). It is observed that the angle is well-preserved. Image from [294]

9.3.2 Constrained Harmonic Map

After we compute brain surface conformal parameterization, we may register 3D surfaces by computing a constrained harmonic map on the parameter domain with the Euclidean metric. Among various rigid and non-rigid surface registration approaches (e.g. [24, 35, 157]), the harmonic map is one of the most broadly applied methods [293, 328]. The advantages of harmonic map computation are: (1) it is physically natural and can be computed efficiently; (2) it measures the elastic energy of the deformation so it has a clear physical interpretation; (3) for a planar convex domain, it is a diffeomorphism; (4) it can be computed by solving an elliptic partial differential equation so its computation is numerically stable; (5) it continuously depends on the boundary condition so it can be controlled by adjusting boundary conditions. In computer vision and medical imaging fields, surface harmonic maps have been used to compute spherical conformal mapping [100], image registration [138], high resolution tracking of non-rigid motion [293], non-rigid surface registration [173], etc.

The basic pipeline of the constrained harmonic map is as follows: on the parameter domain, for each pair of corresponding surfaces, by using a piecewise affine transformation, we map the boundary of the second segment to the boundary of the first one and denote the resulting mapping by ϕ_0. We improve the mapping by using a constrained harmonic map using the heat diffusion method,

$$\frac{\partial \phi(t)}{\partial t} = -\Delta \phi(t), \phi(0) = \phi_0.$$

After that, we resample the meshes using a regular grid in the parameter domain and construct new meshes with the same connectivity for the two segments.

Formally, for two surfaces S_1 and S_2, we would like to find a smooth map $\phi : S_1 \to S_2$, such that $\phi(\partial S_1) = \partial S_2$. To improve the smoothness, we require $\Delta\phi = 0$, namely, ϕ is a harmonic map. It is difficult to find ϕ directly, but instead we can easily find a harmonic map between the parameter domains. Suppose the the conformal parameterization of S_1 is τ_1, conformal parameterization for S_2 is τ_2, then $\tau_1(S_1)$ and $\tau_2(S_2)$ are rectangles in R^2. We want to find a harmonic map $\tau : R^2 \to R^2$, such that

$$\tau \circ \tau_1(\partial S_1) = \tau_2(\partial S_2), \, \Delta\tau = 0,$$

where ∂S_1 and ∂S_2 denote two surface boundaries and Δ is the Laplacian operator defined on the plane. Then the map ϕ can be obtained by the following commutative diagram,

$$\begin{array}{ccc} S_1 & \xrightarrow{\phi} & S_2 \\ \tau_1 \downarrow & & \downarrow \tau_2 \\ R^2 & \xrightarrow[\tau]{} & R^2 \end{array} \tag{9.18}$$

as $\phi = \tau_1 \circ \tau \circ \tau_2^{-1}$. Because both τ_1 and τ_2 are conformal, τ is harmonic, and therefore ϕ is harmonic.

In the implementation, it is equivalent to solving the Laplace's equation with the Dirichlet boundary condition. Specifically, we define the positions of function τ on boundary vertices of surface S_1 parameter domain to be those of boundary vertices of surface S_2 parameter domain. For all interior vertices, their positions are unknown but satisfy $\Delta\tau = 0$. Thus, we define

$$\tau = \begin{cases} \Delta\tau(v) = 0 & \forall v \notin \partial S_1 \\ \tau(\tau_1(\partial S_1)) = \tau_2(\partial S_2) & \forall v \in \partial S_1 \end{cases} \tag{9.19}$$

Then we define the following local stiffness matrix F for all interior vertices v_i on $\tau_1(S_1)$:

$$F_{ij} = \begin{cases} k(v_i, v_j) & [v_i, v_j] \in K \\ 0 & [v_i, v_j] \notin K \end{cases} \tag{9.20}$$

where $k(v_i, v_j)$ is defined in Definition 9.3. Next, we add the contribution of the local stiffness matrix to the global stiffness matrix and define the discrete Laplace–Beltrami operator for triangular mesh as

$$L_p = D - F \tag{9.21}$$

where D, the degree matrix, is the diagonal matrix defined as $D_{ii} = \Sigma_j F_{ij}$.

Further we can construct a linear system to solve the Laplace's equation,

$$A_p \tau_i = B_p \tau_b \tag{9.22}$$

Suppose there are N vertices on the parameter mesh $\tau_1(S_1)$ of surface S_1, m of them are interior vertices and n of them are boundary vertices, i.e. vertices on $\partial\tau_1(S_1)$. Then A_p is an $m \times m$ matrix, B_p is an $m \times n$ matrix and $L_p = [A_p B_p]$; tau_i is an $m \times 1$ vector, representing the unknown position values on interior vertices; τ_b is an $n \times 1$ vector, representing position values on boundary vertices, whose values are fixed. After solving Eq. (9.22), the function τ acquires a position at each interior vertex.

It is well known that if the target mapping domain is convex, then the Euclidean planar harmonic maps are diffeomorphic [232]. Thus constrained harmonic map has been frequently used together brain conformal parameterization for generate point-to-point surface correspondence. By converting landmark curve matching problem to Dirichlet boundary condition, this approach is able to enforce multiple boundary conditions. The harmonic map based solutions are stable and accurate.

The constrained harmonic map brain surface registration has been applied to various parameter domain, including conformal slit map (Fig. 9.2b, [297, 298, 298], punched hole disk (Fig. 9.2c). Besides that, as described below, the constrained harmonic map method was also applied to register general surfaces, i.e. multiply connected surfaces.

The lateral ventricles – fluid-filled structures deep in the brain – are often enlarged in disease, and can provide sensitive measures of disease progression. Surface-based analysis approaches have been applied in many studies to examine ventricular surface morphometry [40, 76, 260, 271, 299]. Ventricular changes typically reflect atrophy in surrounding structures, and ventricular measures and surface-based maps often provide sensitive (albeit indirect) assessments of tissue reduction that correlate with cognitive deterioration in illnesses. Ventricular measures have also recently garnered interest as good biomarkers of progressive brain change in dementia. They can usually be extracted from brain MRI scans with greater precision than hippocampal surfaces or other models [301]. However, it is challenging to apply normal spherical or planar parameter domain as canonical spaces to analyze ventricular surfaces, due to their concave shape, complex branching topology and extreme narrowness of the inferior and occipital horns. Hyperbolic conformal geometry has an important property that it can induce conformal parameterizations on high-genus surfaces or surfaces with negative Euler numbers and the resulting parameterizations have no singularities [246]. Basically, we first automatically locate and introduce three cuts on each ventricular surface, with one cut on the superior horn, one cut on the inferior horn, and one cut on the occipital horn. The locations of the cuts are motivated by examining the topology of the lateral ventricles, in which several horns are joined together at the ventricular "atrium" or "trigone". Meanwhile, we kept the locations of the cuts consistent across subjects. Figure 9.5a show the topology optimization result. After that, a ventricular surface become a genus surface with three open boundaries. We apply hyperbolic Ricci flow method to conformally map the ventricular surface

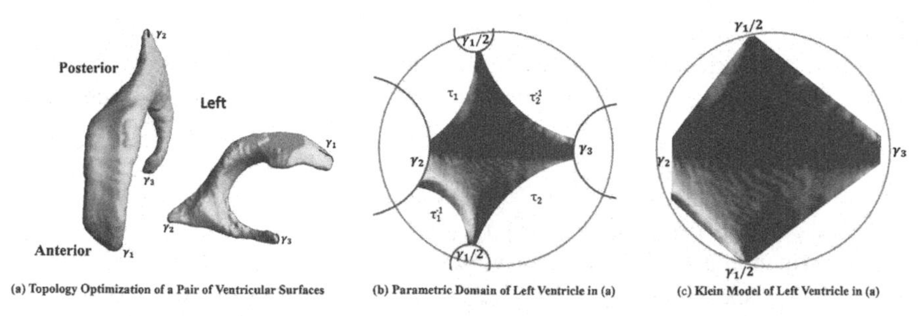

Fig. 9.5 Modeling ventricular surface with hyperbolic geometry. **a** shows three identified open boundaries, $\gamma_1, \gamma_2, \gamma_3$, on the ends of three horns. After that, ventricular surfaces can be conformally mapped to the hyperbolic space. **b**, **c** show the hyperbolic parameter space, where **b** is the Poincaré disk model and **c** is the Klein model. Image from [246]

to the hyperbolic disk. With geodesic lifting, we locate a set of boundaries ($\gamma_1/2$, τ_2, γ_3, τ_2^{-1}, $\gamma_1/2$, τ_1, γ_2 and τ_1^{-1}, shown in Fig. 9.5b), which are consistent across subjects. Furthermore, we convert the Poincaré model to the Klein model with the following transformation [324],

$$z \to \frac{2z}{1 + \bar{z}z}. \tag{9.23}$$

It converts the canonical fundamental domains of the ventricular surfaces to a Euclidean octagon, as shown in Fig. 9.5c. Then we use the Klein disk as the canonical parameter space and apply the constrained harmonic map for ventricular surface registration.

9.3.3 *Hyperbolic Harmonic Map*

The current state-of-the-art surface harmonic map research, including constrained harmonic map, has some limitations. For example, it usually only works with Euclidean metric parameter space and does not directly work on hyperbolic parameter space. Further, a constrained harmonic map combined with landmark matching conditions usually does not guarantee diffeomorphism. All these problems become obstacles using harmonic maps to solve general non-rigid surface matching problems.

Harmonic maps can be defined on general surfaces. Prior work, e.g. [139, 246, 293, 298] mainly computed the harmonic map on Euclidean plane. With our prior work on hyperbolic Ricci flow method [131, 324, 324], here we introduce methods to establish harmonic maps between surfaces with complicated topologies, incorporating landmark curves constraints. we call our method *hyperbolic harmonic map* [251].

Theory
Suppose S is an oriented surface with a Riemannian metric $\mathbf{g}$. One can choose a set of special local coordinates (x, y), the so-called isothermal parameters, such that

$$\mathbf{g} = \sigma(x, y)(dx^2 + dy^2) = \sigma(z)dzd\bar{z},$$

where the complex parameter $z = x + iy$, $dz = dx + idy$.

Given a mapping $f : (M, \mathbf{g}_m) \to (N, \mathbf{g}_n)$, z and w are local isothermal parameters on M and N respectively; $\mathbf{g}_m = \sigma(z)dzd\bar{z}$ and $\mathbf{g}_n = \rho(w)dwd\bar{w}$. Then the mapping has local representation $w = f(z)$ or denoted as $w(z)$. We also name M as the *source surface* and N as the *target surface*.

Definition 9.7 (*Harmonic Map*) The *harmonic energy* of the mapping is defined as

$$E(f) = \int_M \rho(w(z))(|w_z|^2 + |w_{\bar{z}}|^2)dxdy. \tag{9.24}$$

where the complex differential operator is defined as

$$\frac{\partial}{\partial z} = \frac{1}{2}\left(\frac{\partial}{\partial x} + i\frac{\partial}{\partial y}\right), \frac{\partial}{\partial \bar{z}} = \frac{1}{2}\left(\frac{\partial}{\partial x} - i\frac{\partial}{\partial y}\right).$$

If f is a critical point of the harmonic energy, then f is called a *harmonic map*.

Harmonic energy depends on the Riemannian metric on the target surface, and the conformal structure of the source surface. Namely, if the Riemannian metric on the source surface is deformed conformally, the energy does not change. The necessary condition for f to be a harmonic map is the following Euler–Lagrange equation

$$w_{z\bar{z}} + \frac{\rho_w}{\rho} w_z w_{\bar{z}} \equiv 0. \tag{9.25}$$

Theorem 9.8 *Suppose $f : (M, \mathbf{g}_m) \to (N, \mathbf{g}_n)$ is a degree one harmonic map, furthermore the Riemannian metric on N induces negative Gaussian curvature, then for each homotopy class, the harmonic map is unique and diffeomorphic.*

The theories on the existence, uniqueness and regularity of harmonic maps have been thoroughly discussed in [232]. They are the theoretic foundation for the proposed hyperbolic harmonic map computation algorithm.

Definition 9.9 (*Cross Ratio*) Suppose z_1, z_2, z_3, z_4 are points in $\mathbb{C} \cup \{\infty\}$, the complex cross ratio is given by $(z_1, z_2, z_3, z_4) := \frac{z_1 - z_3}{z_1 - z_4} : \frac{z_2 - z_3}{z_2 - z_4}$.

The complex cross ratio is invariant under Möbius transformations. Namely, if φ is a Möbius transformation, then $(\varphi(z_1), \varphi(z_2), \varphi(z_3), \varphi(z_4)) = (z_1, z_2, z_3, z_4)$.

Let z be a point on the Poincaré disk; the tangent space at z is denoted as $T_z\mathbb{H}^2$. Suppose $v \in T_z\mathbb{H}^2$ is a tangent vector at z, and there is a unique geodesic $\gamma(t)$, such that $\gamma(0) = z$, $\dot{\gamma}(0) = v$, then the *exponential map* at z, $\exp(\cdot, z) : T_z\mathbb{H}^2 \to \mathbb{H}^2$ is

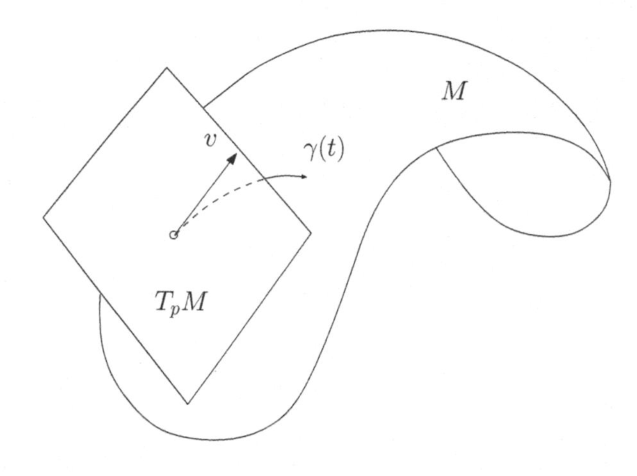

Fig. 9.6 Exponential map. Image from [251]

$$\exp(v, z) := \gamma(1). \tag{9.26}$$

The *logarithm map* is the inverse to the exponential map, the logarithm at the origin 0, $\log(\cdot, 0) : \mathbb{H}^2 \to T_0\mathbb{H}^2$,

$$\log(w, 0) := \rho(w, 0)\frac{w}{|w|}. \tag{9.27}$$

where $\rho(w, 0)$ is the hyperbolic distance between the origin and w. Figure 9.6 illustrates the exponential map on a Riemannian manifold.

Given a surface with a metric $(S, \mathbf{g})$, fix a point $p \in S$. The exponential map at p is diffeomorphic on a small disk at T_pS. The supreme of such a disk radius is called the *injective radius* at p.

Suppose $Q = \{q_1, q_2, \ldots, q_k\} \subset S$ is a point set in a geodesic disk $D(p, r)$, whose radius r is smaller than the injective radius at p. The points are associated with the weights $\Lambda = \{\lambda_1, \lambda_2, \ldots, \lambda_k\}$, each $\lambda_i \geq 0$, then the *weighted geodesic mass center* of (Q, Λ) is defined as $c(Q, \Lambda) := argmin_q \sum_{i=1}^{k} \lambda_j \rho^2(q, q_i)$. From Riemannian geometry, the mass center exists and is unique.

Algorithm

Figure 9.7 illustrates the major steps to compute hyperbolic harmonic maps between surfaces. Algorithm 2 sketches the algorithm pipeline.

In the pipeline, the first four steps, i.e. Step 1. topology optimization; Step 2. hyperbolic Ricci flow; Step 3. hyperbolic pants decomposition; Step 4. constrained harmonic map with Euclidean metric, have been introduced in previous sections. Figure 9.8 illustrates the decomposition of a pair of hyperbolic pants. Here we focus on Step 5, nonlinear heat diffusion.

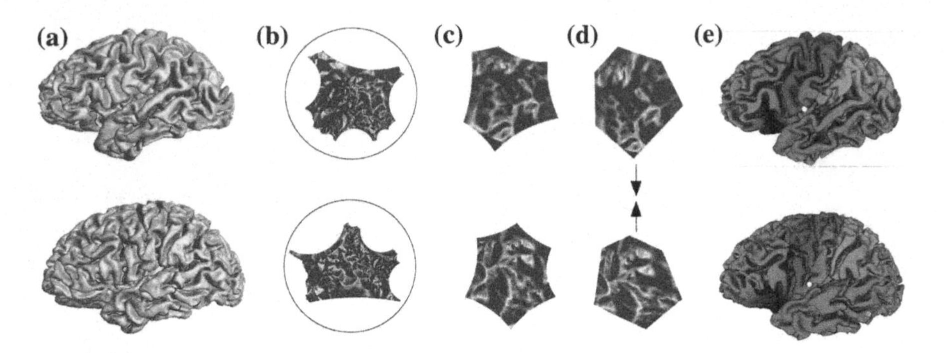

Fig. 9.7 Algorithm Pipeline (suppose we have 2 brain surfaces M and N as input): **a** The input brain models M and N, with landmarks cut open as boundaries. **b** Hyperbolic embedding of M and N on the Poincaré disk. **c** Decompose M and N into multiple pants, and each pant further decomposed to 2 hyperbolic hexagons. **d** Hyperbolic hexagons on the Poincaré disk become convex hexagons under the Klein model, then a one-to-one map between the correspondent parts of M and N can be obtained via the constrained Euclidean harmonic map. Then we can apply our hyperbolic heat diffusion algorithm to obtain a global hyperbolic harmonic diffeomorphism. **e** Color coded registration result of M and N. The colored balls on the models show the detailed correspondence, as the balls with the same color correspond to each other. Image from [251]

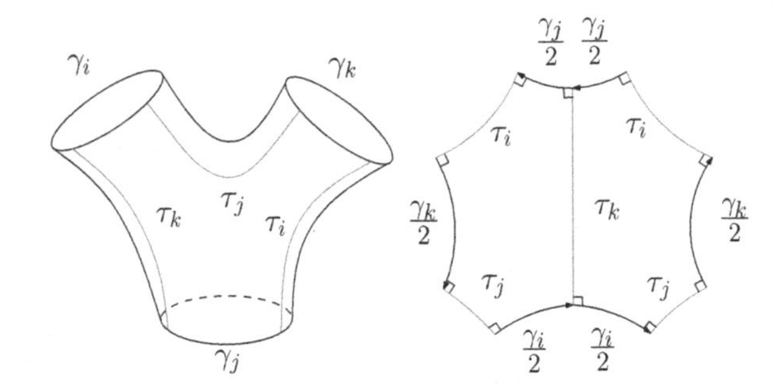

Fig. 9.8 A pair of hyperbolic pants is decomposed to two hyperbolic hexagons. Image from [251]

Algorithm 2 *Surface Hyperbolic Harmonic Mapping Algorithm Pipeline.*

1. *Topology optimization for landmark curve constraints;*
2. *Compute the hyperbolic metric using hyperbolic Ricci flow;*
3. *Hyperbolic pants decomposition, isometrically embed them to the Poincaré disk and then map them to the Klein model;*
4. *Compute harmonic maps using Euclidean metrics between corresponding pairs of pants, with consistent boundary constraints;*
5. *Use nonlinear heat diffusion to improve the mapping to a global hyperbolic harmonic map on the Poincaré disk model.*

Constrained Euclidean harmonic map help generate the initial map, $f : S_1 \rightarrow S_2$. The goal of this step is to diffuse the initial mapping and achieve a global hyperbolic harmonic map which is not restricted on every patch, as used in the prior step. Here a nonlinear heat diffusion method is adopted to compute the harmonic mapping, which is based on a conformal atlas induced by the hyperbolic metric.

Hyperbolic Atlas

Let $(S, \mathbf{g})$ be a dense triangle mesh with hyperbolic metric $\mathbf{g}$. Then for each vertex $v_i \in S$, the one ring neighboring faces form a neighborhood N_i, the union of N_i's covers the whole mesh, $S \subset \bigcup_{v_i \in S} N_i$. Isometrically embed N_i to the Poincaré's disk $\phi_i : N_i \to H^2$, then $\{(N_i, \phi_i)\}$ form a conformal atlas. Furthermore, the chart transitions are Möbius transformations. All the following computations are carried out on local charts of the conformal atlas. The computational result is independent of the choice of local parameters.

Map Representation

Suppose v is a vertex on S_1, with local representation z, its image $w(z)$ is inside a triangular face $t(v)$ of S_2. Suppose the three vertices of $t(v)$ have local representations w_i, w_j, w_k, then we compute the *complex cross ratio*

$$\eta(v) := \left(w(z), w_i, w_j, w_k\right).$$

The image of v is then represented by the pair $[t(v), \eta(v)]$. Note that, all the local coordinates transitions in the conformal chart of S_1 and S_2 are Möbius transformations, and the cross ratio η is invariant under Möbius transformation, therefore, the representation of the mapping $f : v \to [t(v), \eta(v)]$ is independent of the choice of local coordinates.

Hyperbolic Harmonic Map

Let $f : (S_1, \mathbf{g}_1) \to (S_2, \mathbf{g}_2)$ be a mapping, the discrete harmonic energy is similar to the Euclidean one (Eq. (9.2)),

$$E(f) = \frac{1}{2} \sum_{[v_i, v_j] \in S_1} k_{ij} \rho^2(f(v_i), f(v_j)). \tag{9.28}$$

where $\rho(f(v_i), f(v_j))$ is the hyperbolic distance between $f(v_i)$ and $f(v_j)$. By definition, if f is harmonic, then for an arbitrary vertex v_i, fixing all the other vertices, the following energy should be minimized:

$$\min_{f(v_i)} \sum_{[v_i, v_j] \in S_1} k_{ij} \rho^2(f(v_i), f(v_j)), \tag{9.29}$$

therefore, the harmonic map f satisfies the *mean value property*: for each vertex v_i, its image coincides with the weighted geodesic mass center of the images of its neighbors.

Namely, let v_i be a vertex, its one-ring neighbors are $\{v_{i_1}, v_{i_2}, \ldots, v_{i_k}\}$. Let $Q(v_i) = \{f(v_{i_1}), f(v_{i_2}), \ldots, f(v_{i_k})\}$, and weights $\Lambda(v_i) = \{k_{i,i_1}, k_{i,i_2}, \ldots, k_{i,i_k}\}$,

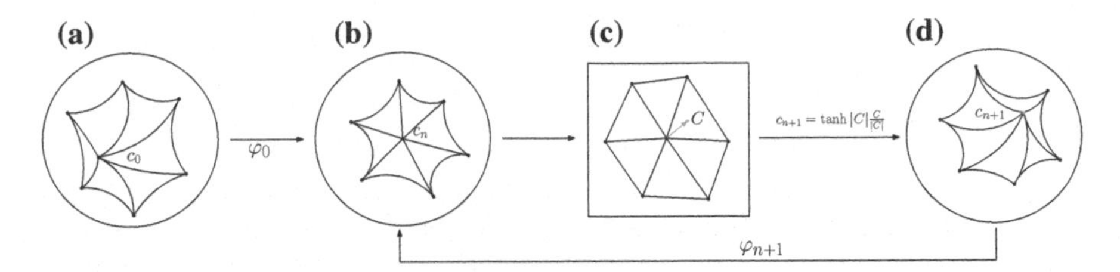

Fig. 9.9 Steps to compute the weighted geodesic mass center. **a** The initial state. **b** Use the Möbius transformation to move c to the center of Poincaré disk. **c** Compute the Euclidean weighted mass center C in the tangent space. **d** Use the exponential map to map C to Poincaré disk. Image from [251]

then $f(v_i) = c(Q(v_i), \Lambda(v_i))$, where the right hand side is the hyperbolic weighted geodesic mass center.

Weighted Geodesic Mass Center

Given a point set $Q = \{z_1, z_2, \ldots, z_k\}$ and weights $\Lambda = \{\lambda_1, \ldots, \lambda_k\}$ on Poincaré's disk, denote the weighted geodesic mass center $c(Q, \Lambda)$ as c. It can be computed by the following iterations and is illustrated in Fig. 9.9.

At the initial step, set $Q_0 = Q$ and set c_0 be any point in the convex hull of Q_0. At the n-th step, apply a Mobiüs transformation to Q_n. $\varphi_n(z) = \frac{z-c_n}{1-\bar{c}_n z}$, Set $z_i^{n+1} = \varphi_n(z_i^n)$, $Q_{n+1} = \{z_1^{n+1}, z_2^{n+1}, \ldots, z_k^{n+1}\}$.

Compute the logarithms of z_i^{n+1}'s using Eq. (9.27). In the tangent plane at the origin, compute the Euclidean weighted mass center $C := \frac{\sum_i \lambda_i \log z_i^{n+1}}{\sum_i \lambda_i}$. Compute the exponential map of C using Eq. (9.26), update the center $c_{n+1} \leftarrow \exp(C, 0)$. Repeat this procedure, then the sequence $\{c_n\}$ converges, $\{z_i^n\}$ converges for all $1 \leq i \leq k$. Then the limit c_∞ is the weighted geodesic mass center of the point set Q_∞ with weights Λ. A detailed algorithm description can be found in Algorithm 3.

Algorithm 3 *Weighted Geodesic Mass Center.*

Input: *A point set $Q = \{z_1, z_2, \cdots, z_k\}$ on the Poincaré disk, the diameter of Q is less than the injective radius of S; A weight set $\Lambda = \{\lambda_1, \lambda_2, \cdots, \lambda_k\}$.*
Output: *The weighted geodesic mass center $c(Q, \Lambda)$.*
1. Initialize $z_i^0 \leftarrow z_i$, $Q_0 \leftarrow \{z_i^0\}$, $c_0 \leftarrow 0$.
repeat
2. Set $\varphi_n \leftarrow (z - c_n)/(1 - \bar{c}_n z)$
3. Set $z_i^{n+1} \leftarrow \varphi_n(z_i^n)$, $Q_{n+1} \leftarrow \{z_i^{n+1}\}$.
4. Compute the logarithm of z_i^{n+1} using Eq. 9.27.
5. Compute the Euclidean weighted mass center in the tangent space at the origin $C \leftarrow (\sum_i \lambda_i \log z_i^{n+1})/(\sum_i \lambda_i)$.
6. Compute the exponential map of C using Eq. 9.26

$$c_{n+1} \leftarrow \tanh|C|\frac{C}{|C|}.$$

until $\rho(c_n, c_{n+1}) < \varepsilon$
7. Construct a Möbius transformation φ, *mapping* $\{z_1^n, z_2^n, z_3^n\}$ *to* $\{z_1, z_2, z_3\}$.
8. Return the weighted geodesic mass center $\varphi(c_n)$.

The computation of the weighted geodesic mass center is guaranteed to converge. Because the surface has a hyperbolic metric, the energy defined in Eq. (9.29) is strictly convex, and the minimizer is unique. The computation on the tangent space can be treated as the first order approximation, thus moving the image to the weighted mass center in the tangent space is equivalent to the gradient descent. The energy decreases monotonically, and the convexity of the energy ensures the process converges to the unique solution.

Non-linear Heat Diffusion

For each vertex v_i, we compute the weighted geodesic mass center of the images of its neighboring vertices $c(Q(v_i), \Lambda(v_i))$, then update the image of v_i to be the mass center, $f(v_i) \leftarrow c(Q(v_i), \Lambda(v_i))$. Update the images of all vertices on S_1. Repeat this process, until for each vertex, the geodesic distance between its image, and the weighted geodesic mass center of the images of its neighbors is less than a given threshold.

We try to solve the optimization problem defined in Eq. (9.29) in the neighborhood of each vertex and move from neighborhood to neighborhood. As a result, although we introduce cuts, e.g. γ_i, γ_j and γ_k in Fig. 9.8, to establish the initial mapping, the non-linear diffusion computation naturally go across these cuts. The diffusion is computed throughout the entire surface and independent of the cutting curves introduced in the initial mapping state. Algorithm 4 illustrates the process step by step. After the hyperbolic heat diffusion converges, we obtain hyperbolic harmonic maps between source and target surfaces.

The non-linear heat diffusion is guaranteed to converge. A brief proof is as follows. In the smooth case, according to the harmonic mapping theory between Riemannian manifolds, if the Riemannian metric of the target induces negative curvature everywhere and the mapping degree is one, then for each homotopy class, the harmonic map is unique and diffeomorphic.

Suppose a degree one mapping between two smooth surfaces is given, $f : (S_1, \mathbf{g}_1) \to (S_2, \mathbf{g}_2)$, where $\mathbf{g}_2$ is a hyperbolic metric. There is a unique harmonic map φ homotopic to f, with harmonic energy $E(\varphi)$. In the discrete approximation, the source surface is approximated by a polyhedral surface $\tilde{S}_1$, and the target is still the smooth hyperbolic surface S_2. The approximated mapping is $\tilde{\varphi} : \tilde{S}_1 \to S_2$. The non-linear heat diffusion method reduces the harmonic energy monotonically; on the other hand, the harmonic energy of the discrete approximation map is always greater than or equal to the smooth energy $E(\tilde{\varphi}) \geq E(\varphi)$ which is non-negative. In particular, the discrete harmonic energy is bounded from below. Therefore, the non-linear heat diffusion process always converges.

Algorithm 4 *Hyperbolic Heat Diffusion Algorithm.*
Input: *Two triangle meshes with hyperbolic metrics* $(S_1, \mathbf{g_1})$ *and* $(S_2, \mathbf{g_2})$*; an initial mapping* $f : S_1 \to S_2$*, represented as* $(t(v_i), \eta(v_i))$, $\forall v_i \in S_1$. *An error tolerance threshold* ε.
Output: *Discrete hyperbolic harmonic map* $h : S_1 \to S_2$.
1. Construct a hyperbolic atlas of the target surface S_2.
repeat
for *each vertex* $v_i \in S_1$ ***do***
2. Collect all its neighboring vertices $N(v_i) = \{v_{i_j}\}$ *and neighboring edge weights* $\Gamma(v_i) = \{k_{i,i_j}\}$.
3. Choose a local chart of S_2*, which covers the images of all neighboring vertices in* $N(v_i)$.
4. Convert the mapping representation of each $v_{i_j} \in N(v_i)$, $(t(v_{i_j}), \eta(v_{i_j}))$ *to local coordinates* w_{i_j}. *Let* $Q(v_i) = \{w_{i_j}\}$.
5. Compute the weighted geodesic mass center $c(Q(v_i), \Gamma(v_i))$*, using Algorithm 3.*
6. Update $f(v_i)$, $f(v_i) \leftarrow c(Q(v_i), \Gamma(v_i))$. *Convert* $f(v_i)$ *to the representation* $(t(v_i), \eta(v_i))$.
end for
until *for all* $v_i \in S_1$, $\rho(f(v_i), c(Q(v_i), \Lambda(v_i))) < \varepsilon$.
8. Output the harmonic map $f : S_1 \to S_2$*, represented as* $(t(v_i), \eta(v_i))$, $\forall v_i \in S_1$.

Experimental Results

General Surface Registration Our method can handle surfaces with general topologies, i.e. surfaces with an arbitrary number of handles and boundaries. For genus zero surfaces with multiple boundaries, the homotopy class of the mapping f is specified by the correspondences between the landmark curves. For high genus surfaces, the homotopy class of the mapping is specified by the correspondences between fundamental group generators (handle loops and tunnel loops) [66], shown in Fig. 9.10. Here we show how it works on the closed surface matching problem. Figure 9.10a, b show two genus-two surface models, the two-holed torus model and the Amphora model. They are marked with a set of canonical fundamental group generators, which can cut the surface into a topological disk with eight sides. (c) and (d) illustrate their embedding of their fundamental domains to the Poincaré disk. Multiple fundamental domains are shown and different domains are color coded differently. By applying Fuchsian transformations which are Möbius transformations, the yellow fundamental polygon is transformed to different fundamental polygons. All the fundamental polygons obtained by Fuchsian transformations tessellate the whole Poincaré disk. As the rigid transformation on the Poincaré disk, the Möbius transformation generates a new fundamental domain across each edge. Cutting along the yellows loops on (e) and (f) may decompose each surface into two pairs of topological pants. On each pair of pants, it can be further cut into two symmetric hyperbolic hexagons, each of which is labeled with a unique color in (g) and (h). The initial mapping is computed between these four hyperbolic polygons.

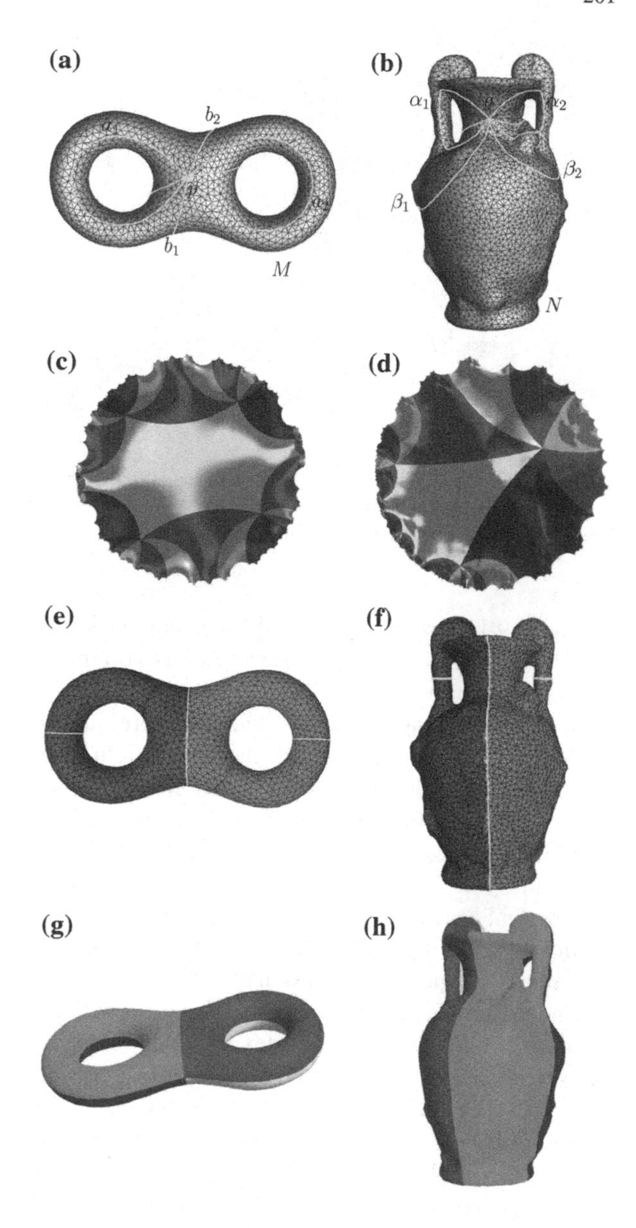

Fig. 9.10 **a**, **b** show the input two-holed torus and Amphora model, as well as their geodesic homotopy bases. **c, d** show their universal covering spaces on Poincaré's disk. **e, f** show their pants decompositions, with each model decomposed into 2 pairs of pants. **g, h** show that each pair of pants can be further decomposed into 2 hyperbolic hexagons. Image from [251]

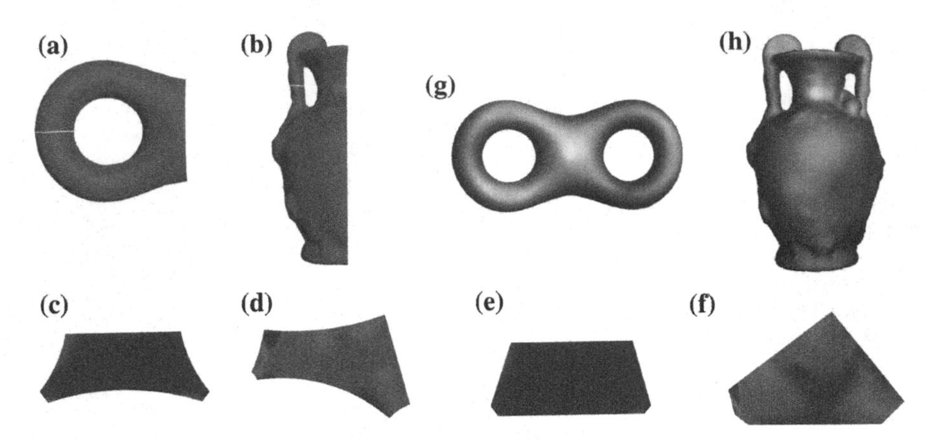

Fig. 9.11 **a**, **b** show one of the hyperbolic hexagons of the two-holed torus and Amphora model. **c**, **d** show the image of **a**, **b** on Poincaré's disk. **e**, **f** show the image of **c**, **d** after converted on the Klein disk. **g**, **h** show the final registration result by color mapping. Image from [251]

Figure 9.11 further illustrates the matching process. Each pair of pants (shown in (a) and (b)) is decomposed into two hexagons on the Poincaré disk. (c) and (d) show two hyperbolic hexagons on the Poincaré disk. With Eq. (9.23), one may convert them to two convex hexagons on the Klein model ((e) and (f)). Note that the geodesics on the Poincaré model are arcs while they are straight lines on the Klein model. Thus the hexagons are convex on the Klein model. Initial mappings are computed between the convex hexagons. Following that, non-linear heat diffusions are computed throughout the entire surfaces and the final hyperbolic harmonic mapping results are color-coded in (g) and (h).

Cortical Surface Registration In this section we apply our method to the brain cortical surface registration problem. Morphometric and functional studies of the human brain require that neuro-anatomical data from a population be normalized to a standard template, so brain cortical surface registration is often needed. Due to the similarity of human cerebral cortex structure, the registration mapping is required to be smooth and bijective, namely, diffeomorphic. Since cytoarchitectural and functional parcellation of the cortex is intimately related the folding of the cortex, it is also important to ensure the alignment of the major anatomic features, such as sulcal landmarks. We used our algorithm on the cortical surface registration problem and compared with [205, 293]. We perform the experiments on 24 brain cortical surfaces reconstructed from MRI images. Each cortical surface has about 150K vertices, 300K faces and was used in some prior research [205]. All these experiments are performed on a laptop computer and the whole pipeline takes no more than 3 min. On each cortical surface, a set of 26 landmark curves was manually drawn and validated by neuroanatomists. Figure 9.12 show the landmark curves and their labels. In our current work, we selected 10 landmark curves, including Central Sulcus, Superior Frontal Sulcus, Inferior Frontal Sulcus, Horizontal Branch of Sylvian Fissure, Cingulate Sulcus, Supraorbital Sulcus, Sup. Temporal with Upper Branch,

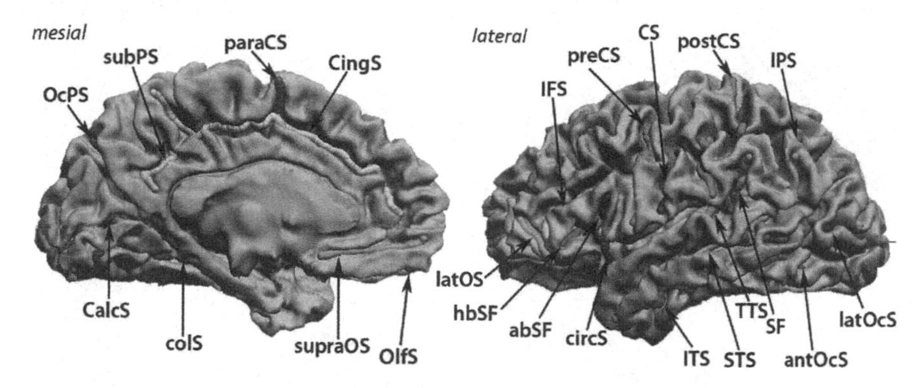

Fig. 9.12 Landmark curves on human cortical surface [205]. Image from [251]

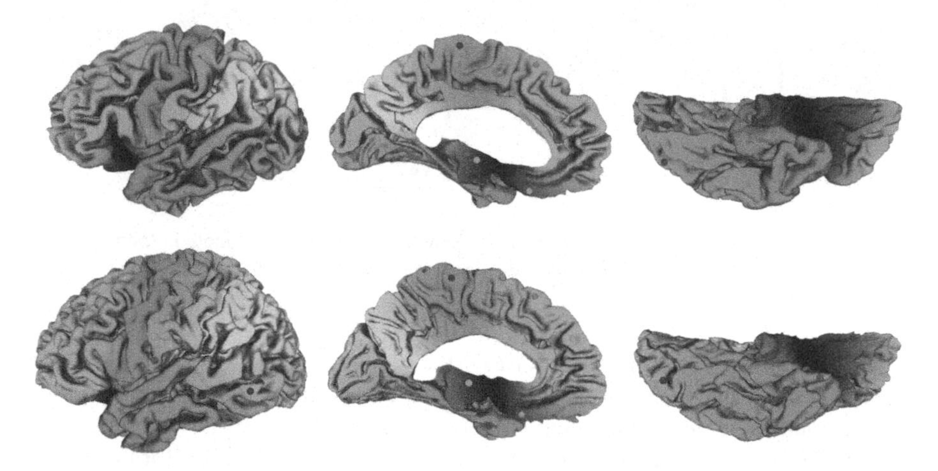

Fig. 9.13 Top row: source brain surface from front, back and bottom view. Bottom row: target brain model. The color on the models shows the correspondence between source and target; the colored balls on the models show the detailed correspondence, as the balls with the same color are correspondent to each other. Image from [251]

Inferior Temporal Sulcus, Lateral Occipital Sulcus and the boundary of Unlabeled Subcortical Region. The hyperbolic Ricci flow takes about 120 s on average, the hyperbolic heat diffusion takes about 100 s on average. The time consumed on pants decomposition and initial map construction depends on the number of landmarks. In the current setting, it takes about 90 s on average.

We show our visualized registration result of 2 brain models in Fig. 9.13, with one as target and the other registered to it. By examining the color consistency between the source and the target surface and comparing the geometric characteristics of the neighborhoods of the corresponding markers, we can see the matching quality is good. The matching results are further validated by experienced neurologists.

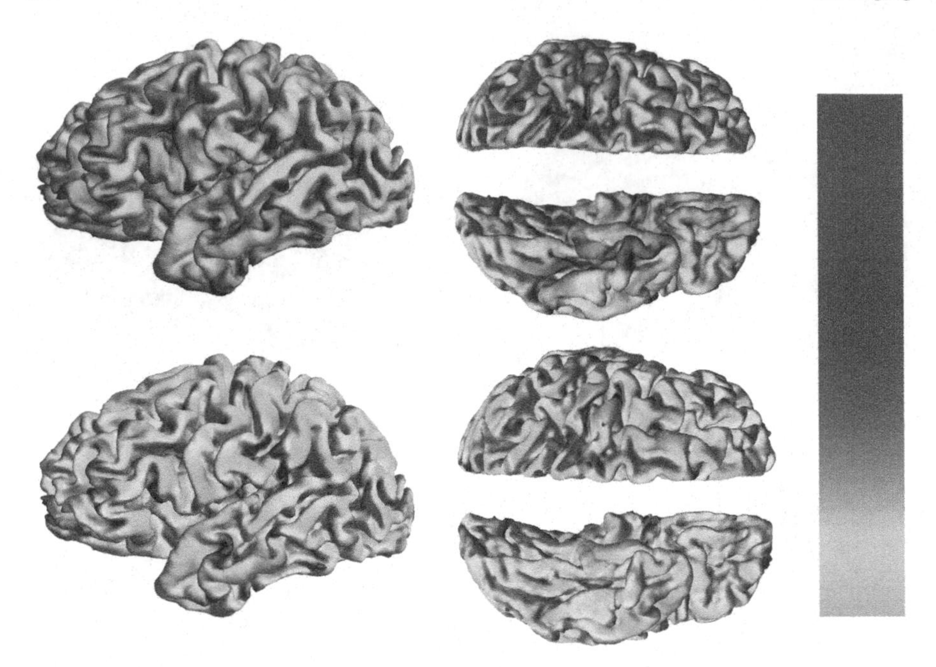

Fig. 9.14 Curvature map difference of previous method (top row) and our method (bottom row). The color goes from green to red with increasing curvature difference. Image from [251]

We measure the curvature distortion as in the previous section. According to [205] which gives the anatomical evidence, the curvature map difference is a standard comparison method in the neroimaging field as similar brain shapes have Riemannian metrics, inducing similar Gaussian curvatures. We use all 24 data sets for the experiment. First, one data set is randomly chosen as the template, then all others are registered to it. The average curvature difference map is color-coded on the template, as shown in Fig. 9.14. The histogram of the average curvature difference map is also computed, as shown in Fig. 9.15.

The reason that our method has smaller curvature distortion is related to the properties of the harmonic map. An isometry preserves the lengths on the surface, which also preserves curvatures; furthermore, it minimizes the elastic stretching energy, and therefore is harmonic. Our method is guaranteed to find the harmonic map between surfaces. Therefore, if there exists an isometry (or a map close to an isometry) between the input brain surfaces, our proposed method will surely find it, which automatically minimizes the curvature difference.

We measure the area distortion according to the previous section. We compute the average of all local area distortion functions induced by the 23 registrations on the template surface. The average local area distortion function on the template is color-coded as shown in Fig. 9.16; the histogram is also computed in Fig. 9.17.

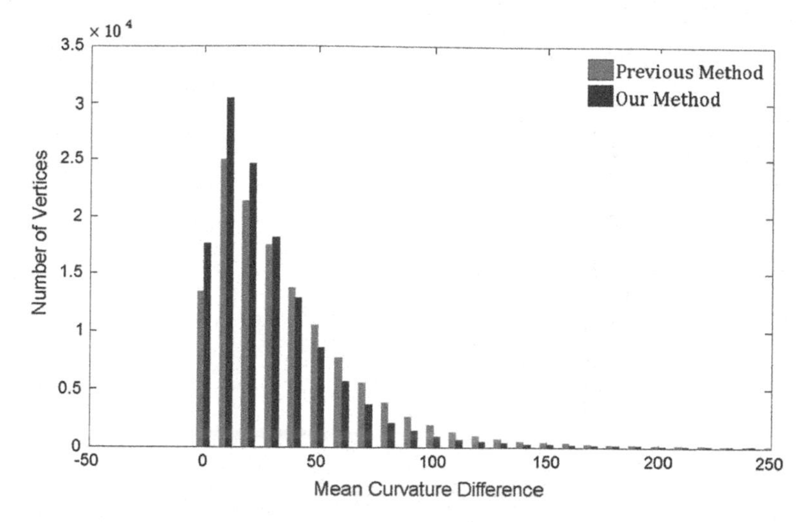

Fig. 9.15 Average curvature map difference. Image from [251]

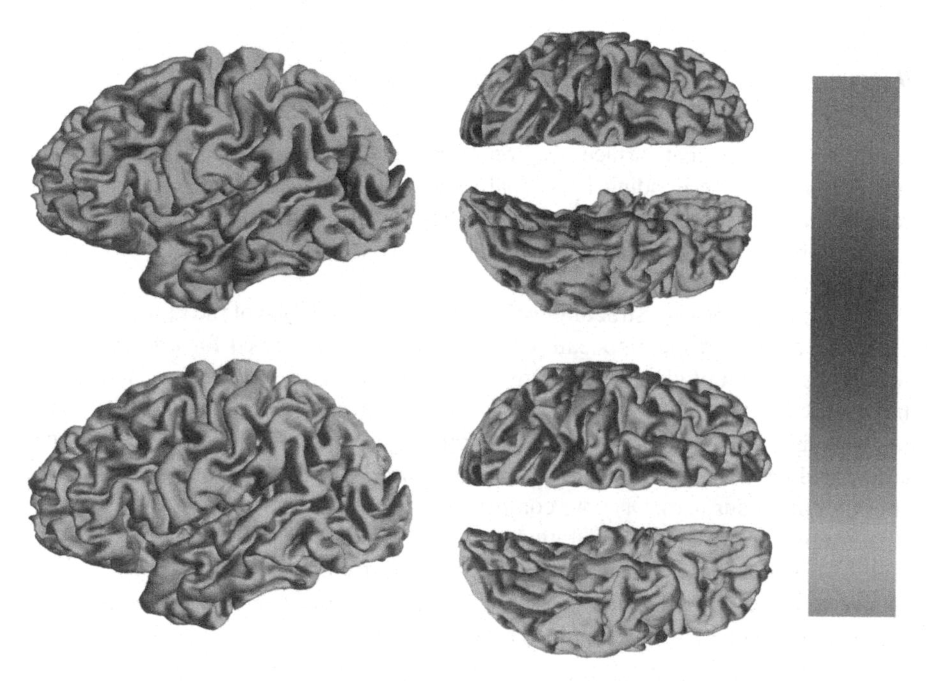

Fig. 9.16 Average area distortion. Comparison between the previous method (top row) and our current method (bottom row). Color goes from green to red with increasing area distortion. Image from [251]

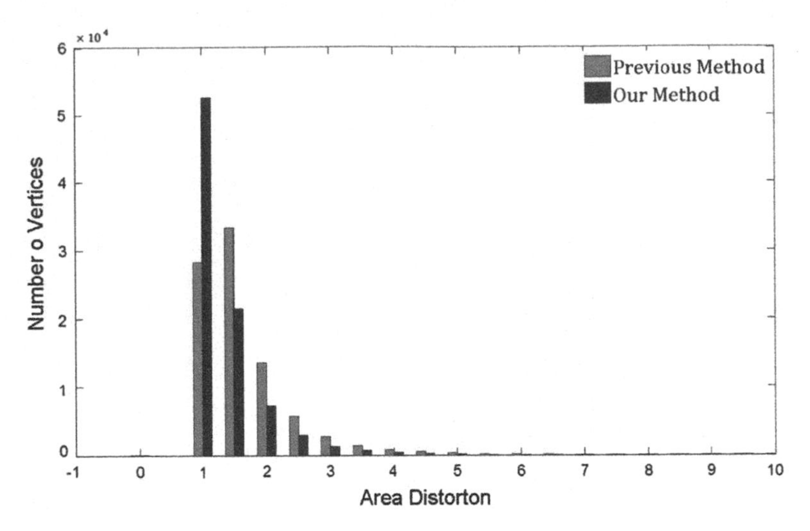

Fig. 9.17 Average area distortion. Image from [251]

9.3.4 *Surface Fluid Registration*

It has been known that surface registration requires defining a lot of landmarks in order to align corresponding functional regions. Labeling features could be accurate but time-consuming. Here we show that surface conformal parameterization could represent surface geometric features, thus avoiding the manual definition of landmarks.

The hippocampus is a structure in the medial temporal lobe of the brain. Parametric shape models of the hippocampus are commonly developed for tracking shape differences or longitudinal atrophy in disease. Many prior studies, e.g. [129, 271], have shown that there is atrophy as the disease progresses and support its shape analysis as a valid brain imaging biomarker. Figure 9.18a–c, e–g shows the planar conformal parameterization of a hippocampal surface (a–c) for source surface and (e–g) for target surface). Once we conformally map the 3D surface to the 2D domain, we can apply surface fluid registration to register surfaces with 3D geometric features and the original surface Riemannian metric by conformal geometry theory [247].

For a general surface and its conformal parameterization $\phi : SR^2$, the conformal factor at a point p can be determined by the formula:

$$\lambda(p) = \frac{Area(B_\varepsilon(p))}{Area(\phi(B_\varepsilon(p)))} \tag{9.30}$$

where $B_\varepsilon(p)$ is an open ball around p with a radius ε. The conformal factor λ encodes a lot of geometric information about the surface and can be used to compute curvatures and geodesic. In our system, we compute the surface mean curvatures only from the derivatives of the conformal factors as proposed in [182], instead of

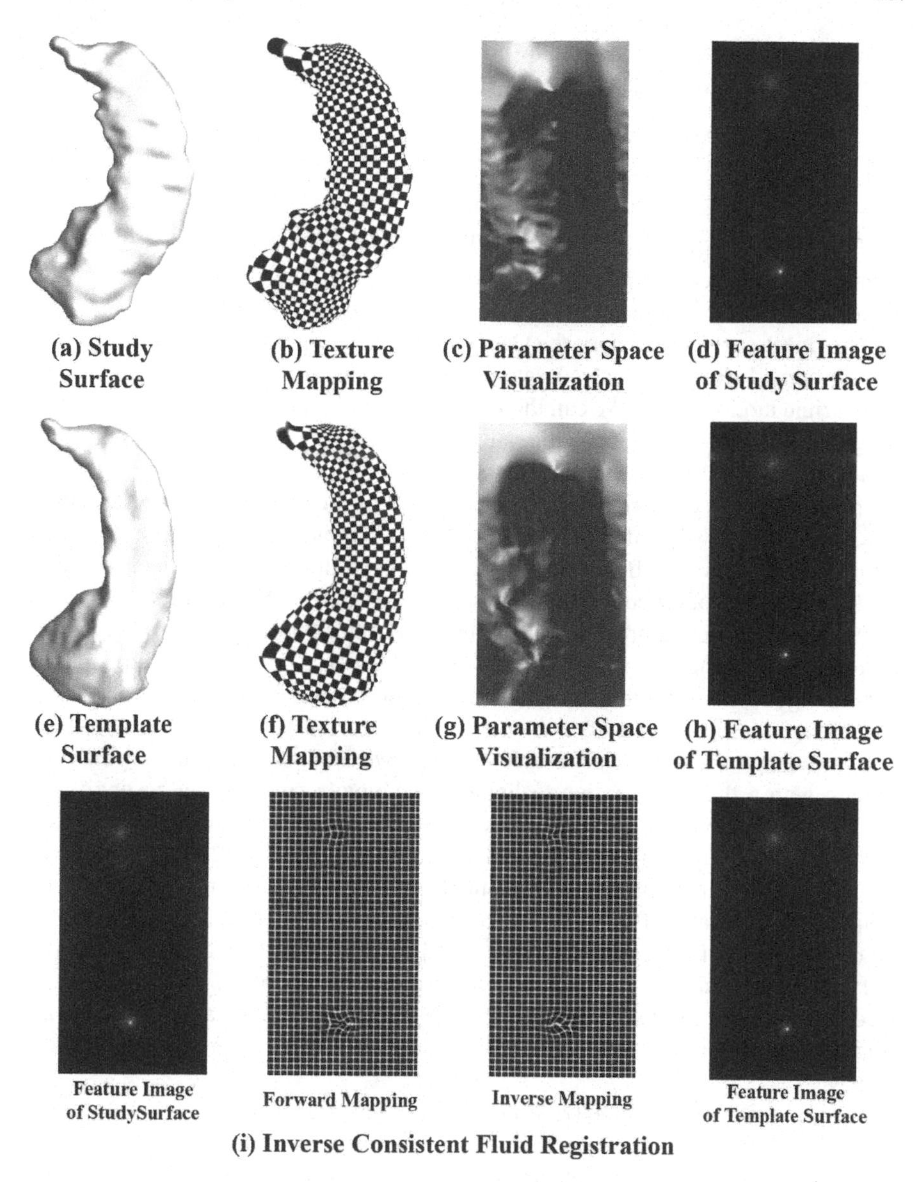

Fig. 9.18 Hippocampal surface registration with inverse consistent surface fluid registration algorithm. Image from [247]

the three coordinate functions and the normal, which are generally more sensitive to digitization errors. Mathematically, the mean curvature is defined as:

$$H = \frac{1}{2\lambda} sign(\phi)|\Delta\phi| \tag{9.31}$$

where $sign(\phi) = \frac{<\Delta\phi, \bar{N}>}{|Delta\phi|}$. Using this formulation of H, we need to use the surface normal $\bar{N}$ only when computing $sign(\phi)$, which takes the value 1 or -1. Thus, the surface normal does not need to be accurately estimated and still we can get more accurate mean curvatures. Using the Gauss and Codazzi equations, one can prove that the conformal factor and mean curvature uniquely determine a closed surface in R^3, up to a rigid motion [101]. We call them the conformal representation of the surface. We choose to encode surface features using a compound scalar function based on the local conformal factor and the mean curvature: $C(u, v) = \beta\lambda(u, v) + H(u, v)$, where (u, v) is the conformal coordinates of the surface and β is a constant scalar to control the ratio of conformal factor and mean curvature. In our work [247], we empirically set β as 7 for both visualization and registration. We then linearly scaled the dynamic range of the conformal representation into [0, 255]. Figure 9.18d, h show the encoded surface features with image intensity set according to the values. Since conformal factor and mean curvature encode both surface intrinsic structure and $3D$ embedding information, they are complete surface features to be used for solving surface registration problems.

After computing surface geometric features, we align surfaces in the parameter domain with a fluid registration technique to maintain smooth, one-to-one topology [51]. Using conformal mapping, we essentially convert the surface registration problem to an image registration problem. In our prior work [289], we proposed an automated surface fluid registration method combining conformal mapping and image fluid registration [62] with mutual information [119, 151, 192, 227, 303] as the driving force of the viscous fluid. In [289], the mutual information between two surface feature images, i.e., the conformal representations of the two surfaces that need to be registered, was maximized by the viscous fluid flow as in [62]. On R^2, fluid flow is governed by the Navier–Stokes equation. For compressible fluid flow, we have

$$\mu\Delta\mathbf{v}(\mathbf{x}) + (\mu + \tau)\nabla(\nabla \cdot \mathbf{v}(\mathbf{x})) = \mathbf{f}(\mathbf{x}, \mathbf{u}(\mathbf{x})). \tag{9.32}$$

Here $\mathbf{v}(\mathbf{x})$ is the deformation velocity, μ and τ are the viscosity constants. $\mathbf{f}(\mathbf{x}, \mathbf{u}(\mathbf{x}))$ is the force field that is used to drive the fluid flow, which was defined as the mutual information in [289]. To simulate fluid flow on Riemann surfaces, we need to extend Eq. (9.32) into surface space by the manifold version of Laplacian and divergence [12, 257, 295].

Theorem 9.10 (Manifold Navier–Stokes Equation) *By covariant derivatives, the Navier–Stokes equation for Riemann surface can be defined as:*

$$\frac{\mu}{\lambda}\Delta \mathbf{v} + \frac{\mu+\tau}{\lambda}\nabla(\nabla\cdot\mathbf{v}) = \mathbf{f}. \tag{9.33}$$

where λ is the conformal factor.

Proof With conformal parameterization, the Riemann metric is defined as:

$$[g_{ij}] = \begin{bmatrix} g_{11} \; g_{12} \\ g_{21} \; g_{22} \end{bmatrix} = \begin{bmatrix} \lambda \; 0 \\ 0 \; \lambda \end{bmatrix}$$

The inverse of $[g_i j]$ is:

$$[g^i j] = \begin{bmatrix} g^{11} \; g^{12} \\ g^{21} \; g^{22} \end{bmatrix} = \begin{bmatrix} 1/\lambda & 0 \\ 0 & 1/\lambda \end{bmatrix}$$

We now provide the expression in general coordinates of the differential operators that appear in Eq. (9.32) [12, 257, 295].

Gradient:

$$\nabla_S \phi = g^{ij}\frac{\partial \phi}{\partial x_j} = g_{i1}\frac{\partial \phi}{\partial x_1} + g^{i2}\frac{\partial \phi}{\partial x_2} = \begin{bmatrix} g^{11}\frac{\partial}{\partial x_1} + g^{12}\frac{\partial}{\partial x_2} \\ g^{21}\frac{\partial}{\partial x_1} + g^{22}\frac{\partial}{\partial x_2} \end{bmatrix}$$

Divergence:

$$\nabla_S \cdot \mathbf{u} = \frac{1}{\sqrt{g}}\frac{\partial}{\partial x_i}(\sqrt{g}u_i) = \frac{1}{\sqrt{g}}\left(\frac{\partial}{\partial x_1}(\sqrt{g}u_1) + \frac{\partial}{\partial x_2}(\sqrt{g}u_2)\right)$$

where $\mathbf{u} = \begin{bmatrix} u_1 \\ u_2 \end{bmatrix}$ and $\sqrt{g} = \sqrt{det([g_{ij}])} = \sqrt{g_{11}g_{12} - g_{12}g_{21}}$.

The Laplacian can be computed by gradient and divergence as:

$$\Delta_S \phi = \nabla_S \cdot (\nabla_S \phi) = \frac{1}{\sqrt{g}}\left(\frac{\partial}{\partial x_1}\left(\sqrt{g}g^{11}\frac{\partial \phi}{\partial x_1} + \sqrt{g}g^{12}\frac{\partial \phi}{\partial x_2}\right)\right)$$
$$+\left(\frac{\partial}{\partial x_2}\left(\sqrt{g}g^{21}\frac{\partial \phi}{\partial x_1} + \sqrt{g}g^{22}\frac{\partial \phi}{\partial x_2}\right)\right)$$

Given conformal parameterization $\phi : SR^2$, where $\sqrt{g} = \lambda, g^{11} = g^{22} = \frac{1}{\lambda}, g^{12} = g^{21} = 0$, we have

$$\Delta_S \mathbf{v} = \frac{1}{\lambda}\Delta \mathbf{v}. \tag{9.34}$$

For a velocity field $\mathbf{v} = \begin{bmatrix} v_1 \\ v_2 \end{bmatrix}$,

$$\nabla_S(\nabla_S \cdot \mathbf{v}) = \nabla(g^{11}\frac{\partial v_1}{\partial x_1} + g^{12}\frac{\partial v_1}{\partial x_2} + g^{21}\frac{\partial v_2}{\partial x_1} + g^{22}\frac{\partial v_2}{\partial x_2}) = \frac{1}{\lambda}\nabla(\nabla\cdot\mathbf{v}). \tag{9.35}$$

By plugging Eq. (9.34), (9.35) into (9.32), one can get Eq. (9.33). □

It is well known that area distortion is an inevitable problem of conformal parameterization. However, considering the definition of conformal factor as Eq. (9.30), we can see that conformal factor is a smooth function which describes the stretching effect of conformal parameterization. In Eq. (9.33), by factoring out the conformal factor λ, the flow induced in the parameter domain is adjusted for the area distortion introduced by the conformal parameterization. As a result, Eq. (9.33) is now governing fluid flow on the manifolds. As pointed out in [166], image registration problem should be symmetric, i.e., the correspondences established between the two images should not depend on the order we use to compare them. However, traditional nonlinear image registration algorithms are not symmetric, thus the deformation field depends on which image is assigned as the deforming image and which image the non-deforming target image. Furthermore, the asymmetric algorithms tend to penalize the expansion of image regions more than the shrinkage [224], making these methods problematic in applications where the Jacobian of the mappings is interpreted as measuring anatomical tissue loss or expansion. Many inverse consistent registration algorithms [50, 223] have been proposed to overcome the shortcomings of conventional inverse non-consistent methods. Leow et al. [166] proposed a novel inverse consistent image registration method. Instead of enforcing inverse consistency using an additional penalty that penalizes inconsistency error as in [50], the method in [166] directly modeled the reverse mapping by inverting the forward mapping. Chiang et al. [45] replaced the linear elastic regularizer in [166] with the fluid regularization to enable large deformations and applied the inverse consistent fluid registration algorithm to diffusion tensor images. Here with the inverse consistent scheme proposed in [45], we extend Eq. (9.33) into an inverse consistent surface fluid registration method.

Let $I_1(\mathbf{x})$, $I_2(\mathbf{x})$ be two images, using the sum of squared intensity differences as the matching cost function, the inverse consistent image registration problem seeks two mappings $\mathbf{h}(\mathbf{x})$ and $\mathbf{g}(\mathbf{x})$ to minimize the following energy function:

$$\begin{aligned} E(I_1(\mathbf{x}), I_2(\mathbf{x})) = \int_\Omega |I_1(\mathbf{h}(\mathbf{x})) - I_2(\mathbf{x})|^2 d\mathbf{x} + \alpha R(\mathbf{h}(\mathbf{x})) + \\ \int_\Omega |I_2(\mathbf{g}(\mathbf{x})) - I_1(\mathbf{x})|^2 d\mathbf{x} + \alpha R(\mathbf{g}(\mathbf{x})) \end{aligned} \tag{9.36}$$

where $\mathbf{h}(\mathbf{x}) = x - u_f(x)$ is the mapping from image I_1 to image I_2 (forward direction) and $\mathbf{u}_f(\mathbf{x})$ is the forward displacement field. $\mathbf{g}(\mathbf{x}) = \mathbf{x} - \mathbf{u}_b(\mathbf{x})$ is the mapping from image I_2 to image I_1 (backward direction) and $\mathbf{u}_b(\mathbf{x})$ is the backward displacement field, $\mathbf{g}(\mathbf{x}) = \mathbf{h}^{(-1)}(\mathbf{x})$. $alpha$ is a positive scalar weighting of the regularization terms applied to the forward and backward mappings. Following prior work in fluid registration [45, 166], we let $\alpha = 1$ to achieve a fast and stable convergence. Equation (9.36) is symmetric and does not depend on the order of I_1 and I_2, i.e., $E(I_1, I_2) = E(I_2, I_1)$. Suppose we have two surfaces S_1, S_2 and their conformal representation I_1, I_2 in R^2. With fluid regularization scheme, $R(\mathbf{h}(\mathbf{x}))$ is defined as

$\int_0^1 \int_\Omega ||L\mathbf{v}_f(\mathbf{x})||^2 d\mathbf{x}dt$ and $R(\mathbf{g}(\mathbf{x})$ is defined as $\int_0^1 \int_\Omega ||L\mathbf{v}_b(\mathbf{x})||2d\mathbf{x}dt$ with the forward and backward velocities $\mathbf{v}_f(\mathbf{x})$ and $\mathbf{v}_b(\mathbf{x})$, respectively. $L = \frac{\mu}{\lambda}\Delta + \frac{\mu+\tau}{\lambda}\nabla(\nabla\cdot)$ is the surface linear operator as in Eq. (9.33). Then the energy function in Eq. (9.36) can be minimized by solving for the velocities $\mathbf{v}_f(\mathbf{x})$ and $\mathbf{v}_b(\mathbf{x})$ in the following general Navier–Stokes equations:

$$\frac{\mu}{\lambda_(f, b)}\Delta\mathbf{v}_{f,b} + \frac{\mu+\tau}{\lambda_(f, b)}\nabla(\nabla\cdot\mathbf{v}_{f,b}) = \mathbf{f}_(f, b) \tag{9.37}$$

where the forward force field $\mathbf{f}_f = -[I_1(x - \mathbf{u}_f(x)) - I_2(x)]\nabla I_1(x - \mathbf{u}_f(\mathbf{x}))$ and backward force field $\mathbf{f}_b = -[I_2(\mathbf{x} - \mathbf{u}_b(\mathbf{x})) - I_1(\mathbf{x})]\nabla I_2(\mathbf{x} - u_b(\mathbf{x}))$. λ_f is the conformal factor of surface S_1 and λ_b is the conformal factor of surface S_2.

With the mappings $\mathbf{h}(\mathbf{x})$, $\mathbf{g}(\mathbf{x})$ initialized as the identical mapping at $t = 0$, the forward and backward mappings at time t are given by the following equations as in [166]:

$$\begin{aligned} \mathbf{h}_t(\mathbf{x}) &= \mathbf{h}_{t-1}(\mathbf{x}) + \varepsilon\eta_1(\mathbf{x}) + \varepsilon\eta_2(\mathbf{x}) \\ \mathbf{g}_t(\mathbf{x}) &= \mathbf{g}_{t-1}(\mathbf{x}) + \varepsilon\xi_1(\mathbf{x}) + \varepsilon\xi_2(\mathbf{x}) \end{aligned} \tag{9.38}$$

Here, ε is an infinitesimally small positive time step. η_1, η_2, ξ_1, ξ_2 are computed as [45]:

$$\begin{aligned} \eta_1(\mathbf{x}) &= -(\nabla\mathbf{h}_{t-1}(\mathbf{x}))\mathbf{v}_f^{t-1}(\mathbf{x}),\ \eta_2(\mathbf{x}) = \mathbf{v}_b^{t-1}(\mathbf{h}_{t-1}(\mathbf{x})) \\ \xi_1(\mathbf{x}) &= v_f^{t-1}(\mathbf{g}_{t-1}(\mathbf{x})),\ \xi_2(\mathbf{x}) = -(\nabla\mathbf{g}_{t-1}(\mathbf{x}))\mathbf{v}_b^{t-1}(\mathbf{x}) \end{aligned} \tag{9.39}$$

The inverse consistent image registration algorithm is used to jointly estimate the forward and inverse transformations between a pair of feature images and to ensure the symmetry of the registration, as shown in Fig. 9.18i. Since conformal mapping and fluid registration generate diffeomorphic mappings, a diffeomorphic surface-to-surface mapping is then recovered that matches surfaces in 3D.

In order to validate the effectiveness of the proposed method, we generated two synthetic surfaces as shown in Fig. 9.19a, b. The two C shapes have different sizes and positions. This can also be seen from the corresponding feature images at the bottom of Fig. 9.19a, b. The feature images were generated by summing up the local conformal factor and the mean curvature, expressed in the conformal parameterization domain. The black lines drawn on the surfaces are used to show equal distances on the surfaces and represent the differences in their shapes. With the inverse consistent fluid registration, in Fig. 9.19c and d, we can see that the feature image of surface 1 was successfully registered to the feature image of surface 2 and the feature image of surface 2 was also registered to the feature image of surface 1. With the forward and backward mappings obtained in the parameter domain, we induced a forward deformation and a backward deformation in surface 1 and surface 2, respectively. As we can see from Fig. 9.19c and d, without changing the shape of the surfaces, the features on them are well aligned to each other.

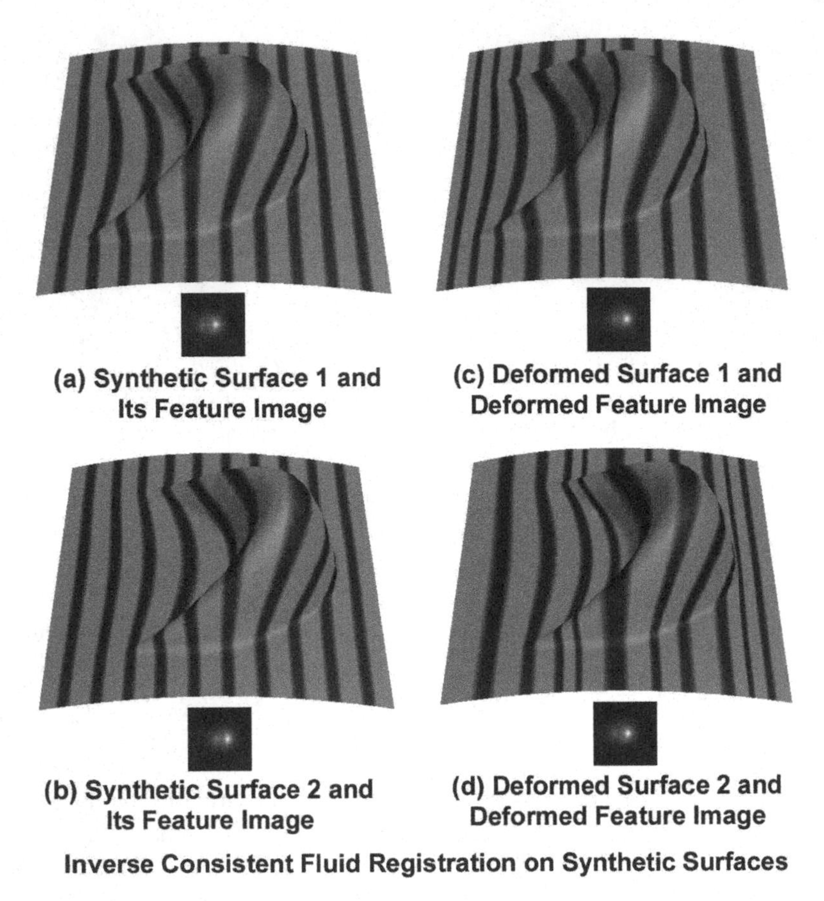

Fig. 9.19 Matching of geometric features in the 2D parameter domain with the inverse consistent fluid registration of two synthetic surfaces. With the forward and backward mappings obtained in the parameter domain, we induced a forward deformation and a backward deformation in surface 1 and surface 2, respectively. As we can see from **c** and **d**, without changing the shape of the surfaces, the features on them are well aligned to each other. Image from [247]

9.4 Global Transformation-Invariant Shape Descriptors

Computational conformal geometry provides powerful tools for us to compute various transformation-invariant shape descriptors [100, 107, 249, 250, 325]. Specifically, computational conformal geometry research may map general surfaces to one of three canonical parameter spaces. The shape descriptors are computed conveniently with the canonical space. Here we briefly introduce four different global shape descriptors, including spherical harmonics based rotation invariant shape descriptor, conformal invariant Teichmüller shape descriptor and Teichmüller shape space coordinates, and isometry invariant Riemannian optimal mass transport map based Riemannian Wasserstein distance.

9.4.1 Spherical Harmonic Analysis Based Rotation Invariant Shape Descriptor

Let $L^2(S^2)$ denote the Hilbert space of square integrable functions on the S^2. In spherical coordinates, θ is taken as the polar (colatitudinal) coordinate with $\theta \in [0, \pi]$, and ϕ as the azimuthal (longitudinal) coordinate with $\phi \in [0, 2\pi)$. The usual inner product is given by

$$< f, h >= \int_0^{\pi} \left[\int_0^{2\pi} f(\theta, \phi)\overline{h(\theta, \phi)} d\phi \right] \sin\theta d\theta. \tag{9.40}$$

A function $f : S^2 \to \mathbb{R}$ is called a *Spherical Harmonic*, if it is an eigenfunction of Laplace–Beltrami operator, namely $\Delta f = \lambda f$, where λ is a constant. There is a countable set of spherical harmonics which form an orthonormal basis for $L^2(S^2)$.

For any nonnegative integer l and integer m with $|m| \le l$, the $(l, m)-$spherical harmonic Y_l^m is a harmonic homogeneous polynomial of degree l. The harmonics of degree l span a subspace of $L^2(S^2)$ of dimension $2l + 1$ which is invariant under the rotations of the sphere. The expansion of any function $f \in L^2(S^2)$ in terms of spherical harmonics can be written

$$f = \sum_{l \ge 0} \sum_{|m| \le l} \hat{f}(l, m) Y_l^m \tag{9.41}$$

and $\hat{f}(l, m)$ denotes the (l, m) *Fourier coefficient*, equal to $< f, Y_l^m >$. Spherical harmonic Y_l^m has an explicit formula

$$Y_l^m(\theta, \phi) = k_{l,m} P_l^m(cos\theta) e^{im\phi}, \tag{9.42}$$

where P_l^m is the *associated Legendre function* of degree l and order m, and $k_{l,m}$ is a normalization factor. The details are explained in [280].

Once the brain surface is conformally mapped to S^2, the surface can be represented as three spherical functions, $x^0(\theta, \phi)$, $x^1(\theta, \phi)$ and $x^2(\theta, \phi)$. The function $x^i(\theta, \phi) \in L^2(S^2)$ is regularly sampled and transformed to $\hat{x^i}(l, m)$ using the Fast Spherical Harmonic Transformation as described in [117].

The geometric representation $(x^1(\theta, \phi), x^2(\theta, \phi), x^3(\theta, \phi))$ depends on the orientation of the brain. Brain registration has to be applied first in order to compare the geometric representations of two different brains. A rotation-invariant shape descriptor can be formulated based on the frequency coefficients. Because the harmonics of degree l span the rotation invariant subspace of $L^2(S^2)$, the following shape descriptor is also rotation invariant

$$s(l) = \sum_{i} \sum_{|m| \le l} ||\hat{x}^i(l, m)||^2. \tag{9.43}$$

Given two brain surfaces, we can compute their shape descriptor from their spherical harmonic spectrum, and compare them directly without any registration.

The brain surface can be represented as a vector valued function defined on the sphere via conformal mapping of its surface to the surface. The brain surface can then be decomposed in terms of linear combination of spherical harmonics. The vector valued spectrum, i.e. the harmonic coefficients expressed as components of a vector, can be used to analyze the shape. The main geometric features are encoded in the low frequency part, while the noise will be in the high frequency part. By filtering out the high frequency coefficients, we can smooth the surface, and compress the geometry. By comparing the low frequency coefficients, we can match surfaces, and compute the similarity of surfaces.

In one prior work [107], 112 1.5T T1-weighted magnetic resonance imaging (MRI) scan images from the Alzheimer's Disease Neuroimaging Initiative (ADNI) database (http://www.adni.loni.usc.edu), with 49 Alzheimer's disease (AD) patients and 63 controls, age and gender-matched (mean age: 76.14, 76.76, $P = 0.609$). Initially, structural MRI images are automatically converted into binary hippocampal masks with the help of the recent Auto Context Model (ACM) [200]. ACM uses a few hand-traced hippocampi as a training set for AdaBoost to create a voxel-level classification function. We then convert the masks to a signed distance function and apply topology-con- strained mean curvature ow following a topology- preserving geometric deformable model algorithm [111]. Following triangle mesh extraction and minor processing [91], a quick visual check is done on each mesh to ensure that the original masks correspond to a hippocampal shape. After that our spherical conformal mapping algorithm [100] was applied to map each hippocampal surface to the sphere where the spherical harmonic coefficients and further the rotation invariant global shape descriptors were computed.

With the global shape descriptors as features, we used support vector machine [233] as the classifier for classification experiments. For each training set in the leave-one-out test, we selected a feature if its t-statistic exceeded a threshold. After testing a few subjects, we noticed that the best overall accuracy is achieved with $6.7 \leq t_{min} \leq 6.9$, and set it globally to 6.8. This yielded between 6 and 14 features, depending on which subject was left out. All selected features $s(l)$ were of order $37 \leq l \leq 58$. Our margin/error coefficient C was set to 1,000. All features were normalized with respect to standard deviation (differently for each excluded subject) and translated so that $min(x) = max(x)$. The transformation was saved and applied to the remaining subject. The result was 75.5% sensitivity and 87.3% specificity for a total correct rate of 82.1% (AD considered positive). By comparison, hippocampal volume gave 67.3% sensitivity and 76.2% specificity in a leave-one-out test, with 72.3% correct overall. The experimental results demonstrated its potential as an effective shape descriptor in medical imaging research.

9.4.2 Conformal Welding Based Teichmüller Shape Descriptor

Definition 9.11 (*Conformal Equivalence*) Suppose $(S_1, \mathbf{g}_1)$ and $(S_2, \mathbf{g}_2)$ are two Riemann surfaces. We say S_1 and S_2 are *conformal equivalent* if there is a conformal diffeomorphism between them.

All Riemann surfaces can be classified by the conformal equivalence relation. Each conformal equivalence class shares the same *conformal invariants*, the so-called *conformal module*. The conformal module is one of the key component for us to define the unique shape signature.

Definition 9.12 (*Teichmüller Space*) Fixing the topology of the surfaces, all the conformal equivalence classes form a manifold, which is called the *Teichmüller space*.

For example, all topological disks (genus zero Riemann surfaces with single boundary) can be conformally mapped to the planar disk. Therefore, the Teichmüller space for topological disks consists of a single point.

All the surfaces in real life are Riemann surfaces, therefore with conformal structures. Two surfaces share the same conformal structure, if there exists a conformal mapping between them. Conformal modules are the complete invariants of conformal structures and intrinsic to surface itself. They can serve as the coordinates in Teichmüller space.

Suppose a genus zero Riemann surface S has b boundary components $\{\gamma_1, \gamma_2, \ldots, \gamma_b\}$, $\partial S = \gamma_1 + \gamma_2 + \cdots + \gamma_b$, $\phi : S \to \mathbb{D}$ is the conformal mapping that maps S to a circle domain $\mathbb{D}$, such that it satisfies the following Möbius normalization conditions,

1. $\phi(\gamma_1)$ is the exterior boundary of the $\mathbb{D}$;
2. $\phi(\gamma_2)$ centers at the origin; and
3. The center of $\phi(\gamma_3)$ is on the imaginary axis.

Definition 9.13 (*Conformal Module*) The conformal module of the surface S (also the circle domain $\mathbb{D}$) is given by

$$Mod(S) = \{(\mathbf{c_i}, r_i) | i = 1, 2, \ldots, b\}, \tag{9.44}$$

where $(\mathbf{c_i} = x_i + iy_i, r_i)$ denotes the center and the radius of circle $\phi(\gamma_i)$.

Due to the Möbius normalization, $(\mathbf{c_1}, r_1) = (0 + i0, 1)$, $(\mathbf{c_2}, r_2) = (0 + i0, r_2)$, $(\mathbf{c_3}, r_3) = (0 + iy_3, r_3)$, then the Teichmüller space of genus zero surfaces with b boundaries is of $3b - 6$ dimensional. For a doubly connected domain, the circle domain by conformal mapping is a unit annulus; its conformal module is of 1 dimensional, defined as

$$\frac{-\log r_2}{2\pi}. \tag{9.45}$$

Theorem 9.14 (Teichmüller Space [239]) *The dimension of the Teichüller space of genus zero surface with b boundaries, $T_{0,b}$, is 1 if $b = 2$, and $3b - 6$ if $b > 2$.*

The Teichmüller space has a so-called Weil–Peterson metric [243], so it is a Riemannian manifold. Furthermore it is with negative sectional curvature, therefore, the geodesic between arbitrary two points is unique.

Suppose $\Gamma = \{\gamma_0, \gamma_1, \dots, \gamma_b\}$ is a family of non-intersecting smooth closed curves on a genus zero closed surface. Γ segments the surface to a set of connected components $\{\Omega_0, \Omega_1, \dots, \Omega_b\}$, each segment Ω_i is a genus zero surface with boundary components. Construct the uniformization mapping $\phi_k : \Omega_k \to \mathbb{D}_k$ to map each segment Ω_k to a circle domain $\mathbb{D}_k$, $0 \le k \le b$. Assume γ_i is the common boundary between Ω_j and Ω_k, then $\phi_j(\gamma_i)$ is a circular boundary on the circle domain $\mathbb{D}_j$, $\phi_k(\gamma_i)$ is another circle on $\mathbb{D}_k$. Let $f_i|_{\mathbb{S}^1} := \phi_j \circ \phi_k^{-1}|_{\mathbb{S}^1} : \mathbb{S}^1 \to \mathbb{S}^1$ be the diffeomorphism from the circle to itself, which is called the *signature of* γ_i. The above construction process is called *conformal welding*. Conformal welding was first discussed in by Sharon and Mumford's seminal work [243]. With computational conformal geometry, we generalize the idea from 2D shape space to 3D shape space.

Definition 9.15 (*Signature of a Family of Loops*) The signature of a family of non-intersecting closed 3D curves $\Gamma = \{\gamma_0, \gamma_1, \dots, \gamma_b\}$ on a genus zero closed surface is defined as the combination of the conformal modules of all the connected components and the diffeomorphisms of all the curves:

$$S(\Gamma) := \{f_0, f_1, \dots, f_b\} \cup \{Mod(\mathbb{D}_0), Mod(\mathbb{D}_1), \dots, Mod(\mathbb{D}_b)\}. \tag{9.46}$$

The following 3D conformal welding theorem plays fundamental role for the current work. Note that if a circle domain $\mathbb{D}_k$ is disk, then its conformal module can be omitted from the signature (Fig. 9.20).

Theorem 9.16 (3D Conformal Welding) *The family of smooth 3D closed curves Γ on a genus zero closed Riemannian surface is determined by its signature $S(\Gamma)$, unique up to a conformal automorphism of the surface $\eta \in Conf(S)$.*

Proof See Fig. 10.3. In the left frame, a family of planar smooth curves $\Gamma = \{\gamma_0, \dots, \gamma_5\}$ divide the plane to segments $\{\Omega_0, \Omega_1, \dots, \Omega_6\}$, where Ω_0 contains the ∞ point. We represent the segments and the curves as a tree in the second frame, where each node represents a segment Ω_k, each link represents a curve γ_i. If Ω_j is included

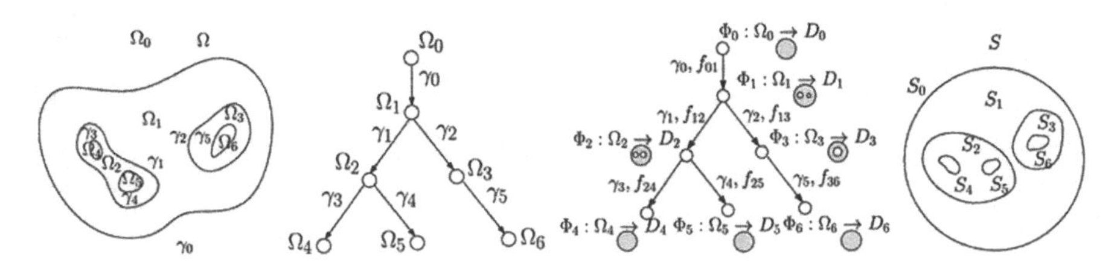

Fig. 9.20 The signature uniquely determines the family of closed curves unique up to a Möbius transformation. Image from [325]

by Ω_i, and Ω_i and Ω_j shares a curve γ_k, then the link γ_k in the tree connects Ω_j to Ω_i, denoted as $\gamma_k : \Omega_i \to \Omega_j$. In the third frame, each segment Ω_k is mapped conformally to a circle domain D_k by Φ_k. The signature for each closed curve γ_k is computed $f_{ij} = \Phi_i \circ \Phi_j^{-1}|_{\gamma_k}$, where $\gamma_k : \Omega_i \to \Omega_j$ in the tree. In the last frame, we construct a Riemann sphere by gluing circle domains D_k's using f_{ij}'s in the following way. The gluing process is of bottom up. We first glue the leaf nodes to their fathers. Let $\gamma_k : D_i \to D_j$, D_j be a leaf of the tree. For each point $z = re^{i\theta}$ in D_j, the *extension map* is

$$G_{ij}(re^{i\theta}) = re^{f_{ij}(\theta)}. \tag{9.47}$$

We denote the image of D_j under G_{ij} as S_j. Then we glue S_j with D_i. By repeating this gluing procedure bottom up, we glue all leafs to their fathers. Then we prune all leaves from the tree, and glue all the leaves of the new tree, and prune again. By repeating this procedure, eventually, we get a tree with only the root node, then we get a Riemann sphere, denoted as S. Each circle domain D_k is mapped to a segment S_k in the last frame, by a sequence of extension maps. Suppose D_k is a circle domain, a path from the root D_0 to D_k is $\{i_0 = 0, i_1, i_2, \ldots, i_n = k\}$, then the map from $G_k : D_k \to S_k$ is given by:

$$G_k = G_{i_0 i_1} \circ G_{i_1 i_2} \circ \cdots \circ G_{i_{n-1} i_n}. \tag{9.48}$$

Note that, G_0 is identity. Then the Beltrami coefficient of $G_k^{-1} : S_k \to D_k$ can be directly computed, denoted as $\mu_k : S_k \to \mathbb{C}$. The composition $\Phi_k \circ G_k^{-1} : S_k \to \Omega_k$ maps S_k to Ω_k, because Φ_k is conformal, therefore the Beltrami coefficient of $\Phi_k \circ G_k^{-1}$ equals to μ_k.

We want to find a map from the Riemann sphere S to the original Riemann sphere Ω, $\Phi : S \to \Omega$. The Beltrami-coefficient $\mu : S \to \mathbb{C}$ is the union of μ_k's each segments: $\mu(z) = \mu_k(z), \forall z \in S_k$. The solution exists and is unique up to a Möbius transformation according to Quasi-conformal Mapping theorem [90]. □

The 3D conformal welding theorem states that the proposed signature determine shapes up to a Möbius transformation. We can further do a normalization that fixes ∞ to ∞ and that the differential carries the real positive axis at ∞ to the real positive axis at ∞, as in Sharon and Mumford's paper [243]. The signature can then determine the shapes uniquely up to translation and scaling.

The shape signature $S(\Gamma)$ gives us a *complete* representation for the space of shapes. It inherits a natural metric. Given two shapes Γ_1 and Γ_2. Let $S(\Gamma_i) := \{f_0^i, f_1^i, \ldots, f_k^i\} \cup \{Mod(\mathbb{D}_0^i), Mod(\mathbb{D}_1^i), \ldots, Mod(\mathbb{D}_k^i)\}$ $(i = 1, 2)$. We can define a metric $d(S(\Gamma_1), S(\Gamma_2))$ between the two shape signatures using the natural metric in the Teichmüller space, such as the Weil–Petersson metric [243]. Our signature is stable under geometric noise. Our algorithm depends on conformal maps from surfaces to circle domains using discrete Ricci flow method (Fig. 9.21).

Figure 10.23 shows the pipeline for computing the conformal module and diffeomorphism signature for a 3D surface with 3 closed contours. Here, we use a human brain hemisphere surface whose functional areas are divided and labeled

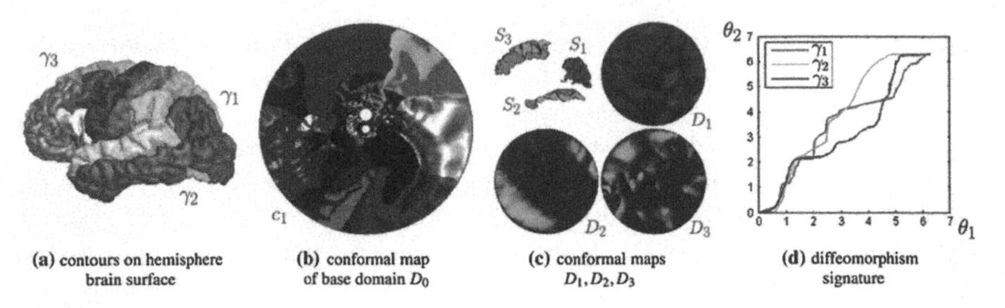

Fig. 9.21 Diffeomorphism signature via uniformization mapping for a genus zero surface with 3 simple closed contours. The curves on surface $\gamma_1, \gamma_2, \gamma_3$ in **a** correspond to the boundaries c_1, c_2, c_3 of the circle domains D_1, D_2, D_3 in **c**, respectively. These three contours are also mapped to the boundaries of the base circle domain D_0 in **b**. The curves in **d** demonstrate the diffeomorphisms for the three contours. Image from [325]

in different color. The contours (simple closed curves) of functional areas can be used to slice the surface open to connected patches. As shown in frames (a-c), three contours $\gamma_1, \gamma_2, \gamma_3$ are used to divide the whole brain (a genus zero surface S) to 4 patches S_0, S_1, S_2, S_3; each of them is conformally mapped to a circle domain (e.g., disk or annuli), D_0, D_1, D_2, D_3. Note that $\gamma_1, \gamma_2, \gamma_3$ are the contours of the inferior parietal area, the fusiform area, and the superior frontal area, respectively. In (b), the base circle domain is normalized by Möbius transformation, such that the circle c_2 is centered at origin, c_3 is centered along imaginary-axis, then conformal module of the base domain is defined as the centers and radii of circles c_2, c_3, i.e., $Mod = (r_2, y_3, r_3) = (0.042263, 0.136767, 0.063546)$, where r_i and $(x_i + iy_i)$ denote the radius and the center of circle c_i, respectively. In the mapping results, one contour is mapped to two circles in two mappings. The representation of the shape according to each contour is a diffeomorphism of the unit circle to itself, defined as the mapping between periodic polar angles (θ_1, θ_2), $\theta_1, \theta_2 \in [0, 2\pi]$. The proper normalization is employed to remove Möbius ambiguity. As shown in (d), the curves demonstrate the diffeomorphisms for three contours. The diffeomorphisms induced by the conformal maps of each curve together with the conformal module form a unique shape signature, which is the Teichmüller coordinates in Teichmüller space and may be used for shape comparison and classification.

The circular uniformization mapping is done by discrete Ricci flow method [131, 296, 324]. After the computation of the conformal mapping, each connected component is mapped to a circle domain. We compute the Teichmüller shape descriptor as in Eq. (9.46).

We define an order for all the non-intersecting closed curves on the surface S, $\{\gamma_0, \gamma_1, \gamma_2, \ldots, \gamma_b\}$, this induces an order for all the boundary components on each segment, $\{S_0, S_1, S_2, \ldots, S_b\}$. By removing all the segments from S, the left segment is denoted as $\bar{S}$, which is a multiple connected domain.

For the multiple connected segments (genus zero surfaces with multiple boundaries), the circle domain is the unit disk with multiple inner holes. Two circle domains are conformally equivalent, if and only if they differ by a Möbius transformation.

Suppose the boundaries of a circle domain D are $\partial D = \gamma_0 - \gamma_1 - \gamma_2 \cdots - \gamma_b$, each γ_k is a circle $(\mathbf{c_k}, r_k)$, where c_k denotes the center, r_k denotes the radius. By the definition for the conformal module of a circle domain, we normalize each circle domain using a Möbius transformation, such that γ_0 becomes the unit exterior circle, $\mathbf{c_1}$ is at the origin, $\mathbf{c_2}$ is on the imaginary axis. Then the normalized circle domain is determined by its conformal module [321], which can be computed directly as in Eq. (9.44),

$$Mod(D) = \{\mathbf{c_k}, k > 1\} \cup \{r_j, j > 0\}. \tag{9.49}$$

For those simply connected segments (genus zero surfaces with only one boundary), the circle domain is the unit disk. We compute its mass center and use a Möbius transformation to map the center to the origin. Their conformal modules can be omitted in the shape signature.

Each closed curve γ_k on the 3D surface becomes the boundary components on two segments, both boundary components are mapped to a circle under the uniformization mapping. Then each boundary component gives a diffeomorphism of the unit circle to itself, defined as the mapping between the radial angles on two circles,

$$Diff(\gamma_k) = (\theta_k^1, \theta_k^2), \theta_k^1, \theta_k^2 \in [0, 2\pi]. \tag{9.50}$$

In order to keep consistency, we define a marker p_k on the boundary as the starting point, i.e., $\theta_k^1(p_k) = \theta_k^2(p_k) = 0$, to compute the radial angles for the whole curve.

Experimental Results

We tested the discrimination ability of the proposed shape descriptor on a set of left brain hemispheres of 152 healthy control (CTL) subjects and 169 AD patients from the ADNI dataset. Each half brain surface mesh has $100K$ triangles. Among 34 cortical functional areas [65] defined by FreeSurfer [78], we selected 3 regions of interest for study, such as superior frontal, fusiform and inferior parietal areas as shown in Fig. 10.23, correspondingly, represented by 3 closed curves, $\gamma_1, \gamma_2, \gamma_3$, on the half brain surfaces. In this work, we used the *left* brain hemisphere surfaces for testing shape descriptors.

These three closed curves segment a brain hemisphere surface to 4 patches; one topological annulus (called the base domain), three topological disks. The base domain with three boundaries is mapped to a circle domain, one boundary to the exterior unit circle, one boundary to the inner concentric circle, the rest one to the inner circle centered at the imaginary axis. The conformal module of the base domain is computed as in Eq. (9.49). In the conformal mapping of each topological disk segment, the mass center is mapped to the origin of the unit disk. In addition, one marker on each curve is extracted as the starting point of computing radial angels. Here, we automatically selected the intersection point of three specified regions along the curve. The diffeomorphism descriptor for each curve, computed by Eq. (9.50), is plotted as a monotonic curve within the square $[0, 2\pi] \times [0, 2\pi]$. We sampled the curve to be 1000 points uniformly. Figure 9.22 illustrates the shape descriptors for 3 CTL brain surfaces and 3 AD brain surfaces.

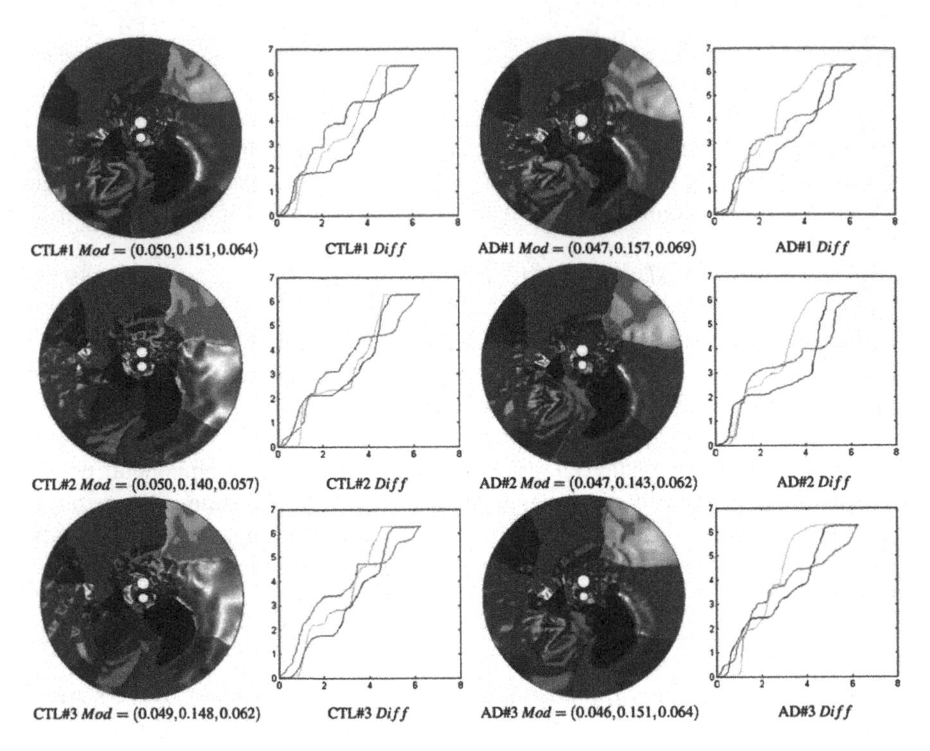

Fig. 9.22 Teichmüller shape descriptor (*Diff*, *Mod*) of 3 healthy control (CTL) brain cortexes and 3 Alzheimer's disease (AD) brain cortexes, both of which are randomly selected from the database. The left half brain with 3 contours is considered. Image from [325]

For our classification purpose, we set 80% of each category to be training samples, the rest 20% testing samples. In order to obtain the fair results, we randomly selected the training set each time and computed the average recognition rate over 1000 times. We applied the support vector machine (SVM) [233] as classifier, where the linear kernel function was employed. Table 9.2 shows that the average recognition rates by the signature (*Diff*, *Mod*) reaches 91.38% under the above experimental setting. We also tested the signatures, diffeomorphism (*Diff*) and conformal module (Mod), separately. The experimental results demonstrate that the recognition rates are much less than the complete signature (*Diff*, *Mod*). That satisfies the fact that (*Diff*) describes the more detailed correlation of each patch to the base domain through the closed curves, while (Mod) captures the global shape information only through the base domain; both together are required to recover the closed curves on 3D surface.

Furthermore, for comparison with the common signatures in literature, we computed the volume for the left brain surfaces as signature, (Vol). The average recognition rate of volume using linear SVM in the above setting is 68.20%. The histogram for volume illustrated in Fig. 9.23 intuitively demonstrates that the volume signature cannot differentiate the AD and CTL subject groups accurately. We also computed the surface area for the base domain and 3 regions as signature,

Table 9.2 Average recognition accuracy rates (%) for applying different signatures among 152 healthy control subjects versus 169 AD subjects, where 80% of the dataset are randomly selected for training and the rest 20% for testing. The average recognition rate is computed over 1000 times. Linear SVM method is used for classification.

Sig.	$(Diff, Mod)$	$(Diff)$	(Mod)	(Vol)	$(Area)$
Rate %	**91.38**	85.71	63.60	68.20	70.23

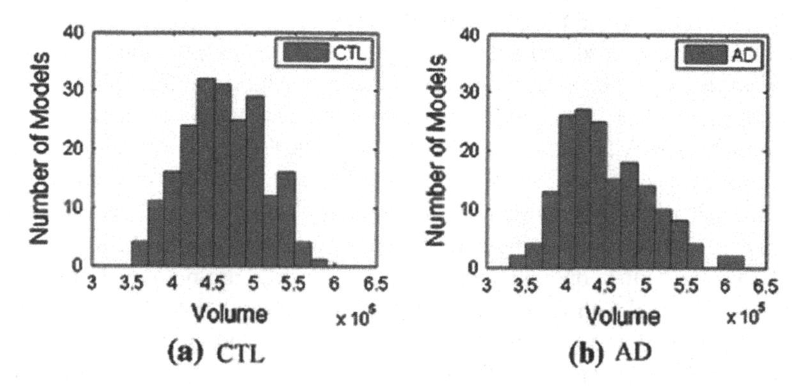

Fig. 9.23 Histogram of volumes for 152 healthy control (CTL) subjects and 169 Alzheimer's disease (AD) patients. Image from [325]

$(Area) = (A_0, A_1, A_2, A_3)$, the average recognition rate is 70.23%. All this demonstrates that the proposed global Teichmüller shape descriptor is very efficient and much more effective to differentiate the shapes within AD and healthy control subject groups.

9.4.3 Teichmüller Space Coordinates for Landmark Curve-Based Brain Morphometry Analysis

Landmark curves were widely adopted in neuroimaging research for surface correspondence computation and quantified morphometry analysis. Most of the landmark based morphometry studies only focused on landmark curve shape difference. With conformal geometry, we develop a set of conformal invariant-based shape indices, which are associated with the landmark curve induced boundary lengths in the hyperbolic parameter domain [249]. Such shape indices may be used to identify which surfaces are conformally equivalent and further quantitatively measure surface deformation. With the surface Ricci flow method, we can conformally map a multiply connected surface to the Poincaré disk. Our algorithm provides a stable method to compute the shape index values in the 2D (Poincaré Disk) parameter domain. The proposed shape indices are succinct, intrinsic and informative.

Genus-zero surfaces with more than two open boundaries are called *multiply connected surfaces*. Multiply connected surfaces have negative Euler characteristics and can be conformally mapped to the hyperbolic space $\mathbb{H}^2$ with their hyperbolic uniformization metric. The lengths of the open boundaries on the multiply connected surfaces under the hyperbolic uniformization metric are the Teichmüller coordinates of the surfaces [185]. The obtained Teichmüller space coordinates, i.e. conformal invariants, are invariants with conformal structures, computationally simple and stable. For a brain cortical surface with certain landmark curves, it will become a multiply connected surface after we cut open the cortical surface along landmark curves. Thus we may use them as such conformal invariants as the shape indices to characterize brain cortical morphometry changes.

We build an integrated and automated framework to compute the shape indices for cortical surfaces. Major steps are summarized in Algorithm 5 and Fig. 9.24.

Similar to prior sections on hyperbolic harmonic map, we cut open brain surfaces along certain landmark curves. After that, we compute the hyperbolic uniformization metric with hyperbolic Ricci flow method [131, 246, 324]. After computing the hyperbolic uniformization metric, we can embed the surface onto $\mathbb{H}^2$. An illustration of the cortical surface embedding onto the Poincaré disk is in Fig. 9.24e.

In the Poincaré disk, open boundaries on the mesh Σ are mapped to hyperbolic geodesics, as shown with red curves in Fig. 9.24e. Assume the edges on boundary b_k are counter-clockwisely ordered, as $(e_i, e_2, \dots, e_m)$, and the two edges adjacent to vertex $v_i \in b_k$ are e_i and e_{i+1}. Then the length of each boundary is computed as

$$\text{len}_{b_k} = \Sigma_{e_i \in b_k} |e_i| \tag{9.51}$$

where $|e_i|$ is the length of edge e_i under the hyperbolic metric. The lengths of the geodesics are invariant to conformal mappings and they form the proposed shape indices for each surface.

Algorithm 5 *Surface Conformal Invariants Computation Pipeline.*

Input:*A structural MR image.*

Output:*Shape indices, which is a set of conformal invariants of the cortical surfaces.*

1. The structural MRI (Fig. 10.23a) is segmented and the cortical surfaces (Fig. 10.23b) are reconstructed automatically using the FreeSurfer software [78].

2. Six landmark curves are automatically traced on each cortical hemisphere surface using the Caret software [278] .

3. With topology optimization, each cortical surface is converted to a multiply connect surface. Open boundaries γ_1, γ_2, γ_3 are shown in Fig. 10.23c.

4. The fundamental group of each cortical surface is computed and the simply connected domain is obtained by slicing each surface along the fundamental group of paths (Fig. 10.23d).

5. Compute the hyperbolic uniformization metric of the multiply connect surface and embed it onto the Poincaré disk with the simply connect domain (Fig. 10.23e).

6. Compute the conformal invariants, i.e. shape indices, with the hyperbolic uniformization metric.

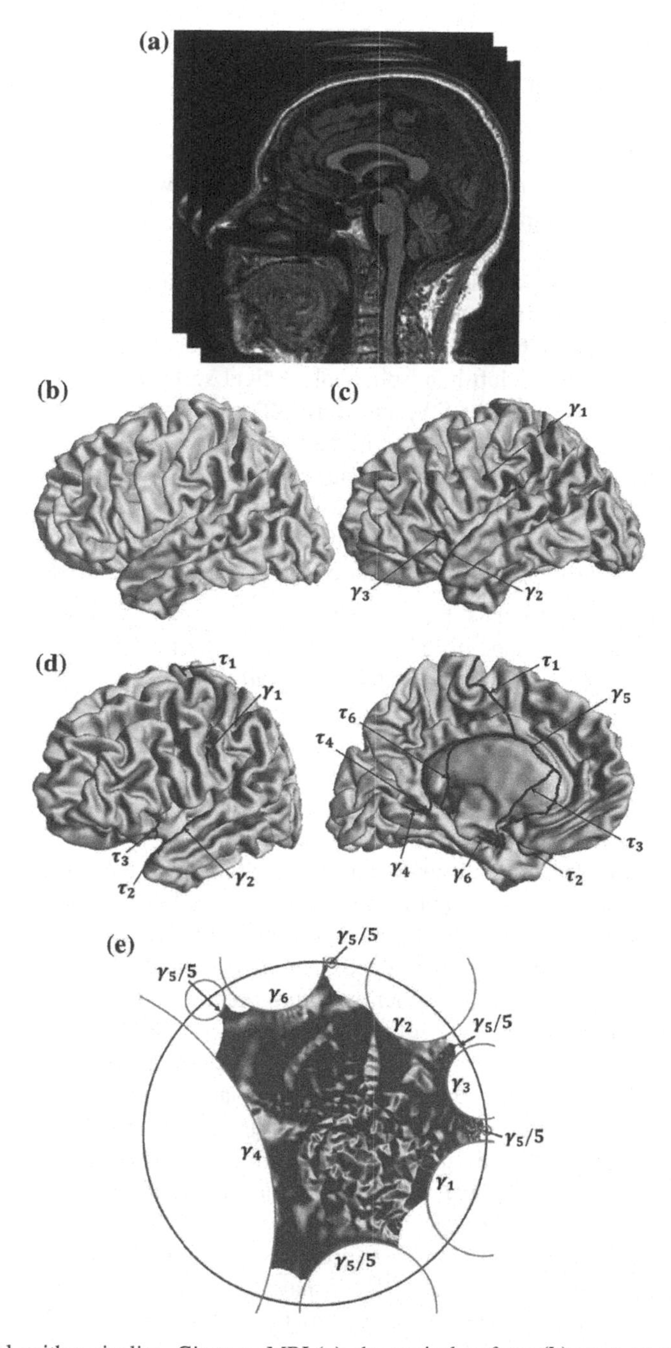

Fig. 9.24 Algorithm pipeline. Given an MRI (**a**), the cortical surfaces (**b**) are reconstructed with the FreeSurfer [78]. Six landmarks are then traced on each cortical surface using the Caret tool [278] (Fig. 9.1). With topology optimization, each cortical surface is converted to a multiply connected surface (**c**). We then compute the fundamental group (**d**) of the multiply connected surface. Finally, with the hyperbolic Ricci flow, we map each cortical surface onto the Poincaré disk (**e**). Image from [249]

Experimental Results

We first tested the feasibility of the proposed shape index in studying longitudinal cortical morphometry associated with AD. We randomly selected two mild cognitive impairment (MCI) patients from the ADNI database [128]. One MCI patient converted to AD 36 months after the baseline screening, which we called the MCI converter subject, the other who did not convert to AD 36 months after the baseline screening was called the MCI stable subject.

In our experiments, we studied their structural MR images at two time points, the baseline and 24 months. Our hypothesis is that we may observe some different atrophy patterns associated with the conversion versus stable progression paths. In this experiment, only the left hemispherical cortical surfaces were used. With the preprocessing with FreeSurfer [78] and Caret [278] software packages, two baseline brain cortical surfaces with the landmark curves are shown in Fig. 9.25a. We computed the lengths of the landmark curves under the hyperbolic metric in the same way as the above section. For each MCI patient, we obtained a shape feature matrix $(B_i, T_i), i = 1, \ldots, 6$, where B represents the baseline cortical surface and T represents the 24-month cortical surface. Then we calculated the L^2 norm of the shape difference for a given subject over time as $d = \sqrt{\sum_{i=1}^{6} (B_i - T_i)^2}$. With the proposed method, the shape difference between the cortical surfaces of two time points is 0.3925 and 1.0438 for the MCI stable and converter subjects, respectively. We see some more differences on the MCI converter subject's longitudinal image data than the MCI stable subject.

Further, to illustrate the difference on each conformal invariant, we plotted the radar chart [42] in Fig. 9.25b. Basically, a radar chart is a graphical method of displaying multivariate data in the form of a two-dimensional chart where three or more quantitative variables are represented on axes starting from the same point. In our case, six landmarks are associated with six corners on a hexagon and six axes connect them to the common origin point. The position on each axis is proportional to the absolute difference between the computed conformal invariants on two time points. The more different, the farther away the point is to the origin. Meanwhile, we used two different color lines to connect the longitudinal differences from two different subjects, the blue color is for the MCI stable subject while the orange color is for the MCI converter subject. From the radar chart, we find that there is a clear difference between these two subjects' radar charts. Interestingly, we also found on the MCI converter radar chart, there are sharp differences associated with two landmark curves on the medial wall ventral segment and medial wall dorsal segment, especially on medial wall ventral segment. Since medial wall ventral segment runs along the medial margin of the hippocampal sulcus and medal wall dorsal segment runs along the corpus callosum (Fig. 9.1). It may be particularly consistent with prior observations that, for AD conversion, grey matter atrophy is more related to medial temporal lobe region and the posterior cingulate [36, 83, 164, 191, 267]. Although multi-subject studies are clearly necessary, this experiment demonstrates that our

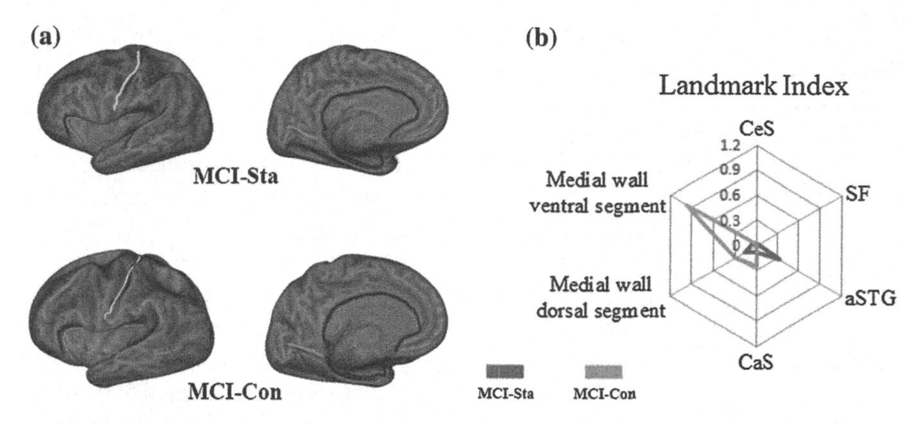

Fig. 9.25 Radar chart showing the variance of each landmark curve on multiply connected cortical surfaces. **a** two baseline flattened cortical surfaces with identified six landmark curves. **b** Radar chart on individual landmark conformal invariants between two MCI subjects. The landmark conformal invariants are computed on their left cortical surfaces. Each landmark is associated with a corner on a hexagon and six axes are connected them to the common origin point. The position on the axis is proportional to the absolute difference between two landmark conformal invariants computed on different time points. By connecting the six points from the same subject, we get two new hexagons. The blue line stands for MCI Stable (MCI-Sta) and the orange for MCI converter (MCI-Con). The radar chart provides a simple way to display multivariate data in the form of a two-dimensional chart. Image from [249]

conformal invariants may potentially be used as a set of shape indices to compare and classify cortical surfaces of AD patients at different stages of disease progression. Our work may also provide a simple way to directly visualize the cortical morphometry changes related to different brain regions.

The second experiment further investigates the variation patterns of the six landmarks to study brain shape morphometry in AD at the group-level population-based study. We hypothesize that the length of those six landmarks under hyperbolic metric may serve as useful shape indices to verify and predict degeneration of AD. In this experiment, we applied a group-wise statistical analysis to our proposed shape indices in left hemisphere cortex between AD and control subjects. Our data set consisted of 60 subjects which were randomly chosen from ADNI database, including 30 AD subjects and 30 matching normal controls. The cortical surface of each subject was computed, from which six landmarks were extracted using the same method as that in the previous experiments. Similarly, we use the lengths of six landmarks in hyperbolic space as the conformal invariants to represent morphometric characteristics of cortical surfaces. Here, conformal invariants for each cortical surface for a given hemisphere was a vector $v = (v_1, \ldots, v_6)^T$, where v_i is the length of a given landmark under hyperbolic metric.

Similar to our prior work [297], we performed the permutation based Hotelling's T^2 test to evaluate group difference. Suppose there are two sets of feature vectors from two different groups, $C = (x_1, x_2, \ldots, x_m)$ and $W = (y_1, y_2, \ldots, y_n)$, where m and n are the number of subjects in AD and the control group respectively. The difference between two groups is measured by the Mahalanobis distance,

$$d(v) = \sqrt{(\mu_C - \mu_W)^T (\Sigma_W + \Sigma_C)^{-1} (\mu_C - \mu_W)} \tag{9.52}$$

where μ_C, μ_W, Σ_C and Σ_W are the mean and covariance of two groups, respectively. We first compute the true group distance based on the true grouped samples. Then we randomly assigned the feature vectors into two groups each with the equal size number of 30 and re-computed group distance. The process was repeated 5000 times with the outcome of 5000 permutation values. Finally, a probability (p value) was computed as the ratio of the number of permutation values greater than the true group distance to the total permutation times. To evaluate if our proposed conformal invariant vector really helped to improve the detection power, we also computed group differences by using cortical surfaces area and cortex volume.

With this protocol, the prominent statistical result of conformal invariants was $p = 0.0133$ between AD and control groups. For comparison purpose, we also computed the natural lengths of the six landmark curves in the ambient space, as they are the counterpart of the proposed shape indices in the Euclidean domain. Similarly, we used Hotelling's T^2 test to perform group comparison. With the same permutation test as above, the Euclidean lengths of the landmarks did not reveal significant difference between the two groups ($p = 0.7844$). Furthermore, we also conducted permutation based t tests using left cortical surface area and left cortex volume, respectively. Similarly, we conducted 5000 times of permutation and estimated the probability of the ground truth data happened in this random permutation results. Neither of the two statistical results reached significant level ($p = 0.5888$ and $p = 0.1152$ for left cortical surface area and left cortex volume, respectively). Our empirical results demonstrated that it may be possible to use our proposed shape indices to detect AD related morphometric variation on group-based studies.

9.4.4 Riemannian Optimal Mass Transport Map and Riemannian Wasserstein Distance

Optimal mass transportation (OMT) problem was first raised by Monge [28] in the 18th century. Given two probability measures on a manifold, the OMT map induces the Wasserstein distance between them, which is a Riemannian metric of the Wasserstein space [281]. Wasserstein distance is continuous and able to measure subtle shape differences. Since OMT considers a transportation between two probability measures, it is robust to noise. Thus Wasserstein space provides suitable mathematical and computational descriptions for both shape representation and comparisons [148].

It has been widely studied and applied for computer vision and shape analysis, due to its significant power to intrinsically compare similarities between shapes. For example, Wang et al. [286] proposed a linear optimal transportation framework for comparing images. Schmitzer et al. [231] proposed Wasserstein based method for joint variational object segmentation and shape matching. Hong et al. [123] introduced a shape feature that characterizes local shape geometry for shape matching based on Wasserstein distance. However, most of these OMT and Wasserstein space computation methods only work for 2D images and the Riemannian OMT and its induced Wasserstein space research is very limited.

With computational conformal geometry, we generalize the OMT map from Euclidean metrics to Riemannian metrics, such that our proposed framework is applicable to any general Riemannian manifolds. We develop the discrete OMT map theory which lays down the theoretic foundation to compute OMT maps with the Riemannian metrics. Further the OMT induced transportation cost defines the Riemannian Wasserstein distance between two general Riemannian manifolds. In this section, we will first introduce the theoretic foundation for our Riemannian OMT work. Later we will use sphere and Poincaré disk as two parameter domains to illustrate how to compute spherical Wasserstein distance [262] and hyperbolic Wasserstein distance [250] and their brain imaging applications.

Problem 9.17 (*Optimal Mass Transport*) Suppose $(M, \mathbf{g})$ is a Riemannian manifold with a Riemannian metric $\mathbf{g}$, let μ and ν be two probability measures on M, which have the same total mass $\int_M d\mu = \int_M d\nu$. A map $T : M \to M$ is *measure preserving*, if for any measurable set $B \subset M$, $\int_{T^{-1}(B)} d\mu = \int_B d\nu$. Namely, the μ pushed forward to ν by T, denoted as $T_{\#}\mu = \nu$. Given a transportation cost function $c : M \times M \to \mathbb{R}$, find the measure preserving map $T : M \to M$ that minimizes the total transportation cost

$$\mathcal{C}(T) := \int_M c(x, T(x))d\mu(x). \tag{9.53}$$

In our current work, the cost function is the squared geodesic distance, $c(x, y) = d^2_{\mathbf{g}}(x, y)$.

In the 1940s, Kantorovich introduced the relaxation of Monge's problem and solved it using the linear programming [145].

Theorem 9.18 (Kantorovich) *Suppose $(M, \mathbf{g})$ is a Riemannian manifold, probability measures μ and ν have the same total mass, μ is absolutely continuous, ν has finite second moment, the cost function is the squared geodesic distance, then the optimal mass transportation map exists and is unique.*

Suppose $(M, \mathbf{g})$ is a Riemannian manifold with a Riemannian metric $\mathbf{g}$.

Definition 9.19 (*Wasserstein Space*) For $p \geq 1$, let $\mathcal{P}_p(M)$ denote the space of all probability measures μ on M with finite pth moment, for some $x_0 \in M$, $\int_M d(x, x_0)^p d\mu(x) < +\infty$, where d is the geodesic distance induced by $\mathbf{g}$.

Given two probability functions μ and ν in $\mathcal{P}_p$, the Wasserstein distance between them is defined as the transportation cost induced by the optimal transportation map $T : M \to M$,

$$W_p(\mu, \nu) := \inf_{T_\# \mu = \nu} \left(\int_M d_{\mathbf{g}}^p(x, T(x)) d\mu(x) \right)^{\frac{1}{p}}.$$

The following theorem plays a fundamental role for the current work.

Theorem 9.20 *The Wasserstein distance W_p is a Riemannian metric of the Wasserstein space $\mathcal{P}_p(M)$.*

Detailed proof can be found in [281].

Suppose μ has compact support on X. Define $\Omega = \text{Supp}\, \mu = \{x \in X | \mu(x) > 0\}$, and assume Ω is a convex domain in X. The space Y is discretized to $Y = \{y_1, y_2, \ldots, y_k\}$ with Dirac measure $\nu = \sum_{j=1}^{k} \nu_j \delta(y - y_j)$. We define a *height vector* $\mathbf{h} = (h_1, h_2, \ldots, h_k) \in \mathbb{R}^k$, consisting of k real numbers. For each $y_i \in Y$, we construct a hyperplane defined on X,

$$\pi_i(\mathbf{h}) : \langle x, y_i \rangle + h_i = 0. \tag{9.54}$$

We define a piece-wise linear convex function

$$u_{\mathbf{h}}(x) = \max_{i=1}^{k} \{\langle x, y_i \rangle + h_i\}, \tag{9.55}$$

We denote its graph by $G(\mathbf{h})$, which is an infinite convex polyhedron with supporting planes $\pi_i(\mathbf{h})$. The projection of $G(\mathbf{h})$ induces a polygonal partition of Ω,

$$\Omega = \bigcup_{i=1}^{k} W_i(\mathbf{h}), \tag{9.56}$$

where each cell $W_i(\mathbf{h})$ is the projection of a facet of the convex polyhedron $G(\mathbf{h})$ onto Ω,

$$W_i(\mathbf{h}) = \{x \in X | u_{\mathbf{h}}(x) = \langle x, y_i \rangle + h_i\} \cap \Omega. \tag{9.57}$$

This partition is equivalent to a power diagram, denoted as $D(\mathbf{h})$, as explained in [98]. The area of $W_i(\mathbf{h})$ is given by

$$w_i(\mathbf{h}) = \int_{W_i(\mathbf{h})} \mu(x) dx. \tag{9.58}$$

The convex function $u_{\mathbf{h}}$ on each cell $W_i(\mathbf{h})$ is a linear function $\pi_i(\mathbf{h})$, therefore, the gradient map

$$\nabla u_{\mathbf{h}} : W_i(\mathbf{h}) \to y_i, i = 1, 2, \ldots, k. \tag{9.59}$$

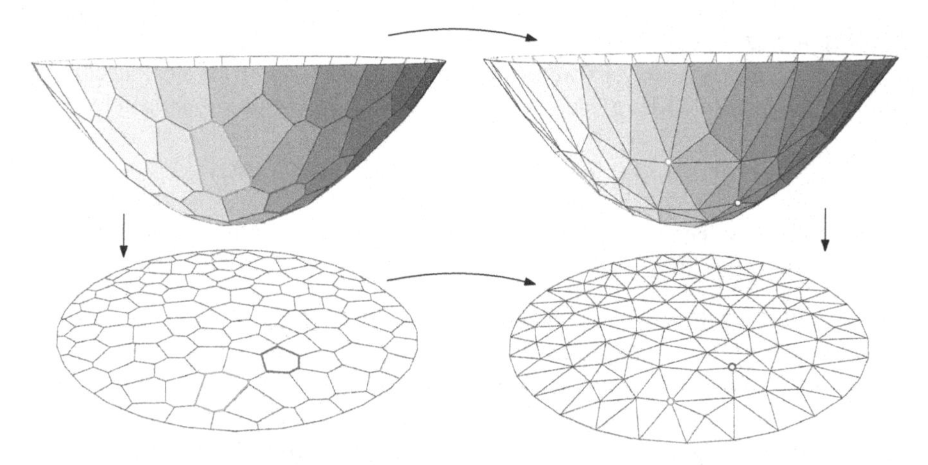

Fig. 9.26 Discrete optimal mass transport map with Brenier's approach. Image from [261]

maps each $W_i(\mathbf{h})$ to a single point y_i. Figure 9.26 shows the cell decomposition induced by a convex function.

The following theorem plays a fundamental role for discrete optimal mass transport theory.

Theorem 9.21 *Given a convex domain* $\Omega \subset \mathbb{R}^n$, *with measure density* $\mu : \Omega \to \mathbb{R}$, *and a discrete point set* $Y = \{y_1, \ldots, y_k\}$ *with discrete measures* $\mathbf{v} = \{\nu_1, \ldots, \nu_k\}$. *Suppose* $\sum_{j=1}^{k} \nu_j = \int_\Omega \mu$, $\nu_j > 0$. *Then there must exist a height vector* $\mathbf{h} = \{h_1, \ldots, h_k\}$ *unique up to translations, such that the convex function Eq.* (9.55) *induces the cell decomposition of Eq.* (9.56). *The following* area-preserving constraints *are satisfied for all cells,*

$$\int_{W_i(\mathbf{h})} \mu(x)dx = \nu_i, i = 1, 2, \ldots, k. \tag{9.60}$$

Furthermore, the gradient map $grad\ u_{\mathbf{h}}$ *optimizes the following transportation cost*

$$\mathcal{C}(T) := \int_\Omega |x - T(x)|^2 \mu(x)dx. \tag{9.61}$$

The existence and uniqueness was first proven by Alexandrov [3] using a topological method; the existence was also proven by Aurenhammer, Hoffmann and Aronov [14], the uniqueness and optimality was proven by Brenier [34]. Recently, Gu et al. [98] gives a novel proof for the existence and uniqueness based on the variational principle, which leads to a simple but elegant optimal mass transport method in Euclidean spaces. We have reported our 2D optimal mass transport method in [261]. The algorithm has the potential to work any dimensional Euclidean spaces. With the same motivation, this paper proposes practical methods to computer OMT on general Riemannian surfaces.

Definition 9.22 (*Geodesic Power Voronoi Diagram*) Given the point set P and the weight $\mathbf{h}$, the geodesic power voronoi diagram induced by $(P, \mathbf{h})$ is a cell decomposition of the manifold $(M, \mathbf{g})$, such that the cell associated with p_i is given by

$$W_i := \{q \in M | d_{\mathbf{g}}^2(p_i, q) + h_i \leq d_{\mathbf{g}}^2(p_j, q) + h_j\}.$$

The following theorem lays down the theoretic foundation of our optimal mass transport map algorithm.

Theorem 9.23 (Discrete Optimal Mass Transportation Map) *Given a Riemannian manifold $(M, \mathbf{g})$, two probability measures μ and ν are of the same total mass. ν is a Dirac measure, with discrete point set support $P = \{p_1, p_2, \ldots, p_k\}$, $\nu(p_i) = \nu_i$. There exists a weight $\mathbf{h} = \{h_1, h_2, \ldots, h_k\}$, unique up to a constant, the geodesic power Voronoi diagram induced by $(P, \mathbf{h})$ gives the optimal mass transportation map,*

$$T : W_i \to p_i, i = 1, 2, \ldots, k,$$

furthermore

$$\int_{W_i} d\mu = \nu_i, \forall i. \tag{9.62}$$

Proof Suppose $M = \cup_{i=1}^k \tilde{W}_i$ is another partition of the manifold, such that $\int_{\tilde{W}_i} d\mu = \nu_i$. The mapping $\tilde{T} : \tilde{W}_i \to p_i$ is another measure-preserving mapping. Then

$$C(\tilde{T}) + \sum \nu_i h_i = \sum_{i=1}^k \int_{\tilde{W}_i} (d_{\mathbf{g}}^2(p, p_i) + h_i) d\mu(p)$$

The right hand side is greater than

$$\sum_{i=1}^k \int_{W_i} (d_{\mathbf{g}}^2(p, p_i) + h_i) d\mu(p) = C(T) + \sum \nu_i h_i.$$

This shows for any measure preserving mapping, $C(\tilde{T}) \geq C(T)$. □

The optimal weight for the geodesic power Voronoi diagram that induces the optimal transportation map can be found by

$$\frac{dh_i}{dt} = \int_{W_i(\mathbf{h})} d\mu - \nu_i. \tag{9.63}$$

According to Poincaré uniformization theorem [131], all shapes can be conformally mapped to one of three canonical spaces: the unit sphere, the Euclidean plane or the hyperbolic plane. The area-distortion factor by the uniformization map defines a probability measure on the canonical uniformization space. All the probability measures on a Riemannian manifold form the Wasserstein space. Given two probability measures, there exists a unique OMT map between them. With the tools of discrete surface Ricci flow and geodesic power Voronoi diagram [287], we develop practical systems that are able to compute Riemannian OMT, spherical Wasserstein distance and hyperbolic Wasserstein distance.

Spherical Wasserstein Distance

Figure 9.27 sketches the outline of the computation of the spherical Wasserstein distance procedure.

First, we apply surface Ricci flow method [131] to conformally map a genus-zero surface to sphere. Using the computed spherical domain as the canonical space, here we propose a general method that compute Riemannian OMT map. On the spherical parameter domain, denote the measure μ and the Dirac measure, $\{(p_1, \nu_1), (p_2, \nu_2), \ldots, (p_k, \nu_k)\}$. From each point p_i in the point set P, we compute the spherical geodesics to reach every other vertex on the mesh, this gives the geodesic distance from every vertex to p_i. The geodesics on the triangle meshes can be efficiently computed using the algorithms in [287]. Repeat this for all vertices in P.

Then, we pursue the optimal weight $\mathbf{h}$ which induces a geodesic power Voronoi diagram which gives the optimal mass transportation map (Theorem 9.23). Specifically, we initialize all the weights to be zeros, then update the weight using the formula

$$\frac{dh_i}{dt} = \nu_i - \int_{W_i(\mathbf{h})} \mu(p)dp.$$

The iteration stops until all differences between the given measures ν_i and the Wasserstein cell areas w_i are less than a given threshold (ε). Details of the algorithm can be found in Algorithm 6. Figure 9.27e shows a Riemannian OMT map between two cortical surfaces.

Algorithm 6 *Riemannian Optimal Mass Transport Map*

Input: *A triangle mesh M, measure μ and Dirac measure $\{(p_1, \nu_1), (p_2, \nu_2), \cdots, (p_k, \nu_k)\}$, $\int_M u(p)dp = \sum_{i=1}^k \nu_i$; a threshold ε.*
Output: *The unique discrete Optimal Mass Transport Map $T : (M, \mu) \to (P, \nu)$.*
Subdivide M for several levels, until each triangle size is small enough.
for all $p_i \in P$ ***do***
  Compute the geodesic from p_i to every other vertex on M,
end for
$\mathbf{h} \leftarrow (0, 0, \cdots, 0)$.
repeat
  for all *vertex v_j on M* ***do***
    Find the minimum weighted squared geodesic distance, decide which Voronoi cell v_i belongs to, $v_i \in W_t(\mathbf{h})$

$$t = argmin_k d_{\mathbf{g}}^2(v_j, p_k) + h_k$$

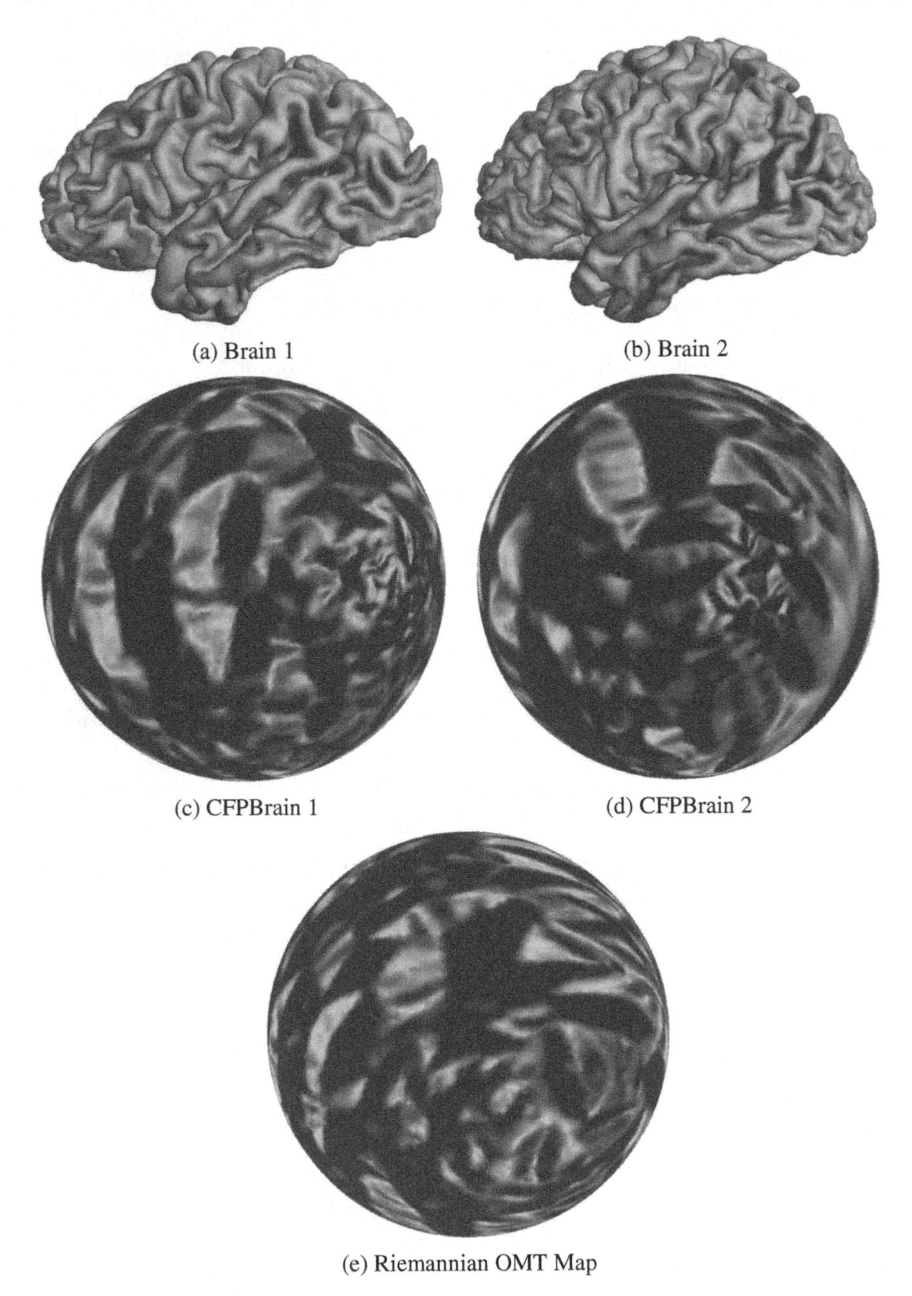

(a) Brain 1 (b) Brain 2

(c) CFPBrain 1 (d) CFPBrain 2

(e) Riemannian OMT Map

Fig. 9.27 Illustration of computation of Wasserstein distance between the left hemisphere brain cortical surfaces. **a** and **b** show two left hemisphere brain cortical surfaces, respectively. **c** and **d** are the spherical conformal parameterization (CFP) of **a** and **b**, respectively. **e** shows the Riemannian optimal mass transport (OMT) map result from **c** to **d**, which induces the Wasserstein distance between **a** and **b**. Image from [262]

end for
for all $p_i \in P$ *do*
Compute the current cell area $w_i = \int_{W_i(\mathbf{h})} d\mu$,
end for
for all $h_i \in \mathbf{h}$ *do*
Update h_i, $h_i = h_i + \delta(\nu_i - w_i)$
end for
until $|\nu_i - w_i| < \varepsilon, \forall i$.
return Power geodesic Voronoi diagram.

Algorithm 7 *Computing Wasserstein Distance*
Input: *Two topological spherical surfaces* $(S_1, \mathbf{g}_1)$, $(S_2, \mathbf{g}_2)$.
Output: *The Wasserstein distance between* S_1 *and* S_2.
1. *Scale and normalize* S_1 *and* S_2 *such that the total area of each surface is* 4π.
2. *Compute the conformal maps (Sect. 9.2.3,* $\phi_1 : S_1 \to \mathbb{S}^2$ *and* $\phi_2 : S_2 \to \mathbb{S}^2$, *where* $\mathbb{S}^2$ *is the unit sphere, and* ϕ_1 *and* ϕ_2 *are with normalization conditions: the mass center of the image points are at the sphere center.*
3. *Compute the conformal factors* λ_1 *and* λ_2. *Construct the measure* $\mu \leftarrow e^{2\lambda_1} dA$.
4. *Discretize* $\mathbb{S}^2$ *into a discrete point set with measure* $(P, \boldsymbol{\nu})$, *where* ν *is computed by Eq.* (9.64).
5. *With* $(\mathbb{S}^2, \mu)$ *and* $(P, \boldsymbol{\nu})$ *as inputs of Algorithm 6, we compute the Optimal Mass Transport map.*
6. *Spherical Wasserstein distance between* S_1 *and* S_2 *can be computed by Eq.* (9.65).

Given two probability measures μ and ν on the unit sphere, there is a unique optimal mass transport map between them, and the transportation cost defines the Wasserstein distance between them. Wasserstein distance gives a Riemannian metric for the Wasserstein space. It intrinsically measures the dissimilarities between shapes. Here we introduce how to compute spherical Wasserstein distance between two topological spherical surfaces by integrating surface Ricci flow method and Algorithm 6. Specifically, Given two topological spherical surfaces $(S_1, \mathbf{g}_1)$, $(S_2, \mathbf{g}_2)$ with total area 4π. We first compute the conformal maps with surface Ricci flow method, $\phi_1 : S_1 \to \mathbb{S}^2$ and $\phi_2 : S_2 \to \mathbb{S}^2$, where $\mathbb{S}^2$ is the unit sphere. The conformal factors, λ_1 and λ_2, define two probability measures on the sphere. With a triangular mesh representation, the discrete conformal factors can be estimated by area ratios of the triangles which are attached to each vertex [101].

Then we discretize $\mathbb{S}^2$ into a discrete point set with measure $(P, \boldsymbol{\nu})$, where ν is computed as follows: first we compute geodesic voronoi diagram induced by P, suppose the Voronoi cell associated with p_i is W_i, then

$$\nu_i := \int_{W_i} e^{2\lambda(p)} dA(p), . \tag{9.64}$$

where dA is the spherical area element. Denote the measure $e^{2\lambda_1}dA$ as μ, use $(\mathbb{S}^2, \mu)$ and $(P, \mathbf{v})$ as inputs of Algorithm 6, we compute the Optimal Mass Transport map $T : \mathbb{S}^2 \to P$, $W_i(\mathbf{h}) \to p_i$, where $p_i \in P, i = 1, 2, \ldots, k$. Therefore, following Theorem 9.23, the Wasserstein distance between S_1 and S_2 can be computed by

$$Wasserstein(\mu, \mathbf{v}) = \sum_{i=1}^{k} \int_{W_i} d_{\mathbf{g}}^2(x, T(x))^2 \mu(x) dx \tag{9.65}$$

An example of the computation process of Wasserstein distance is illustrated in Fig. 9.27. Algorithm 7 gives the implementation details. First, two brain surfaces S_1 and S_2, as shown in Fig. 9.27a and b, are normalized so that each surface has a total area of 4π. Spherical conformal parameterizations ϕ_1 and ϕ_2 of above two brain surfaces are then carried out to obtain two unit spheres which are illustrated in Fig. 9.27c and d. Based on these spherical conformal mapping results and computed measures μ and ν, the Riemannian optimal mass transport map is computed, as shown in Fig. 9.27e. Finally, the Wasserstein distance between two brain surfaces is given by the transportation cost induced by the Riemannian optimal mass transport map.

As an application example, we study how spherical Wasserstein distance help study the correlation between human intelligence and cortical morphometry. Earlier works have studied some significant factors such as cortical surface area, cortical thickness and cortical convolution [127, 179, 180]. To validate the correctness of our framework in some shape comparison applications, we apply our method for the classification problem of brain cortical surfaces with different intelligence quotient (IQ) (based on Ravens Progressive Matrices [221]), and compare it with some other standard brain imaging indices.

The dataset used in our experiments is some brain data from a medical center, it includes 50 male and 50 female, with ages ranging from 18 to 30 years. MRI recording was performed using a standard 12-channel head coil on a Siemens 3T Trio Magnetic Resonance Imaging System with TIM. The brain cortical surfaces are reconstructed from MRI images by FreeSurfer [78]. Among all the brain data, we use the left hemispheres of the brain surfaces for experiments.

The intelligence quotient (IQ) was evaluated by an online version of Ravens Advanced Progressive Matrices (APM) [221]. The test consists of 36 questions and the IQ score is calculated by $N_{correctAnswers}/N_{total} * 100$. The IQ among the data ranges from 0 to 100, which are almost uniformly distributed. Figure 9.27 shows the computation of Wasserstein distance between two brain cortical surfaces. (a) shows an example of a 20-year-old female, with IQ score 88.89; (b) shows an example of a 21-year-old male, with IQ score 33.33.

Instead of claiming whether one human brain is intelligent or not, in our experimental settings we divide the IQ into three classes: A, B, and C, ranging from $A : [0, 33)$, $B : [33, 67)$ and $C : [67, 100]$. The data are uniformly distributed in these three classes. For each gender, we randomly choose 12 examples from each class. Therefore, we create a training set of 72 examples, which is uniformly dis-

Fig. 9.28 Wasserstein distance matrix encoded in a gray image. The distance is normalized from 0 to 1, where 0 indicates black and 1 indicates white. The results show that, mostly, two surfaces in the same class induce smaller Wasserstein distance, yet two surfaces in different classes induce larger Wasserstein distance. Image from [262]

tributed with respect to gender and IQ. And the remaining examples are used as testing data.

For the classification experiments, we first compute the full pair-wise Wasserstein distance matrix based on our method. We index all the data of class A into $i = 1, 2, \ldots, 33$, data of class B into $i = 34, 35, \ldots, 66$ and data of class C into $i = 67, 68, \ldots 100$. Figure 9.28 shows the visualization of the full Wasserstein distance matrix encoded in a gray image. The distance is normalized from 0 to 1, where black color indicates 0 and white color denotes 1. The entry of the matrix $M_{i,j}$ is the Wasserstein distance between brain data i and brain data j. Then we can clearly see that, mostly, two surfaces in the same class induce smaller Wasserstein distance, yet two surfaces in different classes induce larger Wasserstein distance. The results further demonstrate the power of Wasserstein distance for measuring cortical shape similarities.

With the distance matrix, we classified the testing set by k-Nearest Neighbors (k-NN) classifier, where k is chosen to be 11 by running 9-fold cross-validation (we choose 9-fold to make each fold has the same number of examples.). The cross-validation curve is shown in Fig. 9.29. Table 9.3 shows the classification rate of our method is 78.57%.

To demonstrate the efficiency and advantages of our method, we compare our method with some existing popular single brain indices. Previous work [310] shows that cortical surface area and cortical surface mean curvature have significant correlations to human intelligence, since they quantify the complexity of cortical foldings. Thus we compute these two cortical measurements and use surface area, mean curvature, and the combination of these two measurements as three types of features

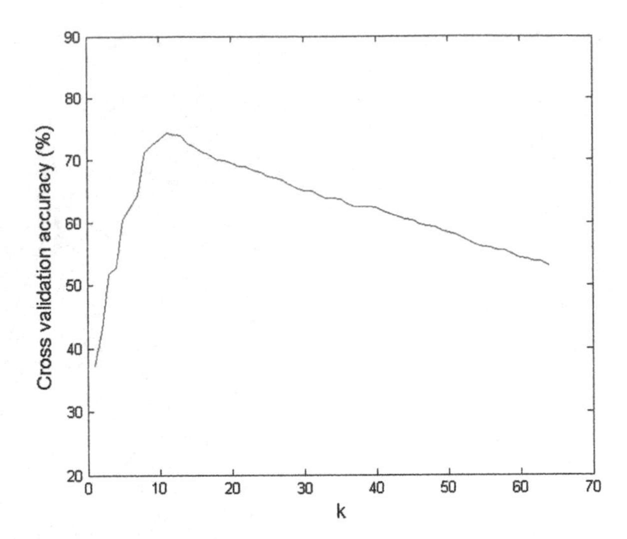

Fig. 9.29 Cross-validation curve. It shows the cross validation accuracy as functions of the parameter k in the k-NN classification. According to the experiments, we chose $k = 11$. Image from [262]

Table 9.3 Classification rate (CR) of our method and previous methods based on cortical surface area, cortical surface mean curvature and combination of previous two cortical measurements. The results demonstrated the accuracy of our method

Method	CR (%)
Our method	78.57
Surface area	53.57
Surface mean curvature	57.14
Combination of area and curvature	67.85

for classification, respectively. We use LIBSVM [233] as the classifier. Linear kernel and regularization parameter $C = 4.5$ are chosen by cross validation. Table 9.3 reports the classification rate of all the three comparison methods. The results indicate that our method outperforms previous methods and show that the proposed univariate Wasserstein distance based shape feature is promising as brain cortical shape biomarkers.

Hyperbolic Wasserstein Distance

A similar framework can also be applied to compute hyperbolic Wasserstein distance. We use genus-0 surfaces with multiple boundaries as examples to illustrate our algorithm. We use the Poincaré disk model to visualize the hyperbolic space. The Poincaré disk is a unit disk on the complex plane $\{z \in \mathbb{C}, |z| < 1\}$ with Riemannian metric $ds^2 = \frac{dzd\bar{z}}{(1-z\bar{z})^2}$. The hyperbolic distance between two points in the Poincaré disk is defined as

$$\mathrm{dist}(z_1, z_2) = \tanh^{-1}\left|\frac{z_1 - z_2}{1 - z_1\bar{z}_2}\right| \tag{9.66}$$

The pipeline is summarized in Algorithm 1 and illustrated in Fig. 9.30.

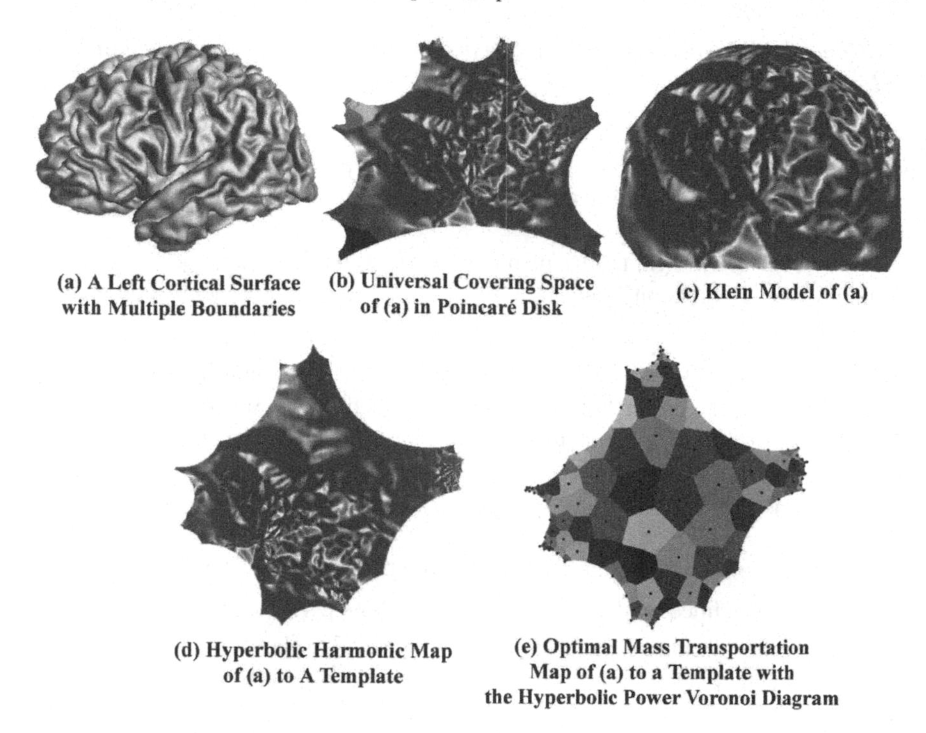

Fig. 9.30 Algorithm pipeline: **a** slice a surface open along landmark curves to generate a genus-0 surface with multiple boundaries; **b** embed the surface onto the Poincaré disk with its hyperbolic uniformization metric, which is computed by the hyperbolic Ricci flow; **c** covert the Poincaré disk to Klein model to construct the initial map between the surface and a template; **d** compute the hyperbolic harmonic map by diffusing the initial map; **e** compute the optimal mass transportation map using hyperbolic power Voronoi diagram, with surface tensor-based morphometry as the probability measure, where the colored regions denote Voronoi cells. Image from [250]

Algorithm 8 *Hyperbolic Wasserstein Distance Computation Pipeline.*

1. Slice the surface open along some delineated landmark curves to generate a surface with multiple boundaries (Fig. 9.30a).

2. Compute the hyperbolic uniformization metric of the surface with hyperbolic Ricci flow.

3. Isometrically embed the surface onto the Poincaré disk and convert it to the Klein model (Fig. 9.30b, c).

4. With the Klein model, construct the initial mapping between the surface and a template surface with the constrained harmonic map.

5. Improve the initial mapping with hyperbolic harmonic map to obtain a global diffeomorphic mapping in the Poincaré disk (Fig. 9.30d).

6. Compute the optimal mass transportation map between the surface and the template surface with the hyperbolic power Voronoi diagram, where the surface tensor-based morphometry of the hyperbolic harmonic map is used as a measure (Fig. 9.30e).

7. Compute the hyperbolic Wasserstein distance between the surface and the template surface.

In Algorithm 7, Step 1–5 apply constrained harmonic map to match two general surfaces with the same topology. We further use the hyperbolic space as the canonical space and TBM as the measure to compute power Voronoi diagram on the Poincarè disk.

Given a surface S with the Riemannian metric $\mathbf{g}$, let $P = \{p_1, p_2, \ldots, p_n\}$ be a set of n discrete points on S and $\mathbf{w} = \{w_1, w_2, \ldots, w_n\}$ be the weights defined on each point.

Definition 2 (Power Voronoi Diagram): Given a point set P and its corresponding weight vector $\mathbf{w}$, the power Voronoi diagram induced by $(P, \mathbf{w})$ is a cell decomposition of the surface $(S, \mathbf{g})$, such that the cell spanned by p_i is given by

$$\text{Cell}_i = \{x \in S | d_{\mathbf{g}}^2(x, p_i) - w_i \leq d_{\mathbf{g}}^2(x, p_j) - w_j\}, j = 1, \ldots, n \text{ and } i \neq j \tag{9.67}$$

In this work, with the Poincaré disk model, the geodesic distance $d_{\mathbf{g}}$ between two points, z_1, z_2 is defined by $\text{dist}(z_1, z_2) = \tanh^{-1}\left|\frac{z_1 - z_2}{1 - z_1 \bar{z}_2}\right|$. The term $d_{\mathbf{g}}^2(x, p_i) - w_i$ is called the *power distance* between x and p_i. Figure 9.31a shows the power distance on the Euclidean plane. Figure 9.31b illustrates the power Voronoi diagram on the Poincaré disk.

The optimal weight for the power Voronoi diagram that induces the optimal mass transportation map can be computed by

$$\frac{dw_i}{dt} = \nu_i - \int_{\text{Cell}_i} \mu(x) dx, x \in S. \tag{9.68}$$

Algorithm 9 gives the details about the optimal mass transportation map computation with hyperbolic metric. Figure 9.30e illustrates the hyperbolic power Voronoi

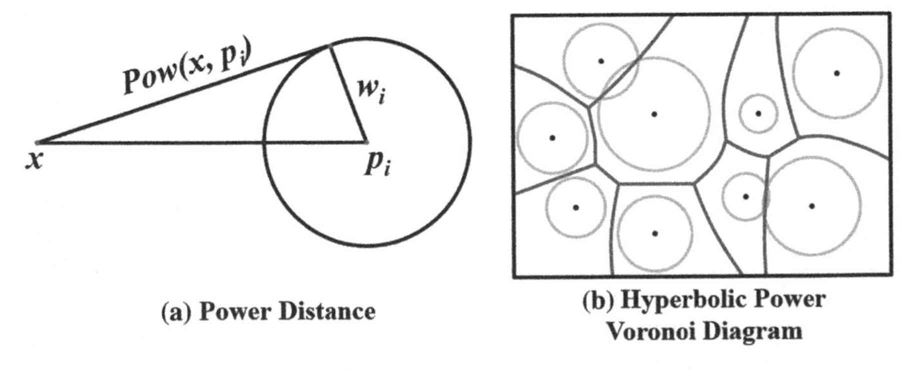

Fig. 9.31 Illustration of the power distance between two points on the Euclidean plane and the power Voronoi diagram on the Poincaré disk. Image from [250]

diagram that results in the optimal mass transportation map between the cortical surface in Fig. 9.30a and a template surface. In Fig. 9.30e, the black points form the discrete point set P. The initial hyperbolic geodesic Voronoi diagram is computed by the method in [213].

Algorithm 9 *Optimal Mass Transportation Map*

1. Given a triangular mesh M with hyperbolic metric $\mathbf{g}$ on the Poincaré disk, define a measure μ and a Dirac measure $(P, \nu) = \{(p_i, \nu_i)\}, i = 1, 2, \ldots, n$, $\int_M \mu(x)dx = \Sigma_{i=1}^n \nu_i$.
2. For each $p_i \in P$, compute it geodesic distances to every other vertex on M with Eq. (9.66).
3. For each vertex $v_i \in M$, determine which Voronoi cell it belongs to with Eq. (9.67).
4. For each $p_i \in P$, compute the total mass of the measures in the cell spanned by it, $\mu_i = \int_{\mathrm{Cell}_i} \mu(x)dx$.
5. Update each weight by $w_i^{t+1} = w_i^t + \varepsilon(\nu_i - \mu_i)$.
6. Repeat steps 3 to 5, until $|\nu_i - \mu_i|$, $\forall i$, is less than a user-specified threshold.

The cost of the optimal mass transportation map computed by Algorithm 9 gives the Wasserstein distance between two measures. With the hyperbolic metric, we define the *hyperbolic Wasserstein distance* between two measures that are defined on the Poincaré disk by

$$\mathrm{HyperbolicWasserstein}(\mu, \nu) = \Sigma_{i=1}^n \int_{\mathrm{Cell}_i} \left(\tanh^{-1}\left|\frac{x - p_i}{1 - x\bar{p}_i}\right|\right)^2 \mu(x)dx. \quad (9.69)$$

We applied our hyperbolic Wasserstein distance to study the classification problem with cortical surfaces between healthy control subjects and Alzheimer's disease (AD) patients. We randomly selected 30 AD patients and 30 healthy controls from the ADNI1 baseline dataset. The data preprocessing is identical with the experiments in Sect. 9.4.2. Only left hemispheric cortices were studied here as some prior research, e.g. [245], which has identified the trend that AD related brain atrophy may starts from left side and subsequently extends to the right. We randomly selected a control subject, which is not in our 60 studied dataset, as the template surface.

With the computed hyperbolic Wasserstein distance (to the common template surface) as the shape features, we applied the complex tree in the Statistics and Machine Learning Toolbox of MATLAB as a classifier. With a 5-fold cross validation, the classification rate of our method is 76.7%. As a comparison, we also computed two other standard cortical surface shape features, the cortical surface area and cortical surface volume. We applied the same classifier on the two measurements with 5-fold cross validation. Their results are summarized in Table 9.4. It can be noticed that our method significantly outperformed them. Our work may build a theoretical foundation to extend other shape space work to general surfaces to further improve AD imaging biomarkers for preclinical AD research.

Table 9.4 Classification rate comparison of our method and two other cortical surface shape features, the cortical surface area and cortical surface volume. The results demonstrated an accuracy rate achieved by the proposed method

Method	Classification rate (%)
Hyperbolic Wasserstein distance	76.7
Surface area	41.7
Surface volume	51.7

9.5 Point-to-Point Local Surface Deformation Measurements

After establishing a one-to-one correspondence map between a pair of surfaces, the Jacobian matrix J of the map is computed as its derivative map between the tangent spaces of the surfaces. Surface tensor-based morphometry (TBM) and its variant, multivariate tensor-based morphometry (mTBM) are defined to measure local surface deformation based on the local surface metric tensor changes. Practically, in the triangle mesh surface, the derivative map is approximated by the linear map from one face $[v_1, v_2, v_3]$ to another $[w_1, w_2, w_3]$. First, we isometrically embed the triangles $[v_1, v_2, v_3]$ and $[w_1, w_2, w_3]$ onto the plane; the planar coordinates of the vertices of v_i, w_j are denoted using the same symbols v_i, w_j. We can explicitly compute the Jacobian matrix for the derivative map

$$J = [w_3 - w_1, w_2 - w_1][v_3 - v_1, v_2 - v_1]^{(-1)}.$$

Then we use multivariate statistics on deformation tensors and adapt the concept to surface tensors. We define the deformation tensors as $S = (J^T J)^(1/2)$. The tensor-based morphometry (TBM) intends to study statistics of Jacobian determinant $det(S)$ or $\log(det(S))$.

For mTBM, we consider a new family of metrics, the "Log-Euclidean metrics" [13]. These metrics make computations on tensors easier to perform, as the transformed values form a vector space, and statistical parameters can then be computed easily using standard formulae for Euclidean spaces. In practice, the matrix logarithm of the surface deformation tensor is a 2×2 symmetric matrix and has two duplicate off-diagonal terms. The mTBM extracts the 3 distinct components of the $log(S)$ and forms a 3×1 vector. The mTBM computes statistics from the Riemannian metric tensors that retain the full information in the deformation tensor fields, thus may be more powerful in detecting surface difference than many other statistics.

The surface-based morphology statistics can be directly used to assess group difference between a clinical population and normal controls, study the simultaneous effects of multiple factors or covariates of interest and evaluate disease burden, progression and response to interventions. Specifically, Students t-test (for univariate statistics) and Hotellings T^2 test together with Mahalanobis distance may be used for group difference study and Pearson correlation for correlation study. A rich set of machine learning algorithms, such as support vector machine (SVM), may use the obtained morphology statistics for disease diagnosis and prognosis research.

Given maps of surface-based morphology statistics, multiple comparisons methods, such as false discovery rate (FDR) methods [22] and permutation methods [84], may be employed to assign overall (corrected) p-values of the map (or the features in the map), corrected for multiple comparisons. As a result, the generated surface p-maps (usually by permutation tests) can help visualize the most statistically significant areas and the corrected p-values quantify the global statistical significance over the whole brain cortical or subcortical surfaces.

Together with efficient brain surface registration, conformal geometry research based point-to-point Surface study is a commonly used population based neuroimaging morphometry scheme and was successfully applied in the many prior work for identifying the relationship of brain structure changes to cognition [160, 246, 247, 297, 298], diseases [159, 161, 199, 244, 248, 299], genetic and environmental factors [168, 245], and the effects of intervention [141, 181]. Here we choose three different applications in which mTBM are studied together with various brain surface parameterization, brain surface registration methods which we discussed in prior part of this chapter. Three applications also involve three different brain structures including hippocampus, lateral ventricle and cerebral cortex, demonstrating the broad impact of computational conformal geometry has made in medical imaging research.

9.5.1 Genetic Influence of APOE4 Genotype on Hippocampal Morphometry

The decline of cognitive skills to a functionally disabling degree is a sign of the clinical onset of AD, but optimizing disease modification strategies requires early intervention against appropriate therapeutic targets that may vary with disease stage. In pre-symptomatic subjects, determining whether AD is present is challenging. The apolipoprotein E (APOE) e4 allele is the most prevalent risk factor for AD. Hippocampal volumes are generally smaller in AD patients carrying the e4 allele compared to e4 non-carriers. However, there still lacks of solid evidence that hippocampal morphometry obtained by analyzing in vivo brain structural MR images is associated with APOE4 dose. We set out to apply computational conformal geometry to study this challenging but important scientific question.

Figure 9.32 shows our overall sequence of processing. First, we segment hippocampal substructures with FIRST [208] and automatically reconstruct hippocampal surfaces based on the segmentations [111]. Second, a conformal grid was generated for each surface with the holomorphic 1-form based surface conformal parameterization (Sect. 9.2.2. With this conformal grid, we register hippocampal surfaces across subjects with inverse consistent fluid registration algorithm (Sect. 9.3.4).

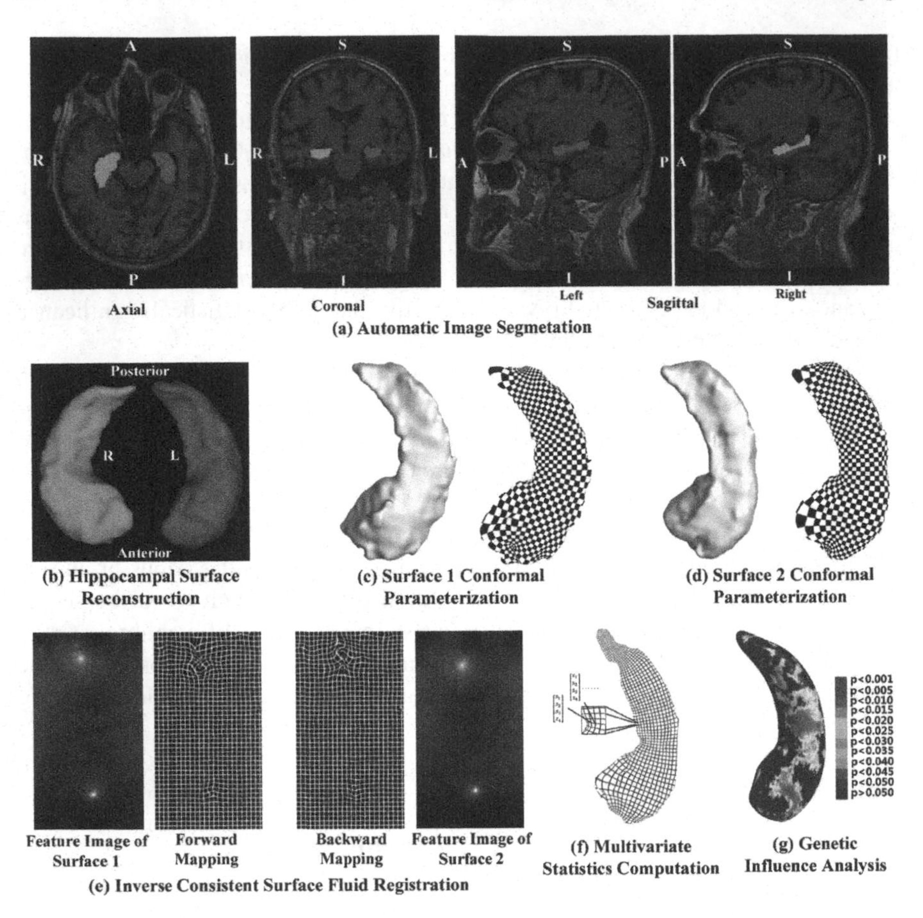

Fig. 9.32 A chart showing the key steps in our system. MR images were automatically segmented by FIRST to extract the hippocampal substructure (**a**). After the hippocampal surfaces were constructed from FIRST segmentations (**b**), we computed their conformal parameterizations with holomorphic 1-forms (**c** and **d**). Then feature images were generated by combining the local conformal factor and mean curvature that were computed from the conformal parameterizations. After the inverse consistent fluid registration was done in the feature image domain, we deformed the surfaces using the obtained displacements (**e**). The new statistics consisting of radial distance and multivariate TBM were computed at each point on the resultant matching surface (**f**). Then the Hotelling T^2 test was applied to study genetic influence of APOE e4 allele (**g**). Image from [245]

Third, we studied the differences between different diagnostic groups with the multivariate tensor-based morphometry (mTBM) statistics, which retain the full tensor information of the deformation Jacobian matrix, together with the radial distance [216], which retains information on the deformation along the surface normal direction. With the surface deformation statistics, we assess group differences with multivariate statistics and Hotelling's T^2 test. Finally, the statistical significance of the area with group difference in surface morphometry are obtained by a permutation test with 10,000 random assignments of subjects to different groups. We also use a pre-defined statistical threshold of $p = 0.05$ at each surface point to estimate the overall significance of the group difference maps by non-parametric permutation testing [122].

We conducted a thorough study to explore a series of questions including (1) whether the presence of APOE e4 allele is associated with greater hippocampal atrophy; (2) whether APOE e4 allele dose affects hippocampal surface morphometry and how this atrophy is related to normal aging; and (3) whether there is an APOE e4 dose effects, i.e. whether morphometric differences would be greater in APOE e4 homozygotes than heterozygotes, who would in turn show greater deformities compared to e4 non-carriers. Detailed experimental results are reported in [245]. Here we briefly report our research results on the first set of questions.

To explore whether the presence of the APOE e4 allele was associated with greater hippocampal atrophy, we conducted two experiments to study the effects of APOE e4 genotype on hippocampal morphometry in two populations:

1. APOE e4 carriers versus non-carriers in the full ADNI cohort;
2. APOE e4 carriers versus non-carriers in the non-demented cohort.

The experiments aimed to determine if the APOE e4 allele was associated with hippocampal atrophy in all subjects or in subjects who have not yet developed AD. Note that the APOE e4 non-carriers are those subjects who are homozygous non-carriers (e3/e3). Subjects with one e2 allele, i.e., e2/e3 and e2/e4 were excluded due to the possible protective effect of e2 allele for AD [200]. In the 725 subjects of known APOE e4 genotype, there were 322 non-carriers (all homozygous for APOE e3) and 343 APOE e4 carriers. The non-demented cohort consisted of 506 subjects who were either MCI or control subjects, including 270 e4 non-carriers and 236 e4 carriers. Figure 9.33a shows the statistical p-map for the full ADNI cohort ($N = 665$; 322 non-carriers and 343 carriers). Non-blue colors show vertices with statistical differences at the nominal 0.05 level, uncorrected for multiple comparisons. As shown in Fig. 9.33a, the APOE e4 carriers differed significantly from the non-carriers ($p < 0.0002$). Figure 9.33b shows the p-map for the non-demented cohort ($N = 506$; 270 non-carriers and 236 carriers). After correcting for multiple comparisons, the difference remained highly significant ($p < 0.0027$).

In our APOE e4 carrier versus non-carriers experiments, comparisons with both the non-demented and the full ADNI cohorts yielded significant differences that were apparently more pronounced on the left hippocampal surface. A prior study [200], which conducted similar experiments with a smaller number of images in ADNI baseline dataset ($N = 490$) was only able to achieve significance for the left hippocampal surfaces on the full ADNI cohort but did not detect significant differences in the non-demented cohort. That aside, our finding of more significant areas on the left than on the right side, agree with [200], despite differences in our image segmentation methods, surface parameterization and registration algorithms, and statistics. Our results also agree with another APOE e4 study with manually segmented hippocampal surfaces [214].

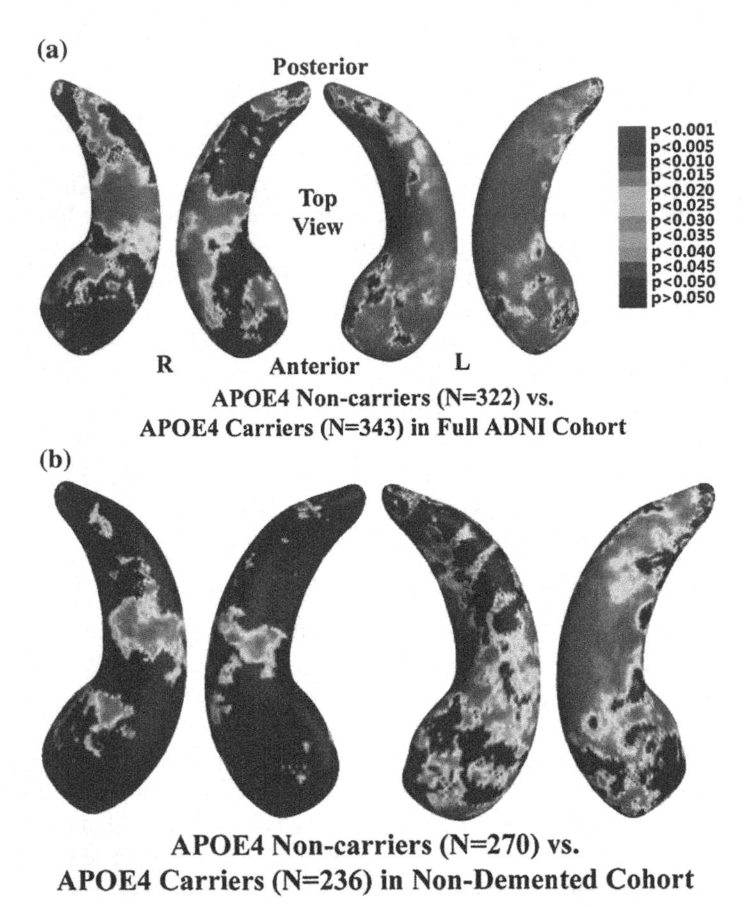

Fig. 9.33 Illustration of local shape differences (P values) between the APOE e4 noncarriers (e3/e3, $N = 322$) and carriers (e3/e4 and e4/e4, $N = 343$) in the full ADNI cohort (**a**) and between the APOE e4 noncarriers (e3/e3, $N = 270$) and carriers (e3/e4 and e4/e4, $N = 236$) in the non-demented cohort (MCI and controls) (**b**). Nonblue colors show vertices with statistical differences, at the nominal 0.05 level, uncorrected. The overall significance after multiple comparisons with permutation test are $P < 0.0002$ (**a**) and $P < 0.0027$ (**b**). Image from [245]

9.5.2 Ventricular Abnormalities in Mild Cognitive Impairment

Mild Cognitive Impairment (MCI) is a transitional stage between normal aging and dementia and people with MCI are at high risk of progression to dementia. MCI is attracting increasing attention, as it offers an opportunity to target the disease process during an early symptomatic stage. Structural magnetic resonance imaging (MRI) measures have been the mainstay of Alzheimers disease (AD) imaging research. Lateral ventricle is one of regional structures frequently studied in structural MRI research. Ventricular enlargement is a highly reproducible measure of AD progression [83], owing to the high contrast between the CSF and surrounding brain tissue on T1-weighted images. The lateral ventricles span a large area within the cerebral hemispheres and abut several substructures including the hippocampus, amygdala and posterior cingulate. Changes in ventricular morphology, such as enlargement, often reflect atrophy of the surrounding cerebral hemisphere which itself may be regionally differentiated (for example, frontotemporal in contrast to posterior cortical atrophy). As regional differences in cerebral atrophy may be reflected in specific patterns of change in ventricular morphology, accurate analysis of ventricular morphology has the potential to both sensitively and specifically characterize a neurodegenerative process. However, the concave shape, complex branching topology and extreme narrowness of the inferior and occipital horns make subregional analysis of ventricular enlargement notoriously difficult to assess, exemplified by the conflicting findings regarding genetic influences on ventricular volumes [46, 154]. Pioneering ventricular morphometry work [260] used spherical harmonics to analyze ventricular morphometry but the underlying spherical mapping may result in significant distortion that underdetermined the analysis as demonstrated previously [299]. Our prior work [299] computed the first global conformal parameterization of lateral ventricular surfaces. However, the holomorphic 1-form based conformal parameterization method always introduces singularity points (zero points, Fig. 9.9a) on lateral ventricular surfaces. To model a topologically complicated ventricular surface, hyperbolic conformal geometry emerges naturally as a candidate method as it can induce conformal parameterization on the surface with a negative Euler number without any singularities.

Figure 9.34 summarizes the overall sequence of steps in our ventricular morphometry system. First, from each MRI scan (a), we automatically segment lateral ventricular volumes with the multi-atlas fluid image alignment (MAFIA) method [47]. The MR image overlaid with the segmented ventricle is shown in (b). A ventricular surface built with marching cube algorithm [178] is shown in (c). After the topology optimization, we apply hyperbolic Ricci flow method on the ventricular surface and conformally map it to the Poincar disk (Sect. 9.2.3. On the Poincar disk, we compute consistent geodesics and project them back to the original ventricular surface (geodesic curve lifting). The results are shown in (d). Further, we convert the Poincar model to the Klein model where the ventricular surfaces are registered by the constrained harmonic map (Sect. 9.3.2. The registration diagram is shown in (e). Next, we compute the TBM features and smooth them with the heat kernel method [53] (f). Finally, the smoothed TBM features are applied to analyze both group difference between the two MCI groups and correlation of ventricular shape morphometry with cognitive test scores and FDG-PET index. Significance p-maps are used to visualize local shape differences or correlations (g). Correction for multiple comparisons is used to estimate the overall significance (corrected p-values).

We perform a group comparison with Students t test on the smoothed TBM features after we register ventricular surfaces with hyperbolic Ricci flow and constrained harmonic map methods. Specifically, for all points on the ventricular surface, we ran a permutation test with 5,000 random assignments of subjects to groups to

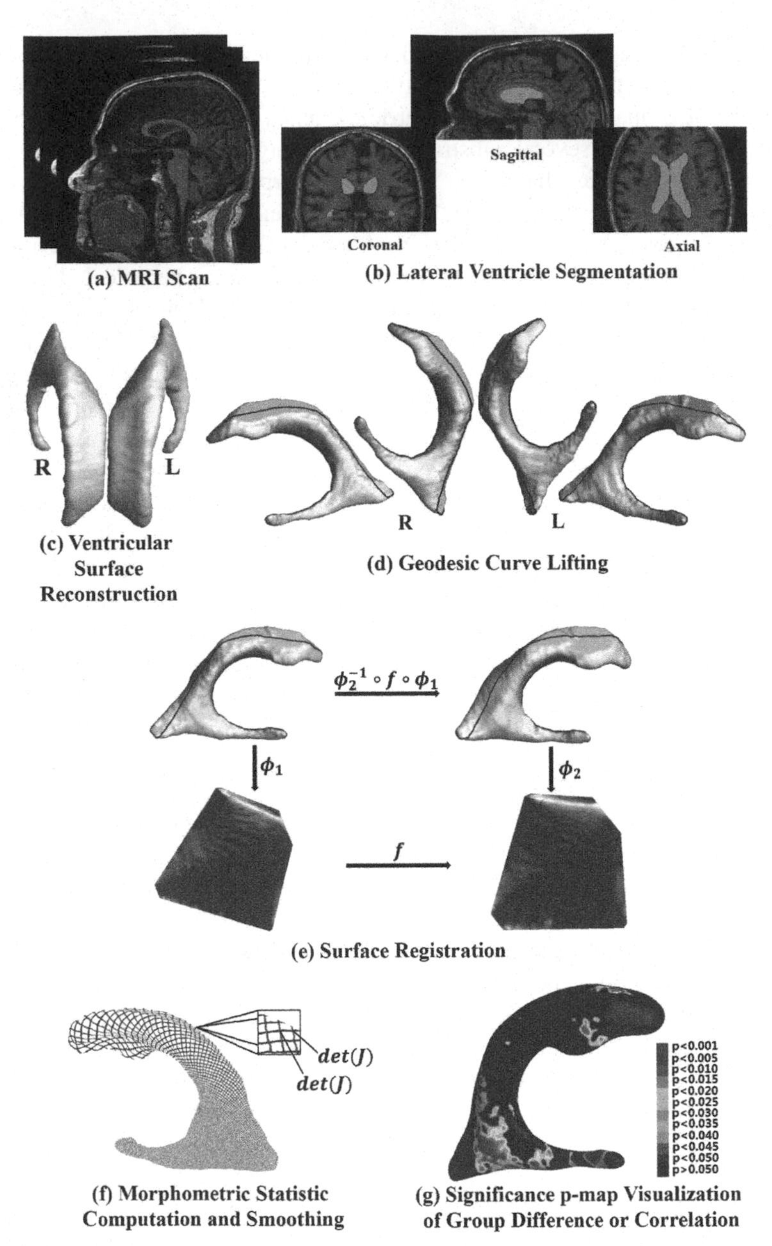

Fig. 9.34 A chart showing the key steps in the ventricular surface registration method. After the lateral ventricles were segmented from MRI scans and surfaces were reconstructed, we computed consistent geodesic curves on each ventricular surface to constrain the registration. Then the constrained harmonic map was used to obtain a correspondence field in the parameter domain represented by the Klein model, which also induced a surface registration in 3D. The statistic of TBM was computed on each point of the resulting matched surfaces. Finally the smoothed TBM features are applied to analyze both group difference between the two MCI groups and correlation of ventricular shape morphometry with cognitive test scores and FDG-PET index. Image from [246]

Fig. 9.35 Illustration of statistical map showing local shape differences (p-values) between MCI converter and MCI stable groups from the ADNI baseline dataset, based on tensor-based morphometry (TBM), which was smoothed by the heat kernel smoothing method [53]. Image from [246]

estimate the statistical significance of the areas with group differences in surface morphometry. The probability was color coded on each surface point as the statistical p-map of group difference. Figure 9.35 shows the p-map of group difference detected between the MCI converter who developed incident AD during the subsequent 36 months ($n = 71$) and stable who did not convert during the same period ($n = 62$) groups, using the smoothed TBM as a measure of local surface area change and the significance level at each surface point as 0.05. In Fig. 9.35, the non-blue color areas denote the statistically significant difference areas between two groups. The overall significance of the map is 0.0172. More experimental results are reported in [246].

For comparison, the analyses of global ventricular volume and surface area did not differentiate MCI converter and MCI stable groups ($p = 0.0803$ and $p = 0.2922$ for volume and area group differences, respectively). Our fine-grained analysis revealed significant differences mostly localized around the subregion of the ventricular body that abuts medial temporal lobe structures. This subregional ventricular enlargement was reported to correlate with atrophy of medial temporal lobe which includes the hippocampal formation. Consistent with prior observations, e.g. [83, 267], our findings suggest that grey matter atrophy starts from the temporal lobe region and then spreads to involve frontal cortices, consistent with Braak staging of neurofibrillary pathology [31]. Importantly, they provide evidence that ventricular subfield analysis provides enhanced statistical power in structural MRI analysis compared with ventricular volume analysis. Our computational conformal geometry based ventricular surface TBM features may enhance the predictive value of MRI-derived data in AD research.

9.5.3 Computer-Assisted Diagnosis with Computational Conformal Geometry

Computer-assisted diagnostic classification is becoming increasingly popular in neuroimaging, especially given the vast number of features available to assist diagnosis in a 3D brain image. Early diagnosis and treatment of degenerative brain diseases, such as AD, depends on the ability to identify disease in its earliest stages, when brain changes may be subtle. In addition, there is interest in understanding which brain imaging features are best for diagnostic classification, as well as biomarkers to measure the severity of disease burden. Over the last decade, many methods have been proposed to study the problem of diagnostic classification based on structural magnetic resonance imaging (MRI) (e.g. [60, 107, 263]), positron emission tomography (PET) [43], or a combination of multi-source datasets [307, 319]. Surface-based modeling is useful in brain imaging to help analyze anatomical shapes, to detect abnormalities in cortical surface folding and thickness, and to statistically combine or compare 3D anatomical models across subjects. Many surface-based morphometry studies describe structural differences at the group level, i.e., between different diagnostic groups. More recently, morphometric maps have also been used to classify individual subjects into diagnostic groups [107, 263, 298]. Overall, a set of surface-based morphometric features combined with a machine learning algorithm may offer a promising way to improve the performance of computer-assisted diagnostic systems.

Figure 9.36 reports our recent efforts [298] to combine computational conformal geometry with sparse learning for computer-assisted diagnostic classification research. It summarizes the steps we used to analyze cortical surface morphometry. The cortical surface data was from a prior study [269]. With 10 selected landmarks on each cortical hemispheric surface, we computed a conformal mapping from a multiply connected mesh to the so-called slit domain, which consists of a canonical rectangle or disk in which 3D curved landmarks on the original surfaces are mapped to parallel lines or concentric slits in the slit domain (Sect. 9.2.2). In this canonical parametric domain, cortical surfaces were matched by a constrained harmonic map (Sect. 9.3.2). Multivariate surface statistics were computed from the registered surfaces. In one experiment, they were applied to identify regions with significant differences between the two groups. In another experiment, cortical features were fed to a sparse learning method to classify each subject into one of two groups by a leave-one-out test. We also tested other possible surface morphometry statistics to compare them with our multivariate surface statistics. Although the method is illustrated on Williams syndrome (WS) data, it is intended to be useful for other disorders as well.

First, we use the Hotelling's T^2 test to identify between-group differences. Specifically, for each point on the cortical surface, given $p = 0.05$ as the significance level, we ran a permutation test with 10,000 random assignments of subjects to groups to estimate the statistical significance of the areas with group differences in surface morphometry. Figure 9.37 shows the significance map of group differences detected

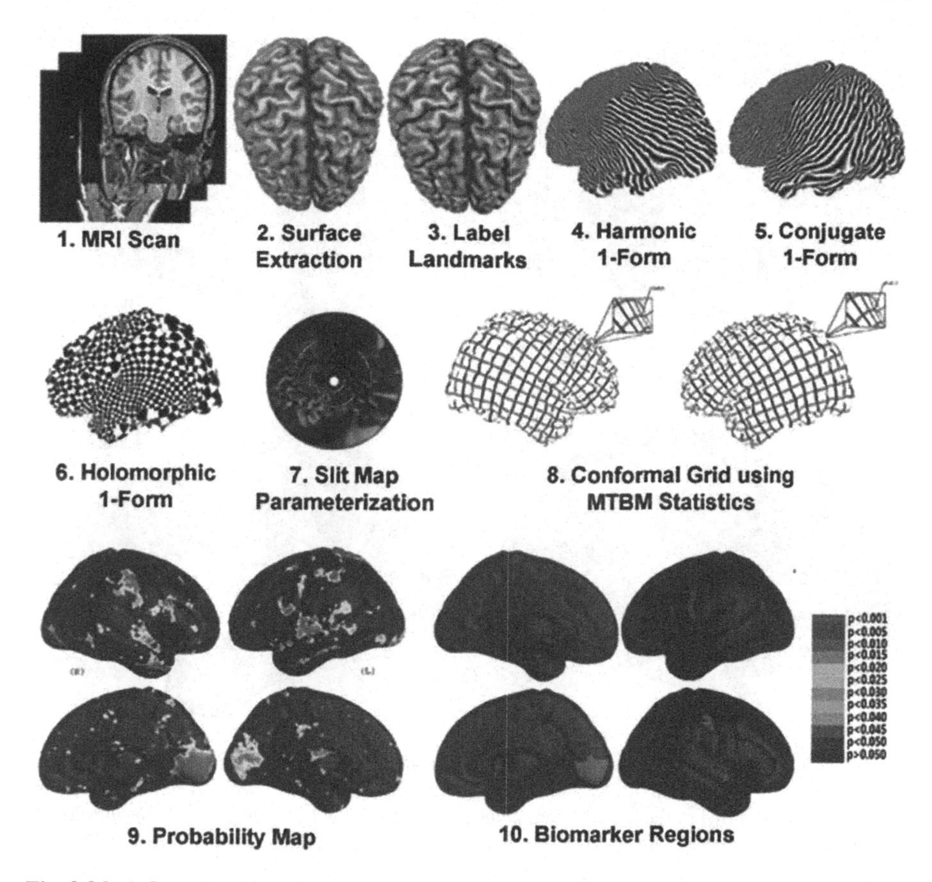

Fig. 9.36 A flow chart shows how circular slit map conformal parameterization is used to model cortical surface shapes. The resulting surfaces are analyzed using multivariate tensor-based morphometry and sparse learning methods. After cortical surfaces are extracted from MRI images and landmark curves are labeled either manually or automatically [269], we compute circular slit map conformal parameterizations for each cortical surface, and register surfaces with a constrained harmonic map. The statistics of multivariate TBM are computed at each point on the resulting matching surfaces, revealing regions with systematic anatomical differences between groups. We also apply a sparse learning algorithm to detect some structural features suitable for classification experiments. Image from [298]

between WS and matched control groups, using mTBM as a measure of local surface area and the significance at each surface point to be $p = 0.05$. In Fig. 9.37, the non-blue colored areas show the areas with (uncorrected) statistically significant differences between the two groups.

The overall significance of the map can be defined as the probability of finding, by chance alone, a statistical map with at least as large a surface area and a statistical threshold more stringent than the predefined level of $p = 0.05$ (note that other methods are also possible, such as those that control the false discovery rate). This omnibus p-value is commonly referred to as the *overall significance of the map* (or of

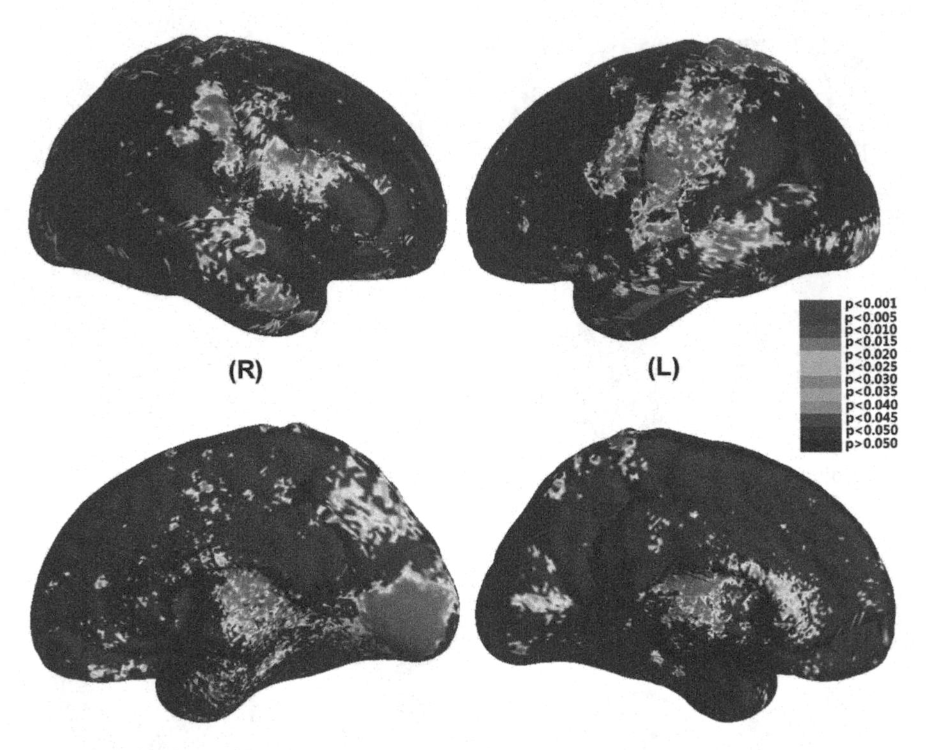

Fig. 9.37 Statistical significance map (uncorrected p-map) shows group differences in regional cortical surface area between 42 WS patients and 40 healthy controls. The local statistic analyzed is the multivariate TBM of the cortical parameterization. On the color-coded scale, non-blue colors denote the vertices where there are significant group differences, at the uncorrected $p = 0.05$ level. Image from [298]

the features in the map), corrected for multiple comparisons. It basically quantifies the level of surprise in seeing a map with this amount of the surface exceeding a predefined threshold, under the null hypothesis of no systematic group differences. We also computed the overall significance p-values, which were $p = 0.0004$ for the left and $p = 0.001$ for the right hemisphere, respectively.

Clearly, our slit map conformal parameterization based surface process identifies the group difference between two group. We first apply the LASSO function [274] to select sparse features and rigid regression classifier to classify. We take a cross validation approach to set up free parameters in the LASSO and rigid regression method. A leave-one-out method is used to evaluate our classifications. We achieve the best specificity (100.00%), sensitivity (59.51%), positive predictive value (100.00%) and negative predictive value (70.18%) are achieved when we used mTBM features from the right cortical hemisphere for the training and testing. Consistent with some data from prior WS studies [269], the right half of the brain may contain more diagnostically useful information relevant to WS classification. This assumption is supported by our classification results.

Overall, it is an exciting new direction to apply brain MRI images for disease diagnosis and prognosis, e.g. [113]. The results, together with our recent work on sparse coding and dictionary learning [329–331], all demonstrate the strong potential to apply computational conformal geometry methods to brain disease diagnosis and prognosis research.

9.6 Summary

Computational conformal geometry has been studied in other medical imaging research fields as well, such as virtual colonoscopy [323] and vestibular system [322]. Here we mainly use human brain mapping research to illustrate the application of computational conformal geometry in medical imaging research. Note similar applications can be also studied in other research fields.

Computational conformal geometry provides new ideas and exciting new opportunities to analyze the medical images accumulated on the daily basis. The fast growing medical images will also push the computational conformal geometry to a even higher level by providing more grand challenges in medical science research.

Chapter 10
Wireless Sensor Networks

Abstract This chapter introduces the applications of computational conformal geometry on wireless sensor network research. Specifically, we focus on large-scale wireless sensor networks where the amount of data generated, stored and transmitted in networks grow proportionally with the network size. The unique and intrinsic challenges in such sensor network design are distributed and scalable computation and communication. Several research topics, including network localization, sensor deployment, greedy routing, in-network data-centric processing, and marching of autonomous networked robots, together with their experimental results, are detailed in this chapter.

10.1 Introduction

A Wireless Sensor Network (WSN) is a self-configuring network of a large number of spatially distributed sensor nodes for monitoring, information sharing, and cooperative processing. The basic components of a sensor node include a low power processor, a modest amount of memory, a wireless network transceiver, a sensor board, and power source. The initial development of WSN was motivated by military applications such as enemy detection, battlefield surveillance, etc. WSNs are used in many other fields, such as agriculture, environmental monitoring including air and water pollution, greenhouse, health monitoring, structural monitoring and more. WSNs has become an appealing technology as a smart infrastructure for building and factory automation, and process control applications.

In comparison with earlier computer communication systems, WSNs form a new kind of wireless networks with a new set of characteristics and challenges. A WSN is designed for unattended operations. The majority of nodes do not communicate directly with the nearest base station, but with their local peers. An individual sensor is highly resource-constrained with extremely limited computing, storage, and communication capacities. However, applications of a WSN often require a large-scale deployment where the amount of data generated, stored, and transmitted in the network grow proportionally with the network size. Therefore, the unique and intrinsic

M. Jin et al., *Conformal Geometry*, https://doi.org/10.1007/978-3-319-75332-4_10

challenges in sensor network design are distributed and scalable computation and communication. The chapter introduces algorithms that apply conformal geometry for WSN operation and design. Specifically, we cover the following research problems in WSN design.

- *Network Localization.* Network localization requires the sensors to self-organize a coordinate system instead of equipping each sensor node with an expensive GPS.
- *Sensor Deployment.* Sensor deployment studies the way to embed an operational sensor network in a real-world environment.
- *Greedy Routing.* Traditional routing table based approaches that each node has to store the whole routing table cannot be directly applied in resource-constrained and highly volatile WSNs. On the contrary, greedy routing uses only local information and greedy decisions to achieve scalable routing for WSNs.
- *In-network Data-centric Processing.* In-network data-centric processing focuses on data rather than individual sensor nodes who collect the data. Data is uniquely named. Data processing is achieved using data names instead of network addresses. In-network data-centric processing aims to establish a self-contained data acquisition, storage, retrieval, and query system.
- *Marching of Autonomous Networked Robots.* A group of autonomous networked mobile sensors coordinates among themselves to complete a task, e.g., to explore or monitor a Field of Interest (FoI).

10.2 Localization

Geographic location information is imperative to a variety of applications in WSNs, ranging from position-aware sensing to geographic routing. While global navigation satellite systems (such as GPS) have been widely employed for localization, integrating a GPS receiver in every sensor of a large-scale sensor network is unrealistic due to the high cost. Moreover, some application scenarios prohibit the reception of satellite signals by part or all of the sensors, rendering it impossible to solely rely on global navigation systems. Sensor network localization refers to the process of estimating the locations of sensor nodes with mere network connectivity or information between neighboring sensor nodes such as local distances and angle measurements.

Even for those ranging information based localization schemes, extra equipments installed to measure the distance or the angle between nodes, can also lead to a dramatically increase of network cost. To this end, many interesting approaches have been proposed for localization with mere connectivity information. Each node only knows which nodes are nearby within its one-hop communication radio range, but does not know how far away and what direction its neighbors are. We introduce a novel mere connectivity-based localization algorithm, well suitable for large-scale planar sensor networks with complex shapes and non-uniform nodal distribution in Sect. 10.2.1. In contrast to current state-of-art connectivity-based localization methods, the algorithm is highly scalable with linear computation and communication costs with respect to the size of a network, and fully distributed where each node only needs the information of its neighbors.

Existing network localization research focuses on sensor networks deployed on two-dimensional (2D) plane or in three-dimensional (3D) space. In real-world applications, many large-scale sensor networks are deployed over complex 3D terrains, such as the volcano monitoring project [302] and ZebraNet [143]. Section 10.2.2 introduces a conformal mapping based localization algorithm for networks deployed over complex 3D terrains.

10.2.1 Planar Sensor Network Localization

Challenges of Previous Approaches

With merely connectivity information available, three major techniques are employed in current state-of-the-art localization schemes: multi-dimensional scaling (MDS), neural networks, and graph embedding.

MDS is a non-linear dimension reduction and data projection technique that transforms distance matrix into a geometric embedding (e.g., a planar embedding for 2D sensor network localization). MDS-based localization is originally proposed in [242]. It constructs a proximity matrix based on the shortest path distance (approximated by hop counts) between all pairs of nodes in the network. The singular value decomposition (SVD) is employed to produce the coordinates matrix that minimizes the least square distance error. Finally, it retains the first 2 (or 3) largest eigenvalues and eigenvectors as 2D (or 3D) coordinates. Subsequent improvements on MDS are made by dividing the graph into patches to enable distributed calculation [241, 282]. In addition [172] proposes to apply SVD to the matrix based on a set of beacon nodes only and thus reduces complexity. A similar idea is adopted in [305], with a simple method (instead of SVD) for error minimization.

One of the major problems for MDS based methods is their low *scalability*. The time complexity for obtaining the distance matrix is $O(nm)$ where n and m are the number of vertices and edges, respectively. The time complexity to compute the two largest eigenvalues and the corresponding eigenvectors is $O(n^2)$. With the increase of a network size n, the computational cost is prohibitive. Another issue is that they are inherently *centralized*. As a sensor network grows large in size, centralized computation has a fundamental bottleneck at nodes near the sink as each sensor node has a limited power and computation capability. So a distributed algorithm is highly preferred especially for a large-scale network. Different algorithms have been proposed to overcome these disadvantages. A basic approach is to partition the network to many subnetworks, and compute the localization of each subnetwork, and then merge these subnetworks together. This method requires delicate strategies and great caution in the merging stage.

Neural networks [93, 170] employ non-linear mapping techniques and neural network models such as self-organizing map (SOM) for dimension reduction of multidimensional data sets, yielding coordinates of sensor nodes that preserve the distances (also approximated by hop counts) between data points of the input and output spaces (i.e., a 2D plane) as much as possible.

For neural network based methods, *stability* is their major problem due to the non-convex shape of their minimized energy. Although several approaches have been proposed to increase the possibility to escape from local minima of the minimized energy, the selection of initial values are still crucial for the final localization results [170].

The localization algorithms based on rigid graph embedding theory [26, 255, 300] aim to create a well-spread and fold-free graph that resembles the given network. They focus on finding a globally rigid graph which can be embedded without ambiguity in plane. While with mere globally rigid structure, like a topological disk triangulation in [300], there exist infinite number of flat metrics that induce different planar embedding as long as the total Gaussian curvatures satisfy the discrete Gauss-Bonnet Theorem as discussed in Sect. 10.2.1. A brute-force way is applied to find one planar embedding of the extracted global structure, which in general cannot be easily guaranteed. So compared with MDS and neural network-based approaches, the graph rigidity-based methods exhibit lower localization accuracy in general.

Motivation

We introduce a localization algorithm that overcomes the major difficulties of conventional MDS and neural network based algorithms and outperform them. We first explain the basic idea using a smooth surface, and then transform it to sensor networks.

Let's consider a smooth surface on plane. It is flat everywhere, so the Gaussian curvature, which measures how much a surface is curved and can be computed based on local distance information, equals to zero at every point of the surface. Assume we only have approximated distance information instead of the exact one of the surface. Such approximated distance generates non-zero Gaussian curvatures and induces a curved surface instead of a planar one. We can distort the approximated distance such that the deformed one generates zero Gaussian curvatures for every point that guarantee a planar surface.

Given a large-scale sensor network deployed on plane, we can extract a triangular mesh structure from the connectivity graph of the network such that the mesh structure approximates well the geographic structure of the sensor network. Specifically, we uniformly select a set of landmark nodes such that any two neighboring landmarks are a fixed k hops away. Landmarks initiate local flooding to build a landmark-based Voronoi diagram of the network such that any non-landmark node is within k hops of some landmark. We then build a triangulation based on the dual of the landmark-based Voronoi diagram. Each vertex of the triangulation is a landmark node. Each edge connecting two neighboring vertices is a shortest path between the two neighboring landmark nodes.

Considering the triangular mesh as a discrete approximation of a smooth surface on plane, the local distance information of the mesh is discrete and represented by the approximated edge lengths (i.e., a fixed K-hop). The Gaussian curvature of the mesh is also discrete and can be computed based on the approximated edge lengths. We apply discrete Ricci flow to deform the length of each edge such that the deformed edge lengths induce zero Gaussian curvature at each vertex of the mesh. The deformed edge lengths guarantee a planar triangular mesh. Denote such edge lengths a flat metric. There exist infinite number of flat metrics. Each one of them can isometrically embed the triangular network into plane, but the optimal flat metric introduces the minimal deformation to the initially approximated edge length. By computing the optimal flat metric, localization (i.e., isometric embedding of the network to plane) is straightforward.

We first introduce the Optimal Flat Metric Theorem, the condition to find the optimal flat metric. Then we explain the localization algorithm with mere connectivity information step by step, followed by part of the simulation results including comparison with previous methods.

Optimal Flat Metric

To find the optimal flat metric, let us denote u_i the distortion of the original metric at each vertex and define the *total distortion energy* as

$$E(\mathbf{K}) = \int_{\mathbf{K_0}}^{\bar{\mathbf{K}}} \sum_{i=1}^{n} u_i dK_i,$$

where $\overline{\mathbf{K}}$ and $\mathbf{K_0}$ represent the set of target and initial vertex Gaussian curvatures respectively, and n is the number of vertices of mesh M. The integration is along an arbitrary path from $\mathbf{K_0}$ to the target curvature $\overline{\mathbf{K}}$. This energy is the Legendre dual to the Ricci energy. Therefore it is also convex with a unique global minimum for a given $\overline{\mathbf{K}}$.

Define

$$\Omega = \bigcap \{\sum K_i = 2\pi\chi(M)\} \bigcap \{K_j = 0, v_j \notin \partial M\}$$

the set of all possible flat metrics of M satisfying the Gauss-Bonnet Theorem.

The problem to find the optimal flat metric can be formulated as:

$$\min_{\mathbf{K}\in\Omega} E(\mathbf{K}). \tag{10.1}$$

Among all possible flat metrics of M satisfying the Gauss-Bonnet Theorem, the following theorem gives a solution of the metric introducing the least distortion from the initially estimated curved metric of M.

Theorem 10.1 (Optimal Flat Metric Theorem) *The solution to the optimization problem (10.1) is unique, and satisfies*

$$u_j = const, \forall v_j \in \partial M. \tag{10.2}$$

Proof The distortion energy $E(\mathbf{K})$ is convex. The domain Ω is a linear subspace of the original domain $\{\mathbf{K} | \sum_i K_i = 2\pi\chi(M)\}$. Therefore the restriction of $E(\mathbf{K})$ on Ω

is still convex, it has a unique global optimum at an interior point. The gradient of the energy is $\nabla E(\mathbf{K}) = (u_1, u_2, \ldots, u_n)$. At the optimal point, the gradient is orthogonal to Ω. Assume $v_i \in \partial M$, $1 \le i \le m$, then the normal vector to Ω is given by $(1, 1, \ldots 1, 0, \ldots, 0)$. Therefore the gradient is along the normal vector. So Eq. 10.2 holds. If we set the constant as 1, the optimal flat metric is the one which satisfies the two conditions: during the process of deforming the estimated metric to a flat one, we only deform the metric of interior vertices; at the end of the process, Gaussian curvatures of all interior vertices are equal to zero.

Computing Optimal Flat Metric

The estimated metric (edge length) of a triangular mesh M can be considered as a set of unit edge length since each edge is approximately a fixed k hops. If the distances between neighboring nodes can be more accurately estimated, the approximated unit edge length can be replaced. An initial circle packing metric (Γ_0, Φ) of M can be easily constructed by assigning each vertex v_i a circle. Its radius equals the unit edge length that forms Γ_0. The intersection angle of circles assigned to v_i and v_j for each edge $[v_i, v_j]$ forms Φ.

The algorithm detects and marks boundary vertices located on the boundary edges of M. Note that a boundary edge is adjacent with only one face. For those non-marked vertices (interior vertices), their target Gaussian curvatures $\bar{K}$ are set to zero. For all vertices of M, the logarithm of the circle radius γ_i assigned to vertex v_i is initialized to zero. In each iteration of discrete Ricci flow, only non-marked vertices are involved. Specifically, an interior vertex v_i collects the u values from its direct neighbors and updates its adjacent edge length with $l_{ij} = e^{(u_i+u_j)}$. For each triangle $[v_i, v_j, v_k]$ adjacent with vertex v_i, v_i can easily compute the corner angle $\angle_i^{jk}$ based on inverse cos law:

$$\angle_i^{jk} = \cos^{-1} \frac{l_{ki}^2 + l_{ij}^2 - l_{jk}^2}{2 l_{ki}^2 l_{ij}^2}.$$

Therefore, v_i computes its current discrete Gaussian curvature K_i as the excess of the total angle sum. If for every interior vertex v_i, the difference between its target Gaussian curvature $\bar{K}_i$ (set to zero) and current Gaussian curvature K_i is less than a threshold (we set it to $1e-5$ in our experiments), the discrete Ricci flow converges. Otherwise, each interior vertex v_i updates its u_i: $u_i = u_i + \delta(\bar{K}_i - K_i)$, where δ is the step length (we set it to 0.1 in our experiments).

When the algorithm stops, all the curvature flux has been absorbed by boundary vertices, such that the interior vertices have zero Gaussian curvature, which induces a flat metric. Since in each step of the algorithm, there is always no deformation of circle radii for boundary vertices (e.g., $u_i - u_i^0 = 0$, $v_i \in \partial M$). According to Theorem 10.1 (the Optimal Flat Metric Theorem), the computed flat metric introduces the least distortion from the estimated one.

Localization

Isometric embedding of the computed flat metric localizes landmark nodes. For a non-landmark node n_i, it first finds its three nearest landmarks, denoted as v_1, v_2, v_3

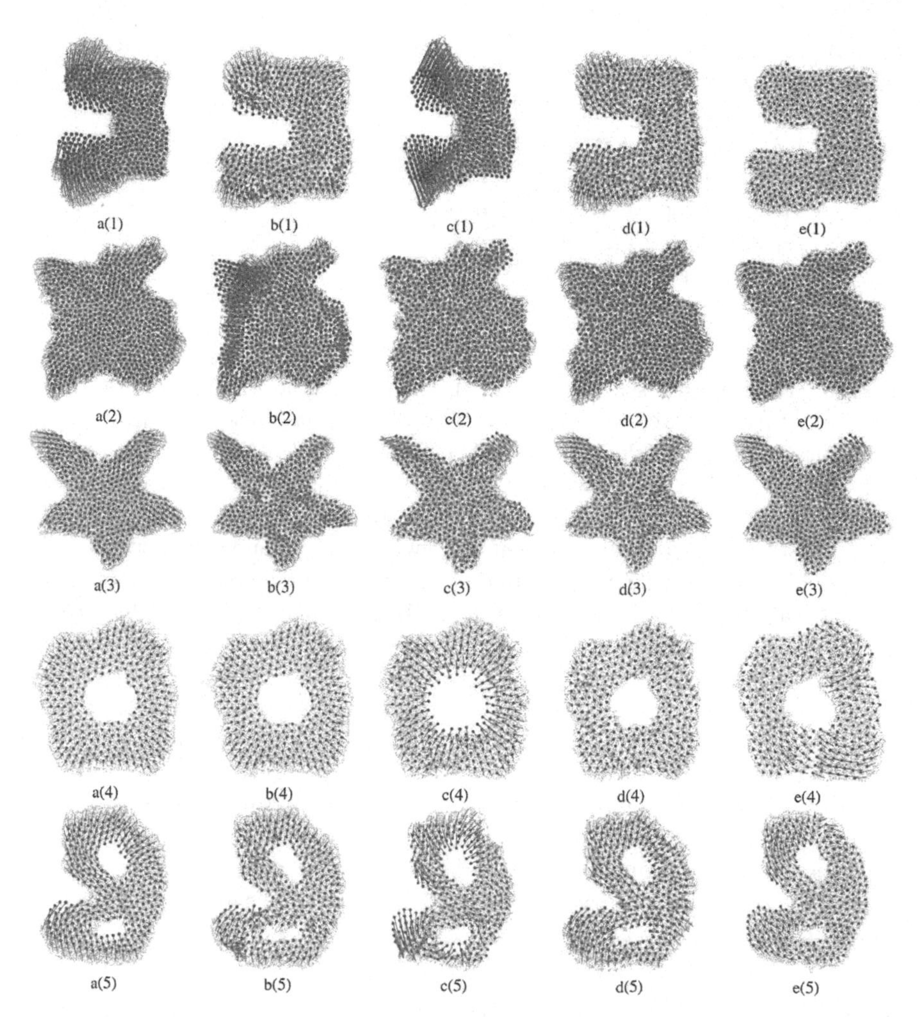

Fig. 10.1 Comparison of different localization approaches on networks with general topologies. **a(1)–a(5)**: C-CCA scheme; **b(1)–b(5)**: D-CCA scheme; **c(1)–c(5)**: MDS-MAP scheme; **d(1)–d(5)**: MDS-MAP(P) scheme; **e(1)–e(5)**: OFM scheme. A red line segment is drawn for each node, starting from its real coordinates marked with red and ending at the computed coordinates marked with grey. Image from [137]

with computed planar coordinates (x_1, y_1),(x_2, y_2), and (x_3, y_3) respectively. Let d_1, d_2, and d_3 be the shortest distances (hop counts) of node n_i to the three landmarks v_1, v_2, v_3 respectively. Then node n_i computes its planar coordinates (x_i, y_i) simply by minimizing the mean square error among the distances:

$$\sum_{j=1}^{3}(\sqrt{(x_i - x_j)^2 + (y_i - y_j)^2} - d_j)^2. \tag{10.3}$$

Table 10.1 Average localization errors with different approaches on network models shown in Fig. 10.1. Image from [137]

Scenario	C-CCA	D-CCA	MDS-MAP	MDS-MAP(P)	OFM
Topology 1 (Figure 10.1 a(1)–e(1))	2.10	0.88	2.52	0.89	0.29
Topology 2 (Figure 10.1 a(2)–e(2))	0.71	0.69	0.56	0.68	0.32
Topology 3 (Figure 10.1 a(3)–e(3))	0.72	0.64	0.62	0.75	0.48
Topology 4 (Figure 10.1 a(4)–e(4))	0.78	0.70	1.18	0.61	0.55
Topology 5 (Figure 10.1 a(5)–e(5))	1.17	0.8	1.27	0.99	0.63

Figure 10.1 shows a set of representative networks and the comparison of the optimal flat metric based localization method (OFM) on the set of networks with those state-of-the-art localization methods including the centralized MDS approach (MDS-MAP) [242], the distributed MDS approach (MDS-MAP(P)) [241], the centralized neural network (C-CCA) and the distributed neural network (D-CCA) approaches [170]. A red line segment is drawn for each landmark node, starting from its real coordinates marked with red and ending at the computed coordinates marked with grey. Clearly, the length of the line segment represents the error of localization at that node. Overall, the more and the longer the red lines are, the worse the performance of the localization is. Table 10.1 gives the localization error computed as the ratio of the average node distance error (all sensors in the network) and the averaged transmission range. The optimal flat metric based localization method achieves the highest localization accuracy.

Figure 10.2a–d gives a series of reversed C-shape networks with mere connectivity information. All the networks have the same communication radio range and under the same transmission model, but the average nodal degree increases from 9 to 18. The localization error of the algorithm decreases with the increase of nodal density as shown in Fig. 10.2e–h correspondingly.

10.2.2 Surface Sensor Network Localization

Localization of a network deployed over a 3D surface generates a unique hardness compared with localization of a network in 2D or 3D space. Specifically, due to limited radio range, the distance between two remote sensors deployed over a 3D surface can only be approximated by their surface distance, the length of the shortest path between them on the surface. Such surface distance is different from the 3D Euclidean distance of two nodes. With precise surface distance between any two points on the surface, a localization algorithm does not exist for a network deployed over a 3D surface. One intuitive example is that a piece of paper shown in Fig. 10.3a can be rolled up to a cylinder illustrated in Fig. 10.3b or to other curved shapes (see Fig. 10.3c). The distance between any pair of points on the paper doesn't change. Such deformation is called isometric deformation that can be applied on closed surfaces (see Fig. 10.3d–f for example) too.

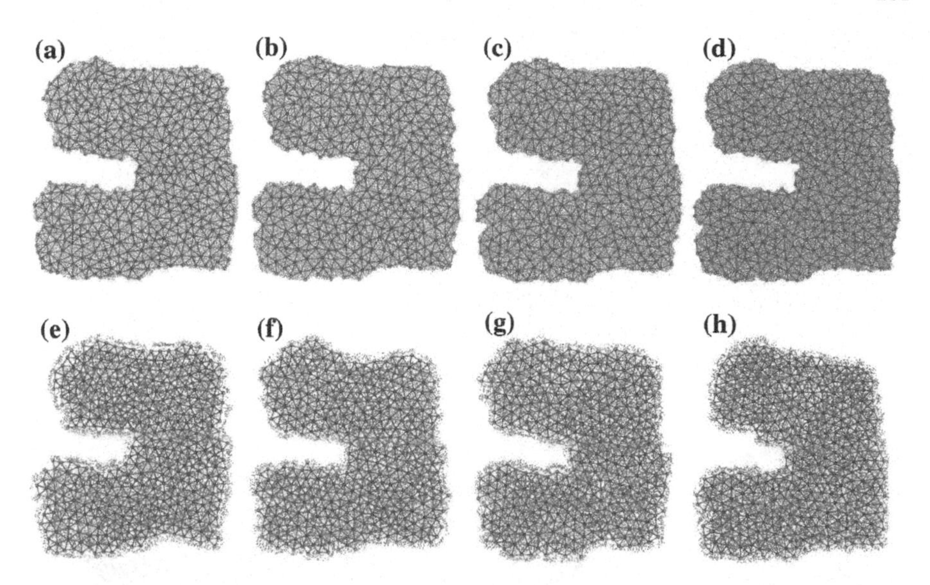

Fig. 10.2 A series of reversed C-shape networks with variant nodal density: **a**–**d** the original network with increased nodal density; **e**–**h** the localization results of our algorithm. All the networks have the same communication radio range and under the same transmission model. **a** Average nodal degree $d = 9.4$, with localization error 0.514 in **e**; **b** Average nodal degree $d = 12.6$, with localization error 0.322 in **f**; **c** Average nodal degree $d = 15.2$, with localization error 0.28 in **g**; **d** Average nodal degree $d = 18.5$, with localization error 0.246 in **h**. Image from [137]

In [334], authors assume each sensor node can measure not only distances between its neighboring nodes but also its own height information. They require a sensor network is deployed on a surface with single-value property - any two points on the surface have different projections on plane. Such property ensures that they can project the network deployed over a 3D surface to 2D plane by removing z coordinate without ambiguity. They apply existing 2D network localization method on the projected one to compute the x and y coordinates of each sensor node, and then add the height information back as the z coordinate.

Later, a cut-and-sew algorithm is proposed in [333] to generalize the localization algorithm introduced in [334] from single-value surfaces to general surfaces. The algorithm takes a divide-and-conquer approach by partitioning a general 3D surface network into a minimal set of single-value patches. Each single-value patch can be localized individually, and then all single-value patches are merged into a unified coordinates system.

However, integrating height measurement into every sensor of a network is not always practical and affordable, especially for a large-scale sensor network. However, a 3D representation of a terrain's surface, called digital terrain model (DTM), is available to public with a variable resolution up to one meter. DTMs are commonly built using remote sensing technology or from land surveying. A DTM is represented as a grid of squares, where the longitude, latitude, and altitude (i.e., 3D coordinates) of all grid points are known. It is straightforward to convert a grid into a triangulation,

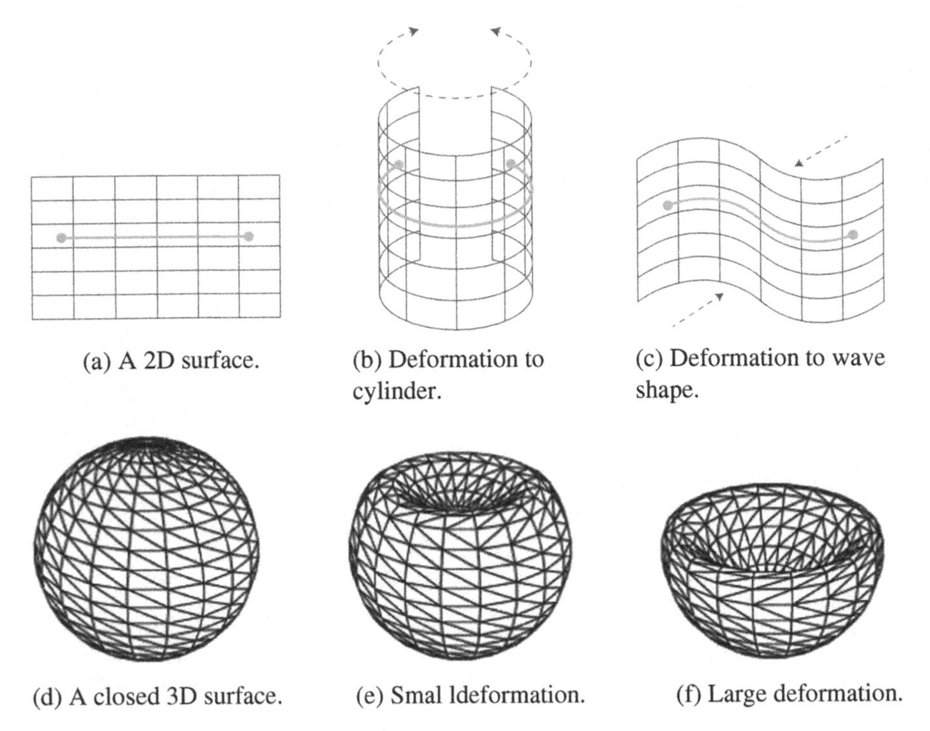

(a) A 2D surface. (b) Deformation to cylinder. (c) Deformation to wave shape.

(d) A closed 3D surface. (e) Smal ldeformation. (f) Large deformation.

Fig. 10.3 Illustration of the non-localizability of general 3D surfaces. A surface can be deformed to another surface without changing the surface distance between any pair of points. Image from [334]

e.g., by simply connecting a diagonal of each square. Therefore a triangular mesh of a terrain surface can be available before we deploy a sensor network on it. On the other hand, given a wireless sensor network deployed on a terrain surface with one-hop distance information available, a simple distributed algorithm can extract a refined triangular mesh from the network connectivity graph [336]. Vertices of the triangular mesh are the set of sensor nodes. An edge between two neighboring vertices indicates the communication link between the two sensors.

The constraint that the sensors must be on the known 3D terrain surface ensures that the triangular meshes of a terrain surface and a network deployed over the terrain surface approximate the same geometric shape. Theoretically, the two triangular meshes share the same conformal structure. We can construct a well-aligned conformal mapping between them. Based on this mapping, each sensor node of the network can easily locate reference grid points of the DTM to calculate its own location.

Figure 10.4 illustrates the basic idea. Figure 10.4a shows the triangular mesh of a terrain surface. Figure 10.4c shows the triangular mesh extracted from the connectivity graph of a network deployed over the terrain surface. We first compute two conformal mappings, denoted as f_1 and f_2 respectively, to map the two triangular meshes to plane as shown in Fig. 10.4b, d respectively. However, the two mapped triangular meshes on plane are not aligned. Three anchor nodes, sensor nodes equipped

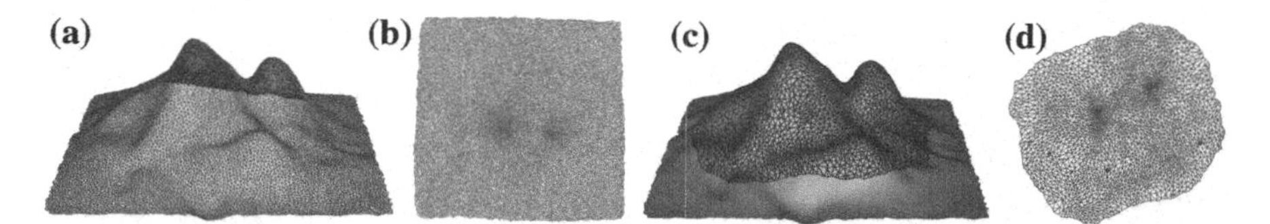

Fig. 10.4 **a** The triangular mesh of the DTM of a terrain surface. **b** The triangular mesh of the DTM is conformally mapped to plane. **c** The triangular mesh extracted from the connectivity graph of a network deployed over the terrain surface. Three randomly deployed anchor nodes are marked with red. **d** The triangular mesh extracted from the network connectivity graph is conformally mapped to plane. Image from [313]

with GPS, marked with red as shown in Fig. 10.4c are deployed with the network to provide the reference for alignment. Based on the positions of the three anchor nodes, we construct a Möbius transformation, denoted as f_3, to align the mapped triangular meshes of the network and the terrain surface on plane. Combining the three mappings, $f_1^{-1} \circ f_3 \circ f_2$, induces a well-aligned conformal mapping between the two triangular meshes shown in Fig. 10.4a, c, respectively. Based on the well-aligned mapping, each sensor node of the network simply locates its nearest grid points, vertices of the triangular mesh of the terrain surface, to calculate its own geographic location.

We briefly explain the alignment of two conformally mapped meshes in plane based on three anchor nodes, followed by discussions including the deployment of anchor nodes and some possible solutions when anchor nodes or one-hop distance information is not available.

Alignment

Denote M_1 the triangular mesh of DTM of a terrain surface and M_2 the triangular mesh extracted from connectivity graph of a network deployed on the surface. Denote f_1 and f_2 the conformal mappings from M_1 and M_2 to planar regions D_1 and D_2, respectively. A Möbius transformation aligns D_2 with D_1 on plane based on anchor nodes.

A Möbius transformation maps four distinct complex numbers z_1, z_2, z_3, z_4 to another four distinct complex numbers w_1, w_2, w_3, w_4, respectively and keeps their cross-ratio invariant, represented as:

$$\frac{(z_1 - z_3)(z_2 - z_4)}{(z_2 - z_3)(z_1 - z_4)} = \frac{(w_1 - w_3)(w_2 - w_4)}{(w_2 - w_3)(w_1 - w_4)}. \tag{10.4}$$

Equation 10.4 provides a natural alignment of two planar regions based on three pairs of anchor points. Denote f a Möbius transformation that maps the planar region D_1 with three distinct points z_1, z_2, z_3 to the planar region D_2 with three distinct points w_1, w_2, w_3. Particularly, z_1, z_2, z_3 are mapped to w_1, w_2, w_3, respectively. We use complex numbers to represent points on plane. Assume we use z_{ij} to denote $z_i - z_j$, and w_{ij} to denote $w_i - w_j$, f can be represented in a closed form from Eq. 10.4,

$$f(z) = \frac{w_2(z - z_1)z_{23}w_{12} - (z - z_2)z_{13}w_{23}w_1}{(z - z_1)z_{23}w_{12} - (z - z_2)z_{13}w_{23}}. \tag{10.5}$$

Again, all the operations in Eq. 10.5 are defined on complex numbers.

Assume three anchor nodes - sensor nodes equipped with GPS - are randomly deployed with other sensors. Each anchor node is assigned planar coordinates, e.g., mapped to plane by f_2. Denote the planar point of an anchor node mapped by f_2 with a complex numbers $z_i (1 \leq i \leq 3)$. Each anchor node then checks its stored M_1 or simply sends a request with its known geographic position to a server to locate three nearest grid points of the DTM, denoted as v_i, v_j, and v_k. Since M_1 and M_2 are not perfectly overlap in general, the anchor node does not necessarily locate inside a face $[v_i, v_j, v_k] \in M_1$. We compute the projection point of the anchor node to $[v_i, v_j, v_k]$. The projection point is the closest point of M_1 to the anchor node. Since f_1 is a continuous and one-to-one mapping, we can compute the planar coordinates of the projection point mapped by f_1 based on the planar coordinates of v_i, v_j, and v_k. Specifically, denote (t_1, t_2, t_3) the Barycentric Coordinates of the projection point on $f = [v_i, v_j, v_k]$, $f_1(v_i)$, $f_1(v_j)$, and $f_1(v_k)$ the planar coordinates of v_i, v_j, and v_k mapped by f_1, the planar coordinates of the projection point mapped by f_1 is: $t_1 f_1(v_i) + t_2 f_1(v_j) + t_3 f_1(v_k)$. Denote the planar point of the projection point mapped by f_1 with a complex number $w_i(1 \leq i \leq 3)$. Each anchor node conducts a flooding to send out its z_i and w_i to the whole network. When receiving the three pairs of planar coordinates, a non-anchor node $v_i \in M_2$ simply plugs them and its planar coordinates into Eq. 10.5. The computed one is the aligned planar coordinates of the sensor node.

Discussion

Deployment of Anchor Nodes

Figure 10.5 shows a set of representative terrain surfaces and their corresponding DTMs, on which wireless sensor networks are randomly deployed (see the black points on these terrain surfaces). Assume sensor nodes with accurate one-hop distance measurement and DTMs with high resolutions. For each network, we randomly deploy three anchor nodes and calculate the localization errors of the network. We repeat eight times for each network. Denote x_i the ith localization error. We compute the arithmetic mean $\mu = \frac{1}{8}\sum_{i=1}^{8} x_i$ and the standard deviation $\sigma = \sqrt{\frac{1}{7}\sum_{i=1}^{8}(x_i - \mu)^2}$. Table 10.2 shows the mean, the median ($\tilde{x}$), and the standard deviation of localization errors under different sets of anchor nodes. The positions of the three anchor nodes affect the performance of the proposed localization algorithm slightly. In general, the more scattered we deploy the three anchor nodes in a network, the lower the localization error is.

The Size of Anchor Nodes

Theoretically speaking, the introduced localization algorithm requires only three anchor nodes to align two triangular meshes on plane, even one triangular mesh is extracted from the connectivity graph of a network with thousands or even tens of

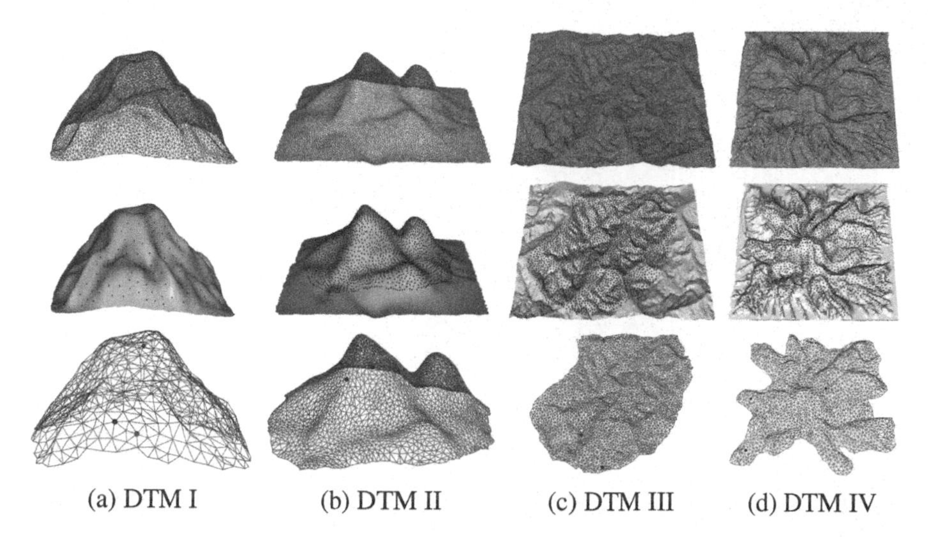

Fig. 10.5 The first row shows a set of DTMs of representative terrain surfaces. The second row shows wireless sensor networks marked with black points randomly deployed on these terrain surfaces. The third row shows the localized sensor networks with anchor nodes marked with red. The given set of anchor nodes provides each network an median localization error of the repeated tests. Image from [313]

Table 10.2 The distribution of localization errors under different sets of anchor nodes

		DTM I	DTM II	DTM III	DTM IV
Error	μ	0.2579	0.1356	0.0951	0.2098
	$\tilde{x}$	0.2306	0.1343	0.0956	0.1512
	σ	0.1089	0.1717	0.0158	0.0352

thousands of sensor nodes. If there are more than three anchor nodes deployed with the network, we can apply the least-square conformal mapping method introduced in [167] instead of Mӧbius transformation to incorporate all anchor nodes into the alignment to improve the localization accuracy.

Figure 10.6 shows one example. For a network with size 2.6k deployed on a 3D surface shown in Fig. 10.4d, the localization error of the network decreases with the increased number of anchor nodes. Compared with Mӧbius transformation based alignment, least-square conformal mapping based alignment is more flexible to take anchor nodes into alignment. But from the other side, least-square conformal mapping method introduced in [167] is centralized with high computational complexity.

Anchor Node Free

Locally conformal mapping introduces no distortion, only scaling. Such scaling is called conformal factor. Conformal factor at v_i can be approximated as the ratio of the triangle areas in $3D$ and mapped in $2D$ plane of all f_{ijk} incident to v_i,

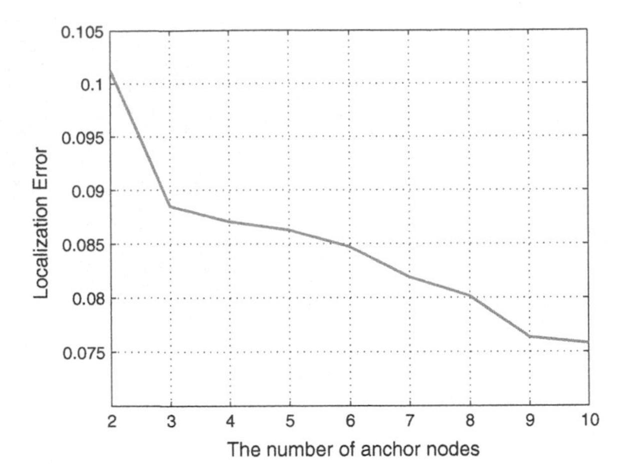

Fig. 10.6 Localization error decreases with the increased number of anchor nodes. Image from [313]

$$cf(v_i) = \frac{\sum_{f_{ijk} \in F} Area_{3D}|f_{ijk}|}{\sum_{f_{ijk} \in F} Area_{2D}|f_{ijk}|}.$$

Conformal factor at the peak of a terrain surface is usually huge. Based on the fact, vertices of M_2, i.e., sensor nodes, with the highest conformal factors are around the peaks of a terrain surface. We can apply them as anchor nodes for alignment. Assume the network shown in Fig. 10.4d is anchor node free. We compute conformal factors of the triangular meshes of the DTM and the network and use colors to encode them at the mapped planar regions shown in Fig. 10.7. It is obvious that areas marked with

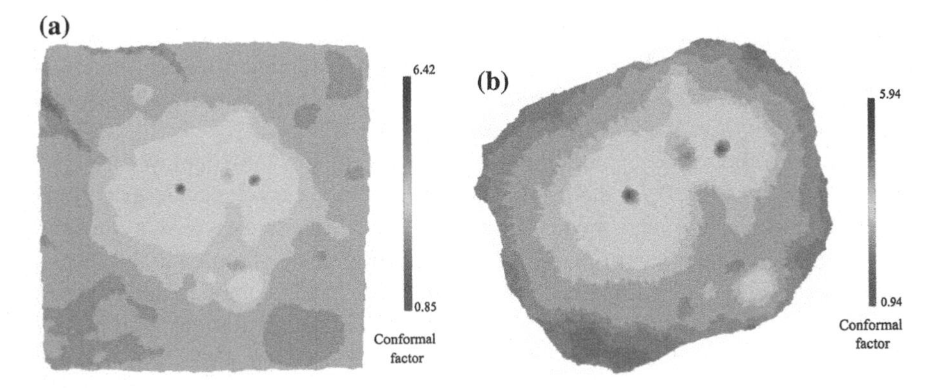

Fig. 10.7 We use colors to encode conformal factors of mapped triangular meshes on plane: **a** The mapped triangular mesh of the terrain surface shown in Fig. 10.4b. **b** The mapped triangular mesh of the network shown in Fig. 10.4d. Image from [313]

red represent the regions with high conformal factors. We pick one vertex with the highest conformal factor for each red marked region. Suppose we pick v_1 and v_2 for the triangular mesh of the network, v_3 and v_4 for the triangular mesh of the DTM. Suppose v_1 shares a similar conformal factor with V_3. v_1 simply determines its 3D coordinates the same as v_3. Similarly, v_2 determines its 3D coordinates the same as v_4.

Note that if the shape of a mountain region is extremely complicated, conformal factors may identify wrong pairs of nodes between M_1 and M_2. The anchor free localization method is not stable in that case.

Connectivity Only

When range distance measurement is not available, a sparse triangular mesh can still be extracted from a network connectivity graph. A simple landmark-based algorithm discussed in [85, 228] uniformly selects a subset of nodes in a distributed way and denotes them as landmarks, such that any two neighboring landmarks are approximately a fixed K hops away ($K \geq 6$). The dual of a discrete Voronoi diagram with generators the set of landmarks forms a triangulation. Vertices of the triangulation is the set of landmarks. Edge between two neighboring vertices is a shortest path between the two landmarks. We simply assume the edge length of the triangulation a unit one, and then apply exactly the same localization algorithm for landmark nodes as discussed in Sect. 10.2.2.

A non-landmark node, denoted as n_i, finds its three nearest landmarks, denoted as v_1, v_2, v_3 with computed 3D coordinates $p(v_1)$,$p(v_2)$, and $p(v_3)$ respectively. Denote d_1, d_2, and d_3 the shortest distances (hop counts) of node n_i to the three landmarks v_1, v_2, v_3 respectively. Then node n_i computes its 3D coordinates $p(n_i)$ simply by minimizing the mean square error among the distances:

$$\sum_{j=1}^{3}(|p(n_i) - p(v_j)| - d_j)^2. \tag{10.6}$$

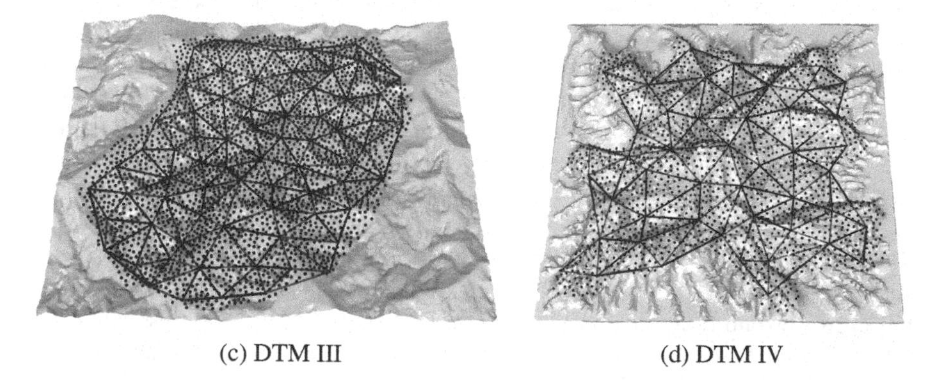

(c) DTM III (d) DTM IV

Fig. 10.8 Networks with Connectivity Information Only. Image from [313]

Figure 10.8 shows the sparse triangular meshes generated from networks with size $3k$ and $2k$ deployed on different terrain surfaces, respectively. The localization errors for landmark nodes of the two networks are 0.2037 and 0.2610 respectively.

10.3 Greedy Routing

Greedy routing, with only local information stored at individual nodes, is known for its scalability to resource-constrained and highly volatile wireless sensor networks. In a basic greedy routing approach, a packet is forwarded at each step to a neighboring node whose position is closest to the destination of the packet based on the standard distance calculation. However, such greedy forwarding is susceptible to problem of local minimum. A node is called a local minimum if it is not the destination, but closer to the destination than all of its neighbors. Clearly, greedy routing fails at a local minimum. Such local minimums may appear at either boundary or internal nodes. A boundary node, especially on a concave boundary, usually becomes a local minimum when the source and destination nodes are located on two sides of the boundary. Although it seems anti-intuitive, an internal node can be a local minimum too, due to local concavity under random deployment of sensor nodes.

Various approaches have been developed to address the problem of local minimum for wireless sensor networks deployed in plane, with primary focus on boundaries. For example, face routing and its alternatives and enhancements [29, 82, 147, 153, 155, 156, 197, 264] exploit the fact that a concave void in a planar network is a face with a simple line boundary. Thus when a packet reaches a local minimum on a boundary, it employs a local deterministic algorithm to search the boundary in either clockwise or counter-clockwise direction as shown in Fig. 10.9a, until greedy forwarding is achievable. However, for wireless sensor networks deployed in 3D, a void is no longer a face. Its boundary becomes a surface, yielding an arbitrarily large number of possible paths to be explored (see Fig. 10.9b) and thus rendering face routing infeasible. On the other hand, greedy embedding [6, 61, 80, 152, 165, 206, 228] provides theoretically sound solutions to ensure the success of greedy routing. Unfortunately, none of the greedy embedding algorithms in literatures can be extended from 2D to 3D general networks.

We introduce a greedy routing approach for wireless sensor networks deployed in 3D volume. We first construct a unit tetrahedron cell (UTC) mesh structure from the connectivity graph of a network. Then a distributed algorithm applies volumetric harmonic mapping to the UTC mesh under spherical boundary condition. The mapping is one-to-one and yields virtual coordinates for each node in the network. Since a boundary has been mapped to a sphere, node-based greedy routing is always successful thereon. At the same time, we exploit the UTC mesh to develop a face-based greedy routing algorithm with guaranteed greedy forwarding success at internal nodes. To route a data packet to its destination, face-based and node-based greedy routing algorithms are employed alternately at internal and boundary UTCs, respectively. For networks with multiple internal holes, a segmentation and *tunnel*-based

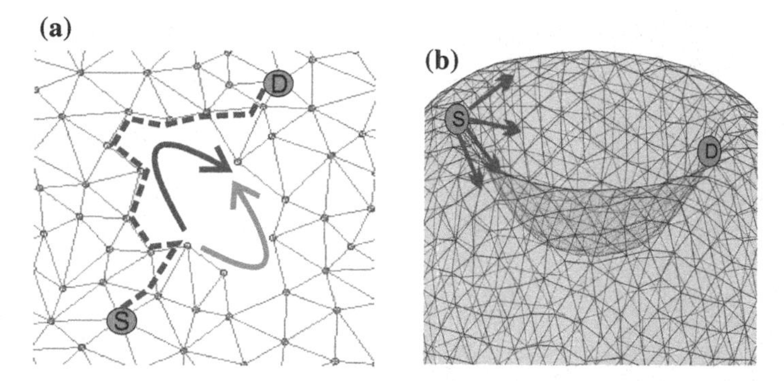

Fig. 10.9 Comparison of face routing in 2D and 3D networks. Node S has a shorter distance to Destination D than all of its neighbors, and thus is a local minimum. **a** Face routing is successful in a 2D planar network because a concave void is a face with a simple line boundary, and thus a local deterministic algorithm can be employed to search the boundary in either clockwise or counter-clockwise direction as shown by the blue and red lines. **b** In a 3D network, a void is no longer a face. Its boundary becomes a surface, yielding an arbitrarily large number of possible paths to be explored (as indicated by the arrows). Thus face routing fails. Image from [306]

routing strategy is proposed to support global end-to-end routing. To make a local routing decision, each node only needs to store virtual coordinates of its neighbors and itself, and a routing table with a size bounded by the number of internal holes.

Figure 10.10 illustrates the basic idea of the approach. Figure 10.10a shows a wireless sensor network deployed in 3D volume with inner boundary nodes marked with red color. Figure 10.10b highlights those local minimum nodes among boundary and internal nodes under greedy routing and marks them with blue squares and red triangles respectively. Figure 10.10c shows a tetrahedral mesh extracted from the connectivity graph of the network. Figure 10.10d shows the mapped tetrahedral mesh using volumetric Harmonic mapping with both outer and inner boundaries mapped to spheres. Such one-to-one map generates virtual coordinates for each node in the network. Since boundaries have been mapped to spheres, greedy routing based on virtual coordinates is always successful at boundaries. Figure 10.10e gives a greedy routing path based on virtual coordinates of the mapped tetrahedral mesh. Figure 10.10f shows the corresponding greedy routing path in the original network.

Sections 10.3.1 and 10.3.2 introduce briefly the routing algorithms for networks without and with multiple internal holes, respectively.

10.3.1 3D Wireless Sensor Networks Without Hole

With a constructed tetrahedral mesh, we first map the boundary of the tetrahedral mesh homeomorphically to a sphere using spherical harmonic map with the heat flow method. Similarly, we apply volumetric harmonic map by minimizing the volumetric

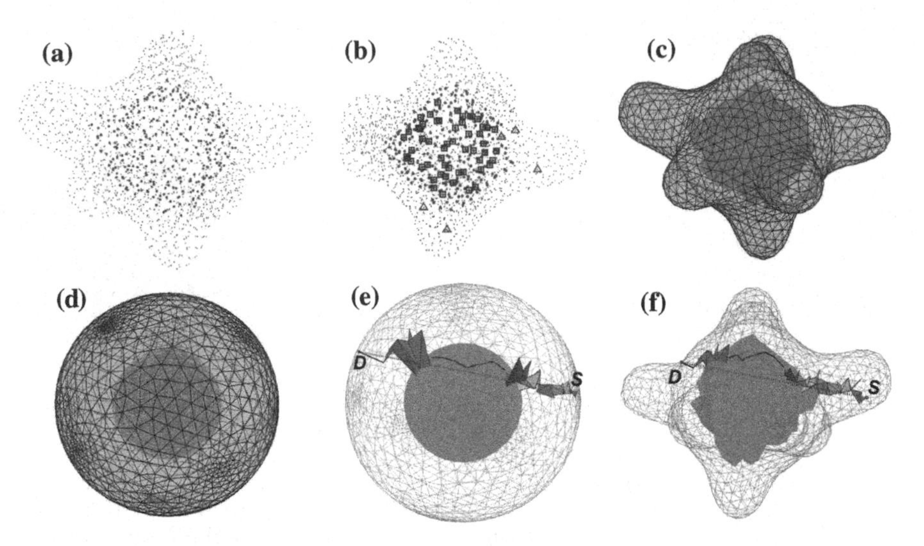

Fig. 10.10 Illustration of the harmonic mapping based greedy routing approach for wireless sensor networks deployed in 3D volume: **a** A wireless sensor network deployed in 3D volume with inner boundary nodes marked with red color. **b** The nodes that are local minimums under greedy routing are highlighted as blue squares and red triangles for boundary nodes and internal nodes, respectively. **c** A tetrahedral mesh extracted from the connectivity graph of the network. **d** The mapped tetrahedral mesh using volumetric Harmonic mapping with both outer and inner boundaries mapped to spheres. **e** A greedy routing path based on virtual coordinates of the mapped tetrahedral mesh. **f** The corresponding greedy routing path shown in the original network. Image from [306]

harmonic energy under the computed spherical boundary condition such that the entire tetrahedral mesh is homeomorphically mapped to a solid tetrahedral ball in $\mathbb{R}^3$. Both algorithms are distributed, where a node only needs to communicate with its one-hop neighbors in each iteration. After mapping, each node has its own virtual coordinates. Since the boundary has been mapped to a unit sphere, greedy forwarding is always successful for boundary nodes.

However, greedy forwarding may get stuck at a local minimum for inner nodes because the mapped tetrahedral mesh is not guaranteed Delaunay. To this end, we introduce a face-based greedy routing strategy for inner nodes. Let us consider a data packet that is to be delivered from Source node denoted as S to Destination node denoted as D. Node S first computes a line segment between S and D, which is denoted by Γ. Clearly, Γ passes through a set of tetrahedrons between S and D and intersects with a sequence of faces, denoted by $\Phi = \{f_i | 1 \leq i \leq k\}$, where k is the total number of intersected faces. The distance from f_i to destination D is calculated as the distance from the intersection point of Γ and f_i to D. Each intermediate node only needs to calculate the next face in Φ. A data packet is then forwarded from f_1 to f_k.

10.3.2 3D Wireless Sensor Networks with Internal Holes

For a network with one internal hole, two boundary surfaces with one outside and the other inside will be detected. The same spherical harmonic mapping algorithm is applied to map them to two unit spheres, respectively. Then the boundary nodes perform simple local calculations to align the inner sphere to the outer one. Specifically, the nodes on the inner boundary scale their coordinates to reduce the radius of inner sphere to r'. It is a constant and less than one. Next, a node on the outer boundary with its virtual coordinates most close to (0,0,1) finds its closest node on the inner boundary. The two nodes and the center of spheres (0,0,0) form an angle, denoted as θ_0, calculated based on virtual coordinates. θ_0 is broadcast to all nodes on the inner boundary. They subsequently apply a rotation matrix with angle θ_0 on their virtual coordinates. Therefore, the inner sphere is aligned with the outer one with respect to this pair of nodes. Then another node on the outer boundary with its virtual coordinates most close to (0,1,0) repeats the above process to initiate the second round of rotation. After two rotations, the inner and outer spheres are approximately aligned. Finally, a volumetric harmonic map with such spherical boundary conditions is applied to generate virtual coordinates for inner nodes in the network.

For a network with multiple internal holes, we first segment the network into a set of clusters such that each contains one internal hole only. Specifically, each inner boundary node initiates a flooding message that includes its boundary ID and a hop counter that records the number of hops the message has traveled. A non-inner-boundary node learns its hop distance to nearby holes according to the received flooding messages. It always keeps the shortest distance and the corresponding boundary ID. When flooding stops, a node checks the shortest distance(s) and the corresponding boundary ID(s) it has learned. If a non-inner-boundary node has equal shortest hop distance to at least two inner boundaries, the node chooses the one with the smaller ID.

Now, the 3D sensor network has been segmented into a set of clusters. For each cluster, we apply the same algorithm as networks with one internal hole to build a local virtual coordinate system. Routing inside a cluster is purely based on the local virtual coordinate system of the cluster. Since we do not require the local virtual coordinate systems of individual clusters aligned with each other, we need a global routing scheme to coordinate these local coordinates to support global routing across clusters.

We use a graph to capture the compact representation of the clusters, where vertices represent individual clusters and edges connecting two vertices indicate adjacent clusters. Each node in the network generates a global routing table based on the graph with size proportional to the number of internal holes.

If source and destination nodes belong to different clusters, the source node checks its global routing table to find the next hop cluster and the gate node to that cluster. A packet is then greedily forwarded to the gate node. Note that the packet is not necessary to actually reach the gate node. Since adjacent clusters share a good amount of intersection nodes, the packet can directly route to the next cluster without passing

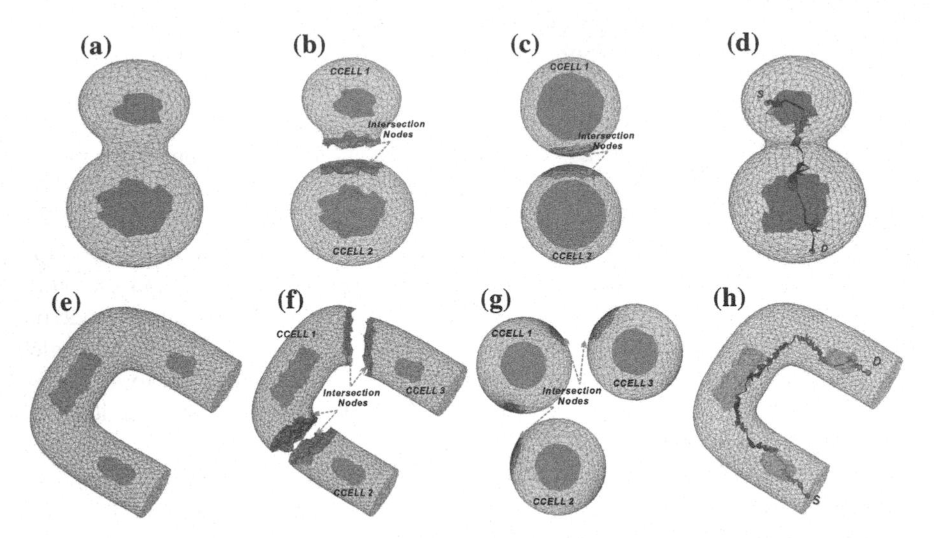

Fig. 10.11 Wireless sensor networks with multiple internal holes. **a** and **e**: Wireless sensor networks deployed in 3D volume with multiple internal holes. The nodes on the inner boundaries are highlighted in purple. **b** and **f**: The networks are decomposed into clusters with each one containing exactly one internal hole. The intersection nodes are highlighted in green and red. **c** and **g**: Each cluster is mapped to a unit sphere. **d** and **h**: Greedy routing paths shown in the original network. Image from [306]

through the gate node, as long as it meets a node on the shared boundary of the adjacent clusters. Thus, the traffic across clusters can be nicely distributed among nodes on the shared boundaries between adjacent clusters. The procedure repeats until the packet enters the cluster containing the destination node.

As shown in Fig. 10.11, we segment wireless sensor networks deployed in 3D volume with multiple internal holes into a set of clusters. Each node stores a small global routing table with a bounded size to route packets across clusters. Figure 10.11 visualizes greedy routing paths with source and destination nodes belonging to different clusters. In summary, the harmonic mapping based greedy routing approach achieves guaranteed delivery with reasonable path stretch and desired scalability for networks in 3D.

10.4 Deployment

Sensor networks offer the ability to monitor real-world phenomena in detail by embedding wireless sensor nodes in a real-world environment. However, a sensor node does not always make perfect measurement. Instead, its reading exhibits inevitable error, which often depends on the distance between the sensor and the target being sensed, as observed in a wide range of applications, e.g., for monitoring pollution, radiation, acoustics, and vibration. With a given set of sensors, a sensor network offers different accuracy in data acquisition when they are deployed in different ways in the Field of Interest (FoI).

There are a handful of works that address the sensing quality of sensor network deployment. Multiple coverage, where each point is covered by at least k different sensors, is proposed to improve the monitoring quality in [16, 253, 288, 312, 337]. Event probability density is considered in [37, 39, 57] to guide mobile sensors to areas with high event density. In [234], each mobile sensor node takes the relative importance of different areas in the FoI and the sensing unreliability of the sensor measurement into a cost function to make moving decision.

Previous research on sensor deployment mainly focuses on networks with FoI on 2D plane or in 3D volume. However, recent study [332] reveals that sensor deployment on 3D terrain surfaces exhibits surprising challenges with provable NP complete property to determine optimal solution with full coverage.

Given a 3D terrain surface with FoI, we introduce a surface network deployment strategy to achieve the best sensing quality for a given set of sensor nodes.

10.4.1 Optimal Surface Network Deployment

We assume the availability of 3D digital terrain models for surfaces with FoI. We also assume the disk topology of the terrain surfaces, although they are not necessary single value or with convex boundary condition. We assume homogeneous sensors with identical communication range and sensing capability. We define the following function to measure data unreliability collected from a single point.

Definition 10.2 Let p_i denote the position of Sensor i on a FoI denoted as A. Given a point q on A, the *sensing unreliability function* $g(||q - p_i||)$ describes how unreliable the measurement of the information at Point q sensed by Sensor i at p_i is, where $||\cdot||$ represents the distance between q and p_i.

The specific form of $g(||q - p_i||)$ is application-dependent. We consider $g(||q - p_i||)$ as a general function in this research, as long as it is continuous and strictly increasing. Note that $g(||q - p_i||)$ will go to infinity if the distance is long enough.

Definition 10.3 Given a set of n sensors deployed on A, let $P = \{p_i|1 \leq i \leq n\}$ be the set of positions of all sensor nodes. A sensing partition of A is defined as $\mathbb{V} = \{\mathbb{V}_i|1 \leq i \leq n\}$ where $\mathbb{V}_i$ represents the sensing region of a sensor at position p_i. Note that the union of $\mathbb{V}_i$ covers the whole A. A point of A may be within sensing ranges of several sensor nodes, but belongs to only one sensing region of a sensor.

The following equation gives the sensing unreliability of a single sensor.

Definition 10.4 Given Sensor i, the sensing unreliability of data collected over its sensing Region $\mathbb{V}_i$ on A is defined as:

$$G_i(p_i, \mathbb{V}_i) = \int_{q \in \mathbb{V}_i} g(||q - p_i||)dq, \tag{10.7}$$

where q is a point inside $\mathbb{V}_i$.

Summing all sensors, we arrive at the overall sensing unreliability of the entire network, given by the following definition.

Definition 10.5 Given a set of n sensors deployed on A, the overall sensing unreliability of data collected by the entire network is defined as:

$$G(P, \mathbb{V}) = \sum_{i=1}^{n} \int_{q \in \mathbb{V}_i} g(||q - p_i||)dq. \tag{10.8}$$

With the above definitions, we can formulate the optimal surface network deployment problem as follows:

Definition 10.6 Given a set of n sensors to be deployed with FoI on a 3D surface, the *optimal surface network deployment problem* is to identify P and $\mathbb{V}$ such that $G(P, \mathbb{V})$ is minimized.

10.4.2 Optimal Solution

To find the solution of optimal surface network deployment problem, we start from a simplified version of this problem, which is defined as:

Definition 10.7 Given a set of n sensors deployed on a field of interest A, the *optimal surface network sensing problem* is to identify $\mathbb{V}$ with fixed P such that $G(P, \mathbb{V})$ is minimized.

With given A and fixed P, there are different ways to assign a sensing region to each sensor that induces infinitely possible solutions for the problem. Lemma 10.9 shows that a Voronoi partition $\mathbb{V}(P, A)$ minimizes the total unreliability of data collected from the sensor network.

Definition 10.8 A *Voronoi partition* is a partition of the space according to the distances to a discrete set of objects (i.e., sensors in our setting), $P = \{p_i | 1 \leq i \leq n\}$, called sites in the space. A Voronoi cell, or a Voronoi region, $\mathbb{V}_i$, of a site p_i is the region of points that are closer to p_i than to any other sites, that is

$$\mathbb{V}_i = \{q \in A \mid ||q - p_i|| \leq ||q - p_j||, \forall j \neq i\}. \tag{10.9}$$

We let $\mathbb{V}(P, A)$ denote the Voronoi partition generated by P on A.

Lemma 10.9 *Denote $\mathbb{V}(P, A)$ a Voronoi partition generated by P on A. For a fixed placement of sensors (i.e., $P = \{p_i | 1 \leq i \leq n\}$) on A, $\mathbb{V}(P, A)$ minimizes the unreliability function $G(P, \mathbb{V})$.*

However, given a set of n sensors, there are infinite ways to place them on A. Each induces a different Voronoi partition. Theorem 10.12 shows that the placement of a set of sensors under a generalized centroidal Voronoi partition is optimal (i.e., with minimum $G(P, \mathbb{V})$) [204].

Definition 10.10 The *mass centroid*, $\widetilde{p_i}$, of the Voronoi region $\mathbb{V}_i$ is defined as:

$$\widetilde{p_i} = \frac{\int_{q \in \mathbb{V}_i} q\rho(q)dq}{\int_{q \in \mathbb{V}_i} \rho(q)dq}, \tag{10.10}$$

where $\rho(q)$ is the density function defined in A and $\rho(q) \geq 0$.

Definition 10.11 Given $P = \{p_i | 1 \leq i \leq n\}$. If the mass centroid $\widetilde{p_i}$ satisfies $\widetilde{p_i} = p_i$ for every $\mathbb{V}_i$ $(1 \leq i \leq n)$, then $\mathbb{V} = \{\mathbb{V}_i | 1 \leq i \leq n\}$ is a *centroidal Voronoi partition.*

In other words, a centroidal Voronoi partition is a Voronoi partition such that each site is located at the mass centroid of its corresponding Voronoi region with respect to a given density function. For any given set of sites, a centroidal Voronoi partition always exists, although not necessarily unique [69].

Theorem 10.12 *For a given set of n sensors on A, the optimal sensor placement $P = \{p_i | 1 \leq i \leq n\}$ and corresponding partition $\mathbb{V} = \{\mathbb{V}_i | 1 \leq i \leq n\}$ that together minimize the unreliability function $G(P, \mathbb{V})$ are achieved when $\mathbb{V}$ is the Voronoi partition of P and p_i is the generalized centroid of $\mathbb{V}_i$, $\forall i$, $1 \leq i \leq n$.*

Figure 10.12 illustrates the basic idea to compute an approximate solution of the optimal surface network deployment problem.

Specifically, Fig. 10.12a shows a set of sensors marked with red randomly deployed on a terrain surface approximated by $5k$ triangles and denoted as M. Figure 10.12b shows a conformal mapping of the triangular surface to a unit planar disk, denoted as $f : M \rightarrow D$, with the set of sensors mapped to the disk accordingly. One appealing property of conformal mapping is that it preserves the surface Riemannian metric (distance) locally up to a scaling factor called conformal factor. The conformal factor denoted as cf at a point p on the surface can be computed as the ratio between the infinitesimal areas around p in the $3D$ surface and $2D$ mapped plane, i.e. $cf(p) = \frac{Area_{3D}(p)}{Area_{2D}(p)}$. For a discrete surface, cf at a vertex v_i can be approximated as the ratio of the triangle areas in $3D$ and $2D$ spaces of all $[v_i, v_j, v_k]$ incident to v_i,

$$cf(v_i) = \frac{\sum_{[v_i, v_j, v_k]} Area_{3D}([v_i, v_j, v_k])}{\sum_{[v_i, v_j, v_k]} Area_{2D}([v_i, v_j, v_k])}.$$

cf of a point q within a triangle $[v_i, v_j, v_k]$ can be approximated by bilinear interpolation,

$$cf(q) = \lambda_1 cf(v_i) + \lambda_2 cf(v_j) + \lambda_3 cf(v_k),$$

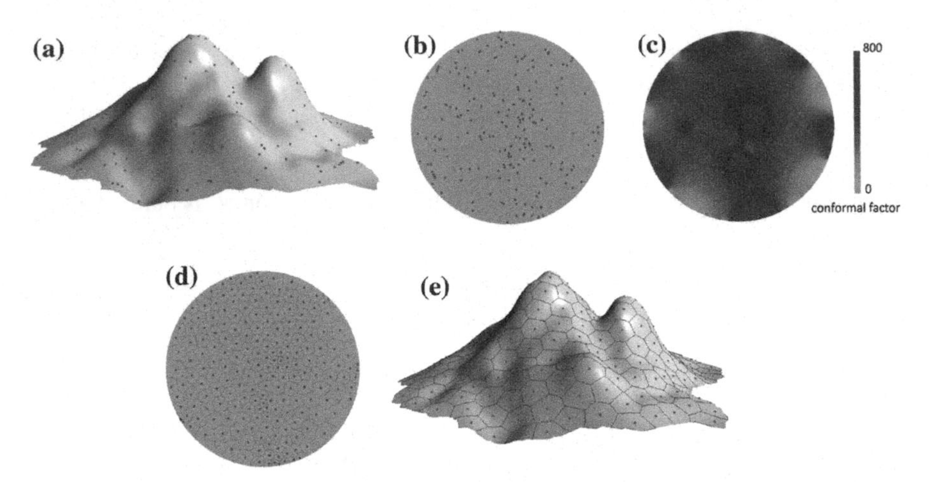

Fig. 10.12 Algorithms: **a** A set of sensors marked with red is randomly deployed on a mountain surface approximated by 5k triangles. **b** The mountain surface is mapped to a planar unit disk based on a conformal parametrization denoted as f, with sensors mapped to the disk accordingly. **c** Metric distortion of the surface on the disk is measured by conformal factor cf with color coded. **d** A generalized centroidal Voronoi partition of the set of sensors is computed on the planar disk based on its compensated metric, where points and polygons representing the computed sensor positions and their sensing regions respectively. **e** The set of sensors and their corresponding sensing regions are projected back to the surface based on f^{-1}. Image from [136]

where λ_1, λ_2, and λ_3 are barycentric coordinates of q inside $[v_i, v_j, v_k]$, satisfying $\lambda_1 + \lambda_2 + \lambda_3 \equiv 1$. Denote the metric of the surface at point p as $d_{3D}(p)$, and the metric of the mapped planar disk at point p as $d_{2D}(p)$, they differ by

$$d_{3D}(p) = cf(p)^2 d_{2D}(p).$$

Figure 10.12c uses color to encode conformal factor cf.

We then sample the mapped unit disk with a regular grid and perform summation over all sampled points to approximate the integration in Eq. 10.10. Note that we add cf to compensate metric distortion.

$$M_{\mathbb{V}_i} = \sum_{q \in \mathbb{V}_i} cf(q)^2 \frac{1}{||q - p_i||} \frac{dg(||q - p_i||)}{d||q - p_i||}, \tag{10.11}$$

and

$$C_{\mathbb{V}_i} = \frac{1}{M_{\mathbb{V}_i}} \sum_{q \in \mathbb{V}_i} cf(q)^2 \frac{q}{||q - p_i||} \frac{dg(||q - p_i||)}{d||q - p_i||}, \tag{10.12}$$

where q is a sampled point inside the Voronoi region $\mathbb{V}_i$, and p_i is the planar position of the ith sensor.

Figure 10.12d shows the computed generalized centroidal Voronoi partition on D based on the compensated metric with red points and marked polygons representing the computed sensor positions and sensing regions respectively. Figure 10.12e depicts the optimal deployment of the set of sensors and their optimal sensing regions on M by projecting the computed generalized centroidal Voronoi partition on D to M based on f^{-1}. In this example, we choose the sensing unreliability increased quadratically along the distance with $g(||q - p_i||) = \frac{1}{2}||q - p_i||^2$.

Let r denote the longest distance between a sensor and a point inside its sensing region, and r_c denote the communication range of the sensor. If $\frac{r_c}{r} \geq 2$, a sensor node has a regular connectivity of six under the discussed optimal surface network deployment. The reason is that centroidal Voronoi partition in $\mathbb{R}^2$ always has congruent regular hexagons as its Voronoi regions [201], and the inverse of conformal mapping keeps the neighboring relations of each Voronoi region.

10.4.3 Decreased Sensing Unreliability

Assume sensors are initially random-deployed on surfaces. The first row of Fig. 10.13 shows that the sensors are randomly deployed on a set of surface models with a random sensing partition, which exhibits a very high sensing unreliability as given in Fig. 10.14. When the sensing area is re-partitioned based on Voronoi partition of the set of sensors, the sensing unreliability decreases 47.58% in average on the

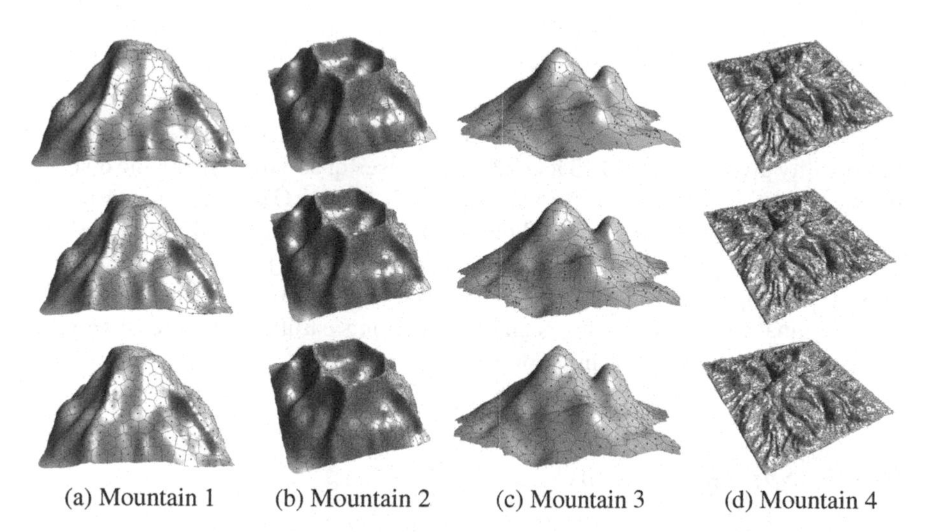

(a) Mountain 1 (b) Mountain 2 (c) Mountain 3 (d) Mountain 4

Fig. 10.13 Given a fixed set of sensors, the first row shows that the sensors are randomly deployed on surfaces with random sensing partition; the second row shows that the sensors are randomly deployed on surfaces with the Voronoi-based sensing partition; the third row shows that the sensors are re-deployed on surfaces with the generalized centroidal Voronoi based sensing partition. Figure 10.14 shows the decreased sensing unreliability of the networks. Image from [136]

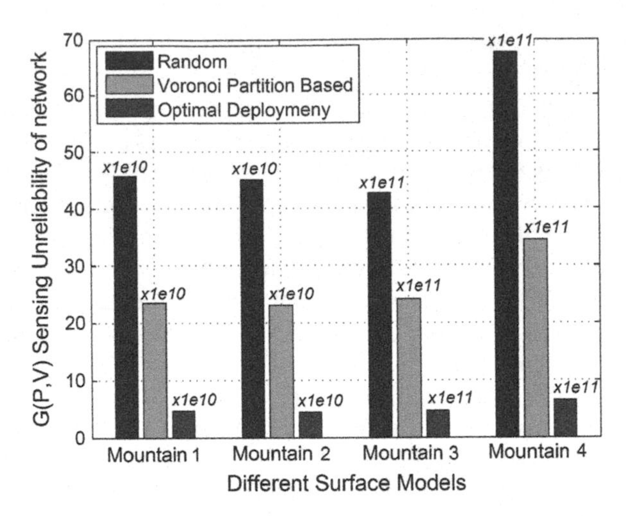

Fig. 10.14 Sensing unreliability with different deployment and sensing partition methods on various surface models shown in Fig. 10.13, where sensing unreliability function $g(||q - p_i||) = \frac{1}{2}||q - p_i||^2$. Image from [136]

testing models. When we re-deploy those sensors such that they satisfy a generalized centroidal Voronoi partition on surface, the total sensing unreliability of a wireless sensor network decreases dramatically 89.94% in average compared with the random sensing deployment. Note that the sensing unreliability function is $g(||q - p_i||) = \frac{1}{2}||q - p_i||^2$. Figure 10.14 experimentally shows the results of the Lemma 10.9 and the Theorem 10.12.

10.4.4 Unreliability Function

Theorem 10.12 is independent of the choice of the unreliability function. The sensing unreliability given in Figs. 10.13 and 10.14 increases quadratically with the distance and the unreliability function $g(||q - p_i||) = \frac{1}{2}||q - p_i||^2$. Given the same set of sensor nodes initially random-deployed on mountain 1 model as shown in Fig. 10.13, the sensing unreliability increases linearly with $g(||q - p_i||) = ||q - p_i||$ and cubically with $g(||q - p_i||) = \frac{1}{3}||q - p_i||^3$, respectively. Figure 10.15 shows the computed optimal positions of sensors and their sensing partitions. Figure 10.16 gives the decreased sensing unreliability of the whole network.

10.4.5 Special Scenarios

In some special scenarios, the sensing reliability of a sensor under its sensing range does not have a noticeable decrease with the distance. The unreliability function can be defined as the following:

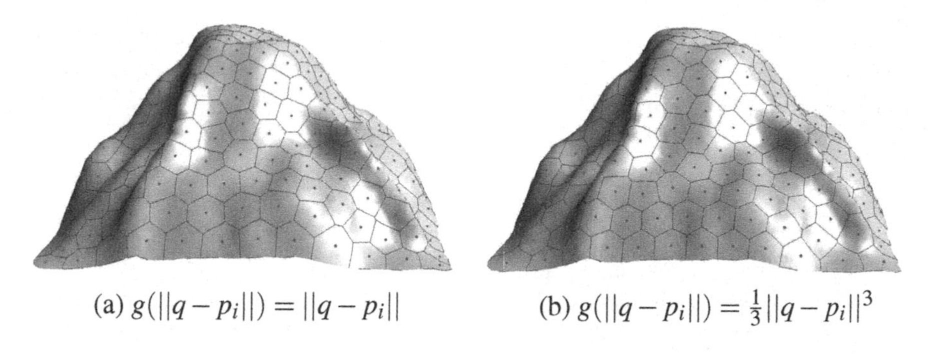
(a) $g(||q-p_i||) = ||q-p_i||$ (b) $g(||q-p_i||) = \frac{1}{3}||q-p_i||^3$

Fig. 10.15 The computed optimal positions of sensors and their sensing partitions on mountain 1 model with the same given set of sensors, but different sensing unreliability functions. Image from [136]

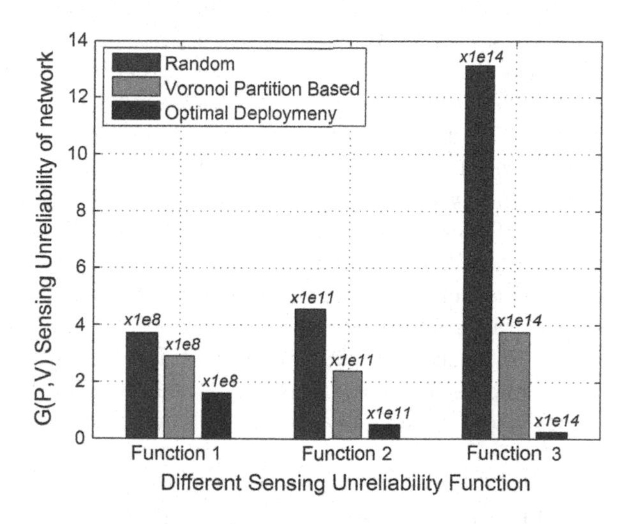

Fig. 10.16 Sensing unreliability of the same set of sensors deployed on mountain 1 surface with different deployment and sensing partition methods and various sensing unreliability functions, specifically, function 1 $g(||q-p_i||) = ||q-p_i||$, function 2 $g(||q-p_i||) = \frac{1}{2}||q-p_i||^2$, and function 3 $g(||q-p_i||) = \frac{1}{3}||q-p_i||^3$. Image from [136]

$$g(||q-p_i||) = \begin{cases} 1, & ||q-p_i|| \le R_s \\ \infty, & ||q-p_i|| > R_s, \end{cases}$$

where R_s is a constant called sensing range. The accuracy of the collected data is guaranteed as long as it is acquired within the sensing range of the collecting sensor. A point q is said covered by a sensor if their distance is less than R_s. Under this special $g(||q-p_i||)$, a full covered sensor network, where every point on the surface is covered by at least one sensor, can obviously provide all the accurate data. To achieve the lowest unreliability in data acquisition, we are actually looking for a surface deployment scheme with full sensor coverage. The optimal surface deployment problem (OSDP) is then converted to the optimal surface coverage problem (OSCP) [332], the minimal number of sensors to fully cover the FoI on a surface. The hardness of the OSCP has been proved in [332]. There are several approximation algorithms proposed in [332] to address this problem in wireless sensor networks. Replacing Eq. 10.12 to the following one:

$$C_{V_i} = \frac{\sum_{q \in V_i} cf(q)^2 q}{\sum_{q \in V_i} cf(q)^2} \tag{10.13}$$

the introduced algorithm can also provide an approximated solution.

We choose the greedy method proposed in [332] to make a comparison because the greedy method performs the best in their simulations. We also compare with the triangle pattern [149], which is the most widely used method to cover FoI on an ideal plane, to illustrate the intrinsic limitation to directly apply planar deployment based method to surface.

Figure 10.17 compares the introduced optimal surface network deployment method with the above two, where the partition of FoI on a surface for the greedy method is $D = 5$. Optimal surface network deployment method achieves the highest coverage ratio under the same set of sensors for the same surface model. Note that Theorem 10.12 can no longer be applied here. But to maximize the coverage ratio of a given set of sensors on a FoI, it is a natural way to consider positioning sensors uniformly on a FoI to minimize the overlapping part of their sensing regions. The Eq. 10.13 is a discrete approximation of Eq. 10.10, the computation of the mass centroid of a Voronoi region with constant mass density ($\rho(q) = 1$). So we actually compute a centroidal Voronoi partition of a given set of sensors with FoI on a $3D$ surface with constant density. Although it is Gershos conjecture [92] that the sites of an Euclidean centroidal Voronoi partition are uniformly distributed in the space for R^n and has been proved by Fejes Tóth only for 2D convex polygon case [75], extensive experiments have verified this conjecture [5, 70].

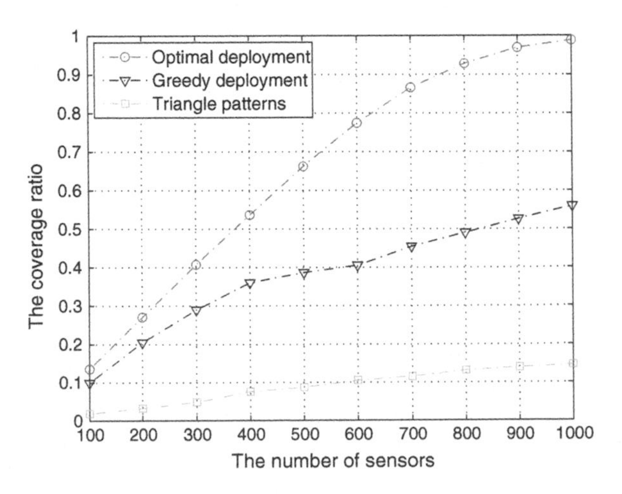

Fig. 10.17 Comparison of our proposed method with the greedy method and the triangle pattern on mountain 4 model. Image from [136]

10.5 In-Network Data-Centric Processing

In comparison with earlier computer communication systems, the unique and intrinsic challenges in sensor networking are distributed and scalable computation and communication. In particular, an individual sensor is highly resource-constrained, with extremely limited computing, storage, and communication capacities. On the other hand, however, the target applications often require a large-scale deployment where the amount of data generated, stored, and transmitted in a network grow proportionally with its size. In-network data-centric processing aims to establish a self-contained data acquisition, storage, retrieval, and query system, where data is uniquely named and data processing is achieved using data names instead of network addresses.

Geographical hash table [2, 10, 25, 220, 229] is a general approach for in-network data-centric processing. A basic geographical hash table based scheme hashes a datum by its type into geographic coordinates and stores at the sensor node geographically nearest to such coordinates. Queries apply the same hash table with the desired type to retrieve data from the storage node. Delivery of the data is implemented by geographic routing, such as GPSR [147]. To reduce bottleneck at the hash nodes and improve data survivability under node failure, a geographical hash table based scheme applies a structured replication with multiple mirrors scattered in the network. Structured replication reduces the cost of storage but increases the cost of queries.

Double-ruling [32, 74, 86, 175, 230, 258, 315] is another approach for in-network data-centric processing. A datum (or a pointer to the datum) is duplicated along a curve called replication curve, and a query travels along another curve called retrieval curve. Successful retrieval is guaranteed if the retrieval curve intersects the replication one. A simple double-ruling scheme on a planar grid is illustrated in Fig. 10.18a where nodes are located at lattice points. The replication and retrieval

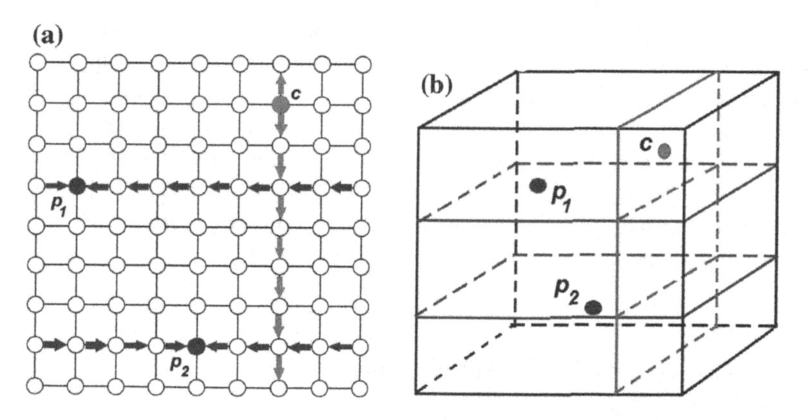

Fig. 10.18 A simple double-ruling scheme on **a** a 2D grid sensor network; **b** a 3D grid sensor network. p_1 and p_2 are two information producers and c is an information consumer. Image from [314]

curves follow the horizontal and vertical lines, respectively. By traveling along a vertical line, a data query, called information consumer, can always find the requested data generated by an information producer. Double-ruling based schemes support efficient data retrieval, since all data with different types generated in a network can be conveniently retrieved along one simple retrieval curve. This is in a sharp contrast to geographical hash table based schemes where an information consumer has to visit multiple nodes scattered in the network to collect data with different types hashed to various locations. Moreover, with modestly increased data replication, a double-ruling based scheme has well balanced load across the network, while nodes near the hashed location suffer much higher traffic load than others in a geographical hash table based scheme. A double-ruling based scheme also has better fault tolerance against geographically concentrated node failure by replicating data on nodes that are uncorrelated with node proximity.

Double-ruling based schemes achieve all the desired properties at the cost of more data duplication and a much stronger geometric constraint on the shape of a sensor network than geographical hash table based schemes. Previous double-ruling based schemes either assume networks with 2D grid shape [175, 258, 315] or with heavy data replication to achieve high probability that the retrieval curve would meet one of the replication curves within the sensor network field [32]. To extend double-ruling scheme to networks with uneven sensor distribution and irregular geometric shapes, landmark-based scheme [74] is proposed to partition the sensor field into tiles. GHT is adopted at the tile level, i.e., a data type is hashed to a tile instead of a single node. Inside each tile, a double-ruling scheme is applied to ensure the intersection of a retrieval path and a replication path. Later, a location-free double-ruling scheme is introduced in [86] based on boundary recognition and the computation of the respective gradient fields. To improve the flexibility of retrieval, a spherical projection-based double-ruling scheme is proposed in [230], where a planar network is mapped to a sphere based on the inverse of stereographic projection. Both the replication and retrieval curves are great circles such that a retrieval curve always intersects all other replication circles.

Although double-ruling has shown highly effective for distributed information storage and retrieval in 2D sensor networks, it cannot be efficiently applied in 3D networks. A naive double-ruling based scheme in 3D sensor networks is shown in Fig. 10.18b. In such a 3D grid-based cube-shape sensor network, data replication and retrieval are along the horizontal and vertical planes respectively, such that a retrieval plane intersects all replication planes. Besides an extremely high cost of data replication, such 3D grid-based double-ruling scheme requires a network with a regular cube shape and uniform node distribution.

10.5.1 Double-Ruling in 3D Network

Two examples given in Fig. 10.21 briefly illustrate the basic idea of the introduced 3D Double-ruling Approach. The first example is a 3D volumetric sensor network model with two handles (see Fig. 10.21a). We first detect the boundary nodes and then build a triangular structure of the identified boundary surface of the network (see Fig. 10.21a). Considering the boundary surface of a 3D volume is always a closed one, we can always cut a closed surface open to a topological disk along an appropriate set of edges called a cut graph. We compute a cut graph of the boundary surface of the network marked with yellow color as shown in Fig. 10.21b. We cut the boundary surface of the network open to a topological disk along the cut graph and then map it to an aligned planar rectangle such that each boundary node of the network is associated with a planar rectangle virtual coordinates (see Fig. 10.21c). Each non-boundary sensor stores the ID of its neighbor nearest to the boundary of the network. A data generator follows the sequence of IDs to the boundary of the network, and then travels along a horizontal line of the virtual planar rectangle and leaves data copies. The two horizontal lines of the virtual planar rectangle marked with blue color shown in Fig. 10.21c correspond to the two real data replication curves of the network shown in Fig. 10.21d marked with the same color. Similarly, a consumer follows the sequence of IDs to the boundary of the network and collects the aggregated data of different types along a vertical line. The vertical line of the virtual planar rectangle marked with red shown in Fig. 10.21c corresponds to the data aggregation curve of the network shown in Fig. 10.21d marked with the same color. The second example is also a 3D sensor network deployed under water (see Fig. 10.21e). We cut the boundary surface of the network open to a topological disk along a computed cut graph (see Fig. 10.21f) and then assign each boundary node of the network a planar rectangle virtual coordinates (see Fig. 10.21g). Similarly, Fig. 10.21h gives a data query example (Fig. 10.19).

Specifically, given a sensor network deployed in 3D volume with distance information within one-hop neighborhood only, we can apply the algorithm proposed in [335] to identify boundary nodes of the network, and then the algorithm proposed

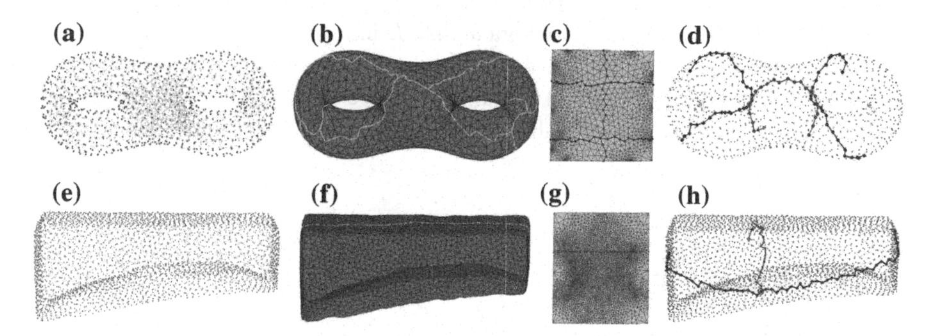

Fig. 10.19 **a**, **d** 3D sensor network models. **b**, **e** Cut graphs of triangulated boundary surfaces marked with yellow. **c**, **f** The boundary surface is cut open to a topological disk along the cut graph and mapped to an aligned planar rectangle. The horizontal lines marked with blue are two data replication curves with different types. The vertical line marked with red is one data retrieval curve. **g** A query of aggregated data corresponding to **c**. **h** A data query corresponding to **f**. Image from [314]

in [336] to extract a triangular structure of the boundary surface. Both algorithms are fully distributed with no constraint on communication models.

The extracted triangular surface, denotes as M, is connected, orientable, and closed. The algorithm to compute a cut graph of M starts from one randomly chosen triangle face $[v_i, v_j, v_k]$ of M or one with the smallest node id. $[v_i, v_j, v_k]$ marks itself and its three edges $[v_i, v_j]$, $[v_j, v_k]$, and $[v_k, v_i]$. Each of the marked edges checks whether it is shared by two marked triangles. For example, edge $[v_i, v_j]$ finds its neighboring triangle $[v_j, v_i, v_l]$ unmarked. $[v_i, v_j]$ then removes mark from itself but adds mark on triangle $[v_j, v_i, v_l]$ and edges $[v_i, v_l]$ and $[v_l, v_j]$. Note that it is possible that $[v_i, v_l]$ or $[v_l, v_j]$ may have been marked already. Such propagation stops when all the triangles of M have been marked.

If we cut M open along the marked edges to a topological disk surface denoted as D, the boundary of D would be extremely zigzagged. The size of the boundary edges of D in the worst case can equal the size of the triangles of M. The reason is that in the worst case the size of the marked edges is increased by one each time when one triangle of M is marked.

To control the size of the cut graph, we can trim away those marked, dangling tree edges. The trimming algorithm starts with each marked edge checking whether both of its two ending vertices connect to marked edges. If only one ending vertex connects to marked edges, the edge is identified as a dangling tree edge and can be unmarked - removed from the cut graph. The unmarked edge will then send messages to its neighboring marked edges through the other ending vertex. Its neighbors then conduct the same checking when receiving the message. The trimming process stops when there is no marked, dangling tree edges.

If M is genus zero, there is no marked edges left after trimming, which provides a convenient way to automatically identify the topology of a 3D sensor network. For a boundary surface of a network detected genus zero, a boundary node of the network with the smallest id conducts a simple flooding to mark the longest shortest path measured by hops to another boundary node along the boundary surface.

We now virtually cut M to a topological disk D along the marked edges and then conformally map D to an aligned planar rectangle. Such planar coordinates are stored at each boundary node as its virtual coordinates. Since a network is location free, we let each non-boundary node store the ID of its neighbor nearest to the boundary of the network.

10.5.2 Data Replication

A producer follows a sequence of nodes to the nearest boundary node of the network denoted as p. The boundary surface of the network has been mapped to a virtual planar rectangle, so each boundary node has a planar rectangle virtual coordinates. We assume a data replication curve is along a horizonal line of the virtual planar rectangle. The horizontal line through p is unique, solely determined by the y coordinate of the planar rectangle virtual coordinates of p. The producer travels and leaves pointers or copies of the data at nodes along the line with two directions - one with the increased and the other with the decreased x values. At each step, the producer simply checks the planar rectangle virtual coordinates of its one range neighbors and chooses the one with the closest distance to the line and along the current direction. Once finishing data replication, the producer turns back and follows the reversed path back.

10.5.3 Data Retrieval

Without awareness of the knowledge of the producer's location and the distance, a consumer follows a sequence of nodes to the nearest boundary node denoted as p. We assume a data retrieval curve is along a vertical line of the virtual planar rectangle. A vertical line passing through p is determined solely by the x coordinate of the planar rectangle virtual coordinates of p. The consumer simply travels along the line with two directions - one with the increased and the other with the decreased y values. At each step, similarly, the producer simply checks the planar rectangle virtual coordinates of its one range neighbors and chooses the one with the closest distance to the line and along the current direction. Note that the boundary surface of the network is only virtually cut open and mapped to a planar rectangle. Once a consumer hits the boundary side of the rectangle, the consumer can cross the boundary side and keep traveling along the line with the same direction. The consumer stops as soon as it hits the replication curve of its desired data. If there are multiple producers and different types of data, the consumer travels along a full vertical line to collect all the aggregated data in the network. Once data has been collected, the consumer turns back and follows the reversed path back.

10.5.4 Delivery of Data and Query

As preprocessing, each boundary node of the network sends messages recording its minimum hop count to boundary (initialized to zero) to its neighbors. A non-boundary node receives a message and compares with its current record (initialized to infinity). If the received count has more than one hop count less, the node updates its current one and records the ID of its neighbor sending this message. The node also updates the count of the message and then sends to its neighbors. Otherwise, the node simply discards the message. When there is no message in the network, each of the non-boundary nodes of the network has recorded the ID of its neighbor nearest to boundary. It is then straightforward for a producer or a consumer to travel along the shortest path to the boundary according to the sequences of IDs.

10.5.5 Storage

We have limited information stored at the nodes of the network. A non-boundary node only stores the ID of its neighbor nearest to boundary. A boundary node stores its computed planar rectangle virtual coordinates. For the data replication, we can leave copies of data on either all the nodes along the replication curve, or just a small portion of nodes sampled along the replication curve, which is a trade off between the storage cost and the retrieval cost.

10.6 Marching of Autonomous Networked Robots

The recent development in sensors, actuators, robots, and mobile and wireless communication technologies has enabled a paradigm shift in robotic systems, named autonomous networked robots (ANRs), where the individual robots coordinate among themselves to complete a task, e.g., to explore or monitor a Field of Interest (FoI). Such an ANR system is extremely valuable in situations where a traditional static sensor network fails or is inapplicable, for example, in disaster areas or toxic urban regions where sensor deployment cannot be performed manually, or hostile environment where sensors can be neither manually deployed nor air-dropped. On the contrary, ANRs can move to correct positions by themselves to provide the required coverage. Compared with the static sensor network that is deployed for a given FoI, an ANR system offers great flexibility to explore different fields as needed. Additionally, an ANR system is more reliable since the failure of an individual robot can be recovered by its peers.

Existing research focuses on improving the coverage performance of a group of ANRs from initially random-deployed positions as an end result of robots movement within a FoI. Virtual-force based algorithms [124, 217, 340] are among the earliest endeavors. The work in [190] enables an ANR system to arrange themselves to a regularly spaced square or rectangle lattice pattern by exploiting a common reference orientation. In [284], algorithms are proposed to discover the existence of coverage holes and then compute the desired target positions to move robots from densely deployed areas to sparsely deployed areas to increase the coverage. Later, in [44, 188], decentralized motion control algorithms are proposed to deploy an ANR system in the so-called triangular lattice pattern, namely a network of equilateral triangles within a given area that is proved optimal in terms of minimum number of sensors required for complete coverage of a bounded area. Centroidal Voronoi Diagram based algorithms [57, 169, 234] can also achieve the layout of an ANR system close to equilateral triangulation, at the same time, easily encoding optimal coverage and sensing policies into the utility function.

Given a group of ANRs with instructions to explore a number of FoIs instead of one. They move to the next FoI after they complete tasks at current one. The new FoI may be far away from the previous one and its shape can also vary dramatically.

Of course, a complete map can be loaded into the memory of each ANR, but the ANRs must be able to redeploy themselves to desired positions in the new FoI based on distributed algorithms. An efficient relocation algorithm will minimize the total moving distance to reduce energy consumption. However, it is more important that the ANRs preserve their local connectivities and organize themselves as a whole network without any isolated nodes during the transition to the new FoI. The global connectivity requirement is mandatory in order to make sure timely coordination among the ANRs. For instance, an unexpected event (such as the change of terrain or weather conditions) may happen during the relocation. As a result, the ANRs must cooperatively determine how to adapt to the event. If an ANR is isolated at this time, it may be excluded from the new plan and thus become permanently lost. The preservation of local connectivity is also highly preferred to reduce the overhead and avoid the delay for pairing the wireless devices. Two ANRs can communicate with each other only if they are paired and have established a secure link. The extensive change of local connectivity may result in significant overhead and delay for re-pairing the wireless links, thus degrading the system performance and even hindering the system function.

The problem is called *optimal marching of autonomous networked robots*. Assume a group of ANRs is initially deployed on a general 2D surface. They are required to redeploy to a new FoI, which is not necessarily close to the current one and may have complicated and concave boundary shapes with inner obstacles or holes. It is easy to prove that it is impossible to preserve local connectivities and at the same time minimize the total moving distance. It is also impossible to maintain all the local connectivities for general cases.

We introduce a harmonic map-based algorithm for a group of ANRs to find paths from current FoI to target one such that both local and global connectivities are maintained. Each ANR then follows a specified rule inside the target FoI to minorly adjust its optimal coverage position. The introduced algorithm guarantees global connectivity and preserves local connectivities as much as possible at negligible cost of the total moving distance.

10.6.1 Optimal Marching Problem

Assume an ANR system consists of a group of identical mobile robots. Each mobile robot has an equipped GPS and capacity to move in a straight line. The sensors mounted on mobile robots are assumed to have disk sensing model, identical sensing range and capability.

Let there be n mobile robots in a known region - the current FoI denoted by M_1. Denote p_i the position of the ith mobile robot in M_1, and denote $P = \{p_1, \ldots, p_n\}$ the positions of the group of mobile robots in M_1. Following an instruction, the group of mobile robots moves to the target FoI denoted by M_2. They automatically redeploy themselves to optimal coverage positions in M_2 denoted by $Q = \{q_1, \ldots, q_n\}$.

Denote d_i the moving distance of the ith mobile robot from p_i in M_1 to q_j in M_2. The total moving distance of the group of mobile robots redeployed from M_1 to M_2 is defined as: $D = \sum_{i=1}^{n} d_i$.

Assume the total transition time from M_1 to M_2 is T. Denote $e_{ij}(t)$ the communication link between the ith and jth mobile robots at time t. We assign a value to $e_{ij}(t)$ such that

$$e_{ij}(t) = \begin{cases} 1 & \text{if the } i\text{th and } j\text{th mobile robots } (i \neq j) \text{ are} \\ & \text{connected at time } t, 0 \leq t \leq T \\ 0 & \text{otherwise.} \end{cases}$$

Denote e_{ij} the communication link between the ith and jth mobile robots during the whole transition time from M_1 to M_2. Similarly, we can assign a value to e_{ij} such that

$$e_{ij} = \begin{cases} 1 & \text{if } e_{ij}(t) = 1, \forall t, 0 \leq t \leq T \\ 0 & \text{otherwise.} \end{cases}$$

Then we can define the total stable link ratio to evaluate the preservation of local connectivity of a transition.

Definition 10.13 *(Total Stable Link Ratio)* Let there be n mobile robots relocating from the current FoI M_1 to the target one M_2. Denote m_i the number of neighbors of the ith mobile robot within its communication range in M_1. The total stable link ratio of the group of mobile robots relocating from M_1 to M_2, denoted by L, is defined as:

$$L = \frac{\sum_{i=1}^{n} \sum_{j=1}^{m_i} e_{ij}}{\sum_{i=1}^{n} m_i}.$$

We now can define the global connectivity of a group of mobile robots during a transition between FoIs.

Definition 10.14 *(Global Connectivity)* Let there be n mobile robots relocating from the current FoI M_1 to the target one M_2. The global connectivity of the group of mobile robots during the transition time T,denoted by C, is defined as:

$$C = \begin{cases} 1 & \text{given any mobile robot, there exists a path} \\ & \text{to network boundary for } \forall t, 0 \leq t \leq T \\ 0 & \text{otherwise} \end{cases}$$

If we consider only the minimization of the total moving distance D during a relocation, we can show that the minimum moving distance marching problem can be converted to the well-known minimum cost bipartite matching problem and solved. Before we continue, we will first introduce some concepts of graph theory.

Definition 10.15 *(Matching)* Given a graph $G = (V, E)$ with a set V of vertices and a set E of edges, a matching $M \in E$ is a collection of edges such that every vertex of V is incident to at most one edge of M. If a vertex v has no edge of M incident to it then v is said to be exposed. A matching is *perfect* if no vertex is exposed.

Definition 10.16 *(Bipartite Graph)* A graph $G = (V, E)$ is bipartite if the vertex set V can be partitioned into two sets V_1 and V_2 such that no edges in E has both endpoints in the same set of the bipartition. A bipartite graph G is balanced if $|V_1| = |V_2| = n$. The bipartite graph G is complete when there are all possible edges between V_1 and V_2.

Definition 10.17 *(Minimum Cost Bipartite Matching Problem)* Given a balanced and complete bipartite graph and a cost c_{ij} for all $e_{(}i, j) \in E$, find a perfect matching with minimum cost where the cost of a matching M is given by $c(M) = \sum_{e(i,j)\in E} c_{ij}$.

Consider the positions of the group of mobile robots $P = \{p_1, \ldots, p_n\}$ in M_1 as the vertex set V_1, and the positions $Q = \{q_1, \ldots, q_n\}$ in M_2 as the vertex set V_2. We can construct a balanced and complete bipartite graph G. The cost associated with each edge is the Euclidean distance between the two incident vertices. The minimum moving distance problem equals to the minimum cost bipartite matching problem of G.

Lemma 10.18 shows the contradiction of minimizing the total moving distance D and maximizing the total stable link ratio L.

Lemma 10.18 *During the process of sensors moving from M_1 to M_2, maximizing the stable link ratio L and minimizing the total moving distance D cannot be achieved at the same time.*

Figure 10.20a gives one example to show the contradiction of maximizing L and minimizing D. Suppose M_1 is a slim rectangle shaped FoI along the x-axis. An optimal deployment of seven mobile robots in M_1 is to form a network of a triangular lattice pattern, namely equilateral triangles [150] as the one shown in the left in Fig. 10.20a. Furthermore, denote r_c the communication range, and r_s the sensing range. If $r_c \geq \sqrt{3}r_s$, every point in the region can be covered by at least one sensor, and every sensor is connected to six neighboring sensors [15]. Suppose the FoI M_2 is also slim rectangle shaped, but along the y-axis. The optimal deployment of the same group of mobile robots in M_2 is the one shown in the right in Fig. 10.20a. To maximize the total stable link ratio, we would choose a moving path with $\{A \to a, B \to b, C \to c, D \to d, E \to e, F \to f, G \to g\}$. However, to minimize the total moving distance, we would choose a totally different moving path with $\{A \to a, B \to g, C \to b, D \to e, E \to c, F \to f, G \to d\}$.

Furthermore, Lemma 10.19 shows that we cannot preserve all the local links during a relocation in general cases.

Lemma 10.19 *The local connectivity cannot be fully preserved during relocation in general cases.*

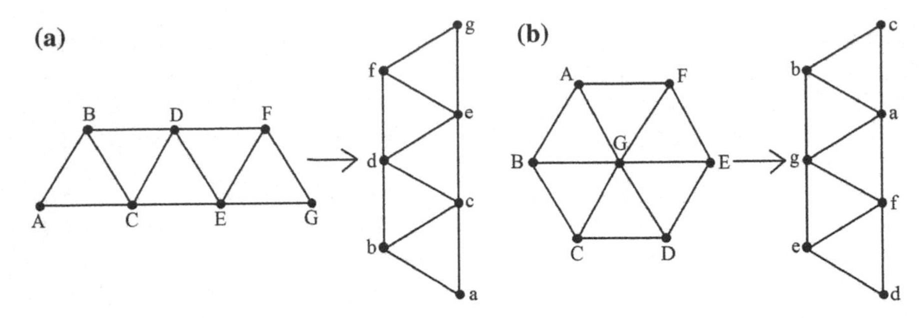

Fig. 10.20 Two examples of a group of mobile robots redeployed from one FoI to the other. Image from [19]

One example is given in Fig. 10.20b. Suppose there are seven mobile robots deployed in M_1, a round shaped FoI. The optimal deployment for the seven mobile robots is one in the central, and the other six circled around to form a network of equilateral triangles as the one shown in the left in Fig. 10.20b. Suppose the FoI M_2 is slim rectangle shaped. The optimal deployment of the group of mobile robots in M_2 is the one shown in the right in Fig. 10.20b. The centered mobile robot and two others circled around in M_1 have to break at least two communication links individually when they redeploy themselves to M_2.

Considering both Lemmas 10.18 and 10.19, we formulate the optimal marching problem as the following one:

Definition 10.20 *(Optimal Marching Problem)* A group of ANRs are instructed to explore a number of FoIs sequentially. These FoIs are not necessarily close and may have complicated and concave boundary shapes with inner obstacles or landscape features that forbid mobile robot placement. The optimal marching problem is to maximize the total stable link ratio L subject to the requirement of the global connectivity $C = 1$ during the transition procedure between FoIs.

10.6.2 An Approximated Solution

We take assumption that the communication range r_c and the sensing range r_s of a mobile robot satisfy $r_c \geq \sqrt{3}r_s$ and the size of a FoI is bounded such that a system of ANRs can achieve full area coverage. We use a graph to model the connectivity of an ANR system. A vertex represents a mobile robot. An edge represents a communication link between two neighboring mobile robots. With position information of each mobile robot, we can easily extract a triangulation from such connectivity graph. Denote G the connectivity graph of a group of mobile robots deployed in M_1, and T the triangulation extracted from G. To redeploy the group of mobile robots from M_1 to M_2 with local connectivities, i.e., edges in T well preserved, basically we want to find a least stretched diffeomorphism of T mapped to M_2. A diffeomorphism is a one-to-one mapping keeping all local neighborhood relationship unchanged. A least stretched diffeomorphism preserves not just local neighborhood relationship but also edge lengths, which are crucial to keep local communication links unbroken during the transition.

Figure 10.21 illustrates the major steps of an approximated solution of redeployment of a group of mobile robots to target FoI M_2 after finishing tasks in current FoI M_1. We first extract a triangulation denoted by T as shown in Fig. 10.21b from the connectivity graph of the group of mobile robots deployed in M_1 as shown in Fig. 10.21a. A vertex represents a mobile robot, and an edge represents a communication link between two neighboring mobile robots. Figure 10.21c shows the computed harmonic map of T to a unit disk. Figure 10.21d shows the surface data of FoI M_2 with a flower-shaped pond inside. Similarly, we can easily grid and triangulate the surface data of M_2, and then harmonic map it to a unit disk. Rotate either of the unit disks with an angle, the two overlapped disks induce a unique harmonic map between T and M_2. We find one optimal rotation angle such that the induced harmonic map gives the maximized total stable link ratio. The group of mobile robots then follow the moving path and redeploy themselves to M_2 as shown in Fig. 10.21e. After a minor local adjustment, each mobile robot moves to the computed optimal coverage position as shown in Fig. 10.21f.

10.6.3 Modified Harmonic Map

We first apply the algorithm [336] to extract a triangulation denoted by T from the connectivity graph of the group of mobile robots deployed in M_1. With position information available at each mobile robot, each edge computes a weight that measures the number of triangles shared by the edge and the local neighbor sets of the edge. An iterative algorithm keeps removing edges based on current edge weight and local neighbor set information. The algorithm is fully distributed with computational complexity linear to the size of the edges.

We then compute the harmonic map of T to a unit disk. Since a boundary edge incidents with only one triangle, we can easily identify the boundary edges of T. A boundary vertex with the smallest ID (a unique ID assigned to each mobile robot) initiates a message with a counter that records how many hops the message has traveled along the boundary. The starting vertex sends the message to one of its neighboring boundary vertices. The receiver updates the counter and records the number, and then forwards the message to its next neighboring boundary vertex. The message will come back to the starting vertex as the boundary vertices form a closed loop. The starting vertex notifies other boundary vertices the size of the boundary. Based on the recorded hop number and the size of the boundary vertices, each boundary vertex then computes a position along the boundary of a unit disk such that the boundary vertices are uniformly and sequentially distributed along the boundary. Inner vertices, i.e., non-boundary vertices, initiate their positions at the center of the unit disk. Then at each step, an inner vertex computes its position as

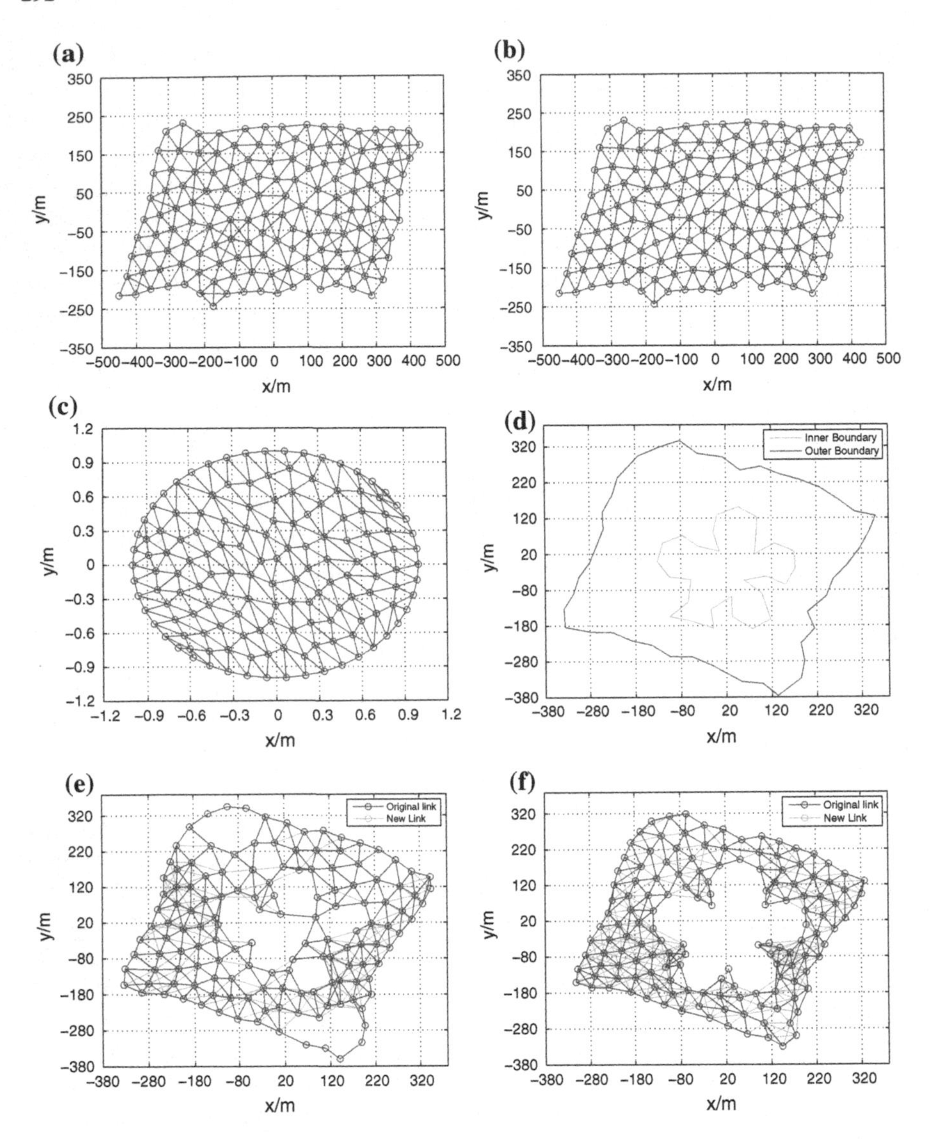

Fig. 10.21 Algorithm pipeline: **a** The connectivity graph of a group of mobile robots deployed in FoI denoted by M_1. **b** A triangulation denoted by T extracted from the connectivity graph. **c** The computed harmonic map of T to a unit disk. **d** The surface data of FoI denoted by M_2 with a flower-shaped pond inside. **e** The group of mobile robots follow the moving path of the harmonic map and redeploy themselves to M_2. **f** After a minor adjustment, each mobile robot moves to the optimal coverage position. Note that blue color marked edges represent local communication links preserved during the transition from M_1 to M_2. Red color marked edges, on the contrary, are new communication links. Image from [19]

the average of the positions of its neighboring vertices. Note that only inner vertices update their positions. We map T to the unit disk when no inner vertex updates its position and each vertex has a unique position in the unit disk. Similarly, we can add grid points and triangulate the surface data of FoI M_2, and then harmonic map it to a unit disk. The computation of the harmonic map of M_2 to a unit disk can be done by each mobile robot individually.

When T and M_2 are both mapped to unit disks, we can rotate either of the disks with any angle and the two overlapped disks induces a unique harmonic map between T and M_2. We want to find the optimal rotation angle such that the induced harmonic map gives the maximized total stable link ratio. However, it is a non-linear problem. To avoid complicated computation, each mobile robot applies a simple binary search method to find the desired rotation angle with a pre-defined search depth. At each step, a mobile robot divides current search interval of angle into two and rotates its mapped position in unit disk with the midpoint angle of the interval. The mobile robot computes its mapped position in M_2 and exchanges the position with its one-range neighbors. After calculating its own stable link ratio, the mobile robot then floods the information to other mobile robots.

After we rotate one mapped disk with the computed optimal rotation angle, the two overlapped unit disks naturally induce a harmonic map between T and M_2. Specifically, a vertex of T, denoted by v, locates three nearest grid points of M_2 in the overlapped unit disks. Denote g_i, g_j, and g_k the three nearest grid points, and $q(g_i)$, $q(g_j)$, and $q(g_k)$ the original geographic coordinates of g_i, g_j, and g_k in M_2. Denote (t_1, t_2, t_3) the Barycentric Coordinates of v in the triangle formed by g_i, g_j, and g_k when mapped to unit disk. The linear combination

$$q(v) = t_1 q(g_i) + t_2 q(g_j) + t_3 q(g_k), \tag{10.14}$$

gives the geographic coordinates where the mobile robot, represented by v, should deploy itself in M_2. Denote $p(v)$ the geographic coordinates of v in M_1, T the transition time of the group of mobile robots from M_1 to M_2, and t a time parameter. A linear combination of $p(v)$ and $q(v)$ with t:

$$\frac{T-t}{T} p(v) + \frac{t}{T} q(v), t \in [0, T] \tag{10.15}$$

gives the mobile robot both the moving path and speed to M_2.

10.6.4 Minor Local Adjustment

After the group of mobile robots redeploy themselves to M_2, they only need a minor local adjustment to optimal coverage positions. We adopt *centroidal Voronoi diagram* based algorithms [57, 136, 169, 234] to compute optimal coverage position. Informally speaking, a Voronoi diagram is a partition of the space according to the distances to a discrete set of objects, denoted by sites such that a Voronoi region of a site is the region of points that are closer to the site than to any other sites. A centroidal Voronoi diagram is a Voronoi diagram such that each site is located at exactly the mass centroid of its corresponding Voronoi region with respect to a given density function. As proved in [201], centroidal Voronoi diagram always has congruent regular hexagons as its Voronoi regions in $R2$ space, which induces a layout of sites forming equilateral triangulation.

Considering the group of mobile robots as sites, and the FoI M_2 as the partition space. We apply Lloyd algorithm [177] to compute the centroidal Voronoi diagram. Lloyd algorithm is an iterative method. At each step, a mobile robot collects the position information of its two-range neighbors, computing its corresponding Voronoi region and the centroid of the Voronoi region. The mobile robot then moves to the centroid position. Since mobile robots have already been very close to the optimal coverage positions after redeploying to M_2, Lloyd algorithm only needs a few steps to converge when no mobile robot needs to update its position.

10.6.5 Global Connectivity

If the shapes of the two FoIs M_1 and M_2 differ too much, some edges of T will be largely stretched when T is mapped to M_2 even though harmonic map is already a least stretched diffeomorphism. Such a largely stretched edge means a broken communication link between the two mobile robots represented by the two ending vertices of the edge. It is possible that all communication links of a vertex will be broken and the vertex will be isolated from the network.

To guarantee global connectivity during the transition procedure, we need a few modifications of the proposed modified harmonic map algorithm in Sect. 10.6.3. Right after computing the harmonic map of T to M_2, a straightforward solution is that each mobile robot exchanges its mapped position with its one-range neighbors and checks whether all of its communication links will be broken. For an isolated vertex, it chooses the closest one-range neighbor as reference and adjusts its moving path to M_2 parallel to the path of the neighbor and with the same speed.

However, in extreme case, a subgroup of mobile robots instead of a single one will get disconnected from the network. Considering that boundary vertices of T are mapped to the boundary of M_2 and form a closed loop, it is easy to check and require that the boundary vertices of T are still connected when mapped to M_2. A boundary vertex of T compares the mapped positions of its one-range neighbors with itself and initiates a packet with a counter set to zero to its one-range neighbors with communication links still preserved in M_2. If there exists at least a communication path connecting a vertex and a boundary vertex when mapped to M_2, the vertex will receive such a packet. When a vertex receives a packet from a boundary vertex that is further away from its current nearest boundary vertex, it stops forwarding this packet. Otherwise, the vertex updates the counter and record the number. Similarly, the vertex forwards the package to its neighbors with communication links still preserved in M_2. Assume the boundary vertices initiate their packages at approximately the same time, and each packet travels at approximately the same speed, the flooding of such packets in the network will then stop quickly depending on the diameter of the network. As a result, we can identify the isolated subgroups vertices.

For each isolated subgroup vertices, we choose a vertex with one of its one-range neighbors not just connecting but also nearest to a boundary vertex and set the vertex as root of the isolated subgroup vertices. The root vertex will then choose the neighbor as reference and adjusts its moving path to M_2 parallel to the path of the neighbor and with the same speed. The root will broadcast its moving path and speed to the subgroup vertices. Each vertex inside the subgroup will then adjust its moving path parallel to the path of the root one and with the same speed.

After the group of mobile robots redeploy themselves to M_2, we also need to guarantee global connectivity at each step of Lloyd algorithm when a mobile robot moves to the updated centroid position. At each step, a mobile robot collects the computed centroid positions of its one-range neighbors and compares with its own. If no mobile robot will disconnect from the network, every robot simply moves to its centroid position; otherwise, each robot checks whether it is safe to move to half of the distance to the centroid position and so on.

10.6.6 Moving Distance

We can modify the introduced algorithm slightly to find transition paths with less total moving distance with the cost of a little bit lower of the total stable link ratio. When T and M_2 are both mapped to unit disks, we want to find an rotation angle such that the induced harmonic map gives the moving path of a group of mobile robots with the minimized total moving distance. Similarly, each mobile robot divides current search interval of angle into two and rotates its mapped position in unit disk with the midpoint angle of the interval. The mobile robot computes its mapped position in M_2 and the moving distance to the mapped position, and then floods the distance information to other mobile robots. We can also set the search depth to 4.

10.6.7 Holes

A FoI can have complex shape. It may also contain obstacles or landscape features that forbid mobile robot placement. For simplicity, we call them holes of a FoI.

Harmonic map assumes not just convex boundary condition of the target FoI, but also topological disk shape for both current and target FoIs. For a FoI with holes, we cannot directly compute its harmonic map to a unit disk. A solution is to add a virtual vertex for each hole and fill all holes with virtual triangulations. Specifically, we apply the rule that a boundary edge incidents with only one triangle to detect boundary vertices along an inner hole. The position of an virtual vertex assigned to the hole is computed as average of the positions of boundary vertices along the hole. Each boundary vertex along the inner hole stores the position of the virtual vertex and adds one virtual edge connecting to it. At each step of harmonic map, boundary vertices along the inner hole exchange information to compute updated position of the virtual vertex.

With all holes filled with virtual triangulation, we can construct the harmonic map from T to M_2 as introduced in Sect. 10.6.3. Note that it is possible that a mobile robot in T is mapped to a hole in M_2. The robot can simply choose the nearest grid point in M_2 as the mapped position. It is also possible that the moving path of a mobile robot computed by Eq. 10.15 passes through a hole. When the mobile robot hits the boundary of the hole, the robot goes along the boundary until it can follow its computed moving path again.

A FoI with holes can also affect the computation of the centroidal Voronoi diagram. At each step, if the computed centroid falls into a hole, we choose the nearest grid point along the hole boundary as the centroid. It is also possible that a mobile robot hits a hole when moving to the computed centroid. Similarly, the robot goes along the boundary of the hole until it can follow the straight line to the computed centroid.

10.7 Summary and Further Reading

The fast growing applications of WSNs continuously generate new challenges. Computational conformal geometry provides models and efficient tools to tackle challenges in WSNs. Computational conformal geometry has been applied in other research topics in WSNs design as well, such as resilient routing [326], load-balanced routing [95], and greedy routing in 2D WSNs [228]. We refer readers to [89] for a summary of computational conformal geometry applied in geometric routing field.

References

1. L.V. Ahlfors, *Complex Analysis* (McGraw-Hill Education, Maidenheach, 1979)
2. M. Albano, S. Chessa, F. Nidito, S. Pelagatti, Data-centric storage in non-uniform sensor networks, in *Prooceedings of the 2nd International Workshop on Distributed Cooperative Laboratories: Instrumenting the Grid (INGRID 2007)* (2007), pp. 407–413
3. A.D. Alexandrov, *Convex Polyhedra* (Springer, Berlin, 2005)
4. P. Alliez, M. Meyer, M. Desbrun, Interactive geomety remeshing, in *SIGGRAPH* (2002), pp. 347–354
5. P. Alliez, É. Colin de Verdière, O. Devillers, M. Isenburg, Centroidal Voronoi diagrams for isotropic surface remeshing. Gr. Models **67**(3), 204–231 (2005)
6. P. Angelini, F. Frati, L. Grilli, An algorithm to construct greedy drawings of triangulations, in *Proceedings of the 16th international symposium on graph drawing* (2008), pp. 26–37
7. S. Angenent, S. Haker, A. Tannenbaum, R. Kikinis, Conformal geometry and brain flattening, in *Proceedings of the Medical Image Computing and Computer-Assisted Intervention* (1999), pp. 271–278
8. S. Angenent, S. Haker, A. Tannenbaum, R. Kikinis, On the Laplace-Beltrami operator and brain surface flattening. IEEE Trans. Med. Imaging **18**(8), 700–711 (1999)
9. S. Angenent, S. Haker, A. Tannenbaum, R. Kikinis, Conformal geometry and brain flattening, in *MICCAI* (1999), pp. 271–278
10. F. Araújo, J. Kaiser, C. Mitidieri, L. Rodrigues, C. Liu, CHR: a distributed hash table for wireless ad hoc networks. International Conference on Distributed Computing Systems Workshops **4**, 407–413 (2005)
11. B.D. Argall, Z.S. Saad, M.S. Beauchamp, Simplified intersubject averaging on the cortical surface using SUMA. Hum. Brain Mapp. **27**(1), 14–27 (2006)
12. R. Aris, *Vectors, Tensors and the Basic Equations of Fluid Mechanics* (Dover, New York, 1989)
13. V. Arsigny, P. Fillard, X. Pennec, N. Ayache, Log-Euclidean metrics for fast and simple calculus on diffusion tensors. Magn. Reson. Med. **56**(2), 411–421 (2006)
14. F. Aurenhammer, Power diagrams: properties, algorithms and applications. SIAM J. Comput. **16**(1), 78–96 (1987)
15. X. Bai, S. Kumar, D. Xuan, Z. Yun, T. Lai, Deploying wireless sensors to achieve both coverage and connectivity, in *Proceedings of the 7th ACM international symposium on mobile ad hoc networking and computing* (2006), pp. 131 – 142
16. X. Bai, Z. Yun, D. Xuan, B. Chen, W. Zhao, Optimal multiple-coverage of sensor networks, in *Proceedings of the INFOCOM* (2011), pp. 2498–2506

M. Jin et al., *Conformal Geometry*, https://doi.org/10.1007/978-3-319-75332-4

17. M. Bakircioglu, S. Joshi, M.I. Miller, Landmark matching on brain surfaces via large deformation diffeomorphisms on the sphere. Proceedings of the SPIE Medical Imaging **3661**, 710–715 (1999)
18. M. Balasubramanian, J.R. Polimeni, E.L. Schwartz, Exact geodesics and shortest paths on polyhedral surfaces. IEEE Trans. Pattern Anal. Mach. Intell. **31**(6), 1006–1016 (2009)
19. B. Ban, M. Jin, H. Wu, Optimal marching of autonomous networked robots, in *Proceedings of the 36th International Conference on Distributed Computing Systems*, ICDCS'16 (2016), pp. 149–158
20. A.F. Beardon, *A Primer on Riemann Surfaces* (Cambridge University Press, Cambridge, 1984)
21. S. Belongie, J. Malik, J. Puzicha, Shape matching and object recognition using shape contexts. IEEE Trans. Pattern Anal. Mach. Intell. **24**, 509–522 (2002)
22. Y. Benjamini, Y. Hochberg, Controlling the false discovery rate: a practical and powerful approach to multiple testing. J. R. Stat. Soc. Ser. B (Methodol.) **57**, 289–300 (1995)
23. J.P. Benzécri, Variétés localement affines. *Sem. Topologie et Géom. Diff., Ch. Ehresmann (1958-1960)*, (7) (1959)
24. P.J. Besl, N.D. Mckay, A Method for Registration of 3D Shapes. IEEE Trans. Pattern Anal. Mach. Intell. **14**(2) (1992)
25. F. Bian, R. Govindan, S. Shenker, X. Li, Using hierarchical location names for scalable routing and rendezvous in wireless sensor networks, in *Proceedings of the 2nd international conference on Embedded networked sensor systems* (2004) pp. 305–306
26. P. Biswas, Y.Ye, Semidefinite programming for ad hoc wireless sensor network localization, in *Proceedings of IPSN* (2004), pp. 46–54
27. A.I. Bobenko, I. Izmestiev, Alexandrov's theorem, weighted delaunay triangulations, and mixed volumes (2007), arXiv:math0609447v1
28. N. Bonnotte, From Knothe's rearrangement to Brenier's optimal transport map (2012), pp. 1–29, arXiv:1205.1099 [math.OC]
29. P. Bose, P. Morin, I. Stojmenovic, J. Urrutia, Routing with guaranteed delivery in ad hoc wireless networks, in *Proceedings of Third Workshop Discrete Algorithms and Methods for Mobile Computing and Communications* (1999), pp. 48–55
30. M. Botsch, L. Kobbelt, M. Pauly, P. Alliez, B. Levy, in *Polygon Mesh Processing*, ed. by A.K. Peters (CRC Press, Boca Raton, 2010)
31. H. Braak, I. Alafuzoff, T. Arzberger, H. Kretzschmar, K. Del Tredici, Staging of Alzheimer disease-associated neurofibrillary pathology using paraffin sections and immunocytochemistry. Acta Neuropathol. **112**(4), 389–404 (2006)
32. D. Braginsky, D. Estrin, Rumor routing algorthim for sensor networks, in *Proceedings of the 1st ACM International Workshop on Wireless Sensor Networks and Applications* (2002), pp. 22–31
33. Ch. Brechbühler, G. Gerig, O. Kübler, Parametrization of closed surfaces for 3-D shape description. Comput. Vis. Image Underst. **61**(2), 154–170 (1995)
34. Y. Brenier, Polar factorization and monotone rearrangement of vector-valued functions. Commun. Pure Appl. Math. **64**, 375–417 (1991)
35. A.M. Bronstein, M.M. Bronstein, R. Kimmel, Generalized multidimensional scaling: a framework for isometry-invariant partial surface matching. Proc. Natl. Acad. Sci. U.S.A. **103**(5), 1168–1172 (2006)
36. R.L. Buckner, J.R. Andrews-Hanna, D.L. Schacter, The brain's default network: anatomy, function, and relevance to disease. Ann. N. Y. Acad. Sci. **1124**, 1–38 (2008)
37. F. Bullo, J. Corts, Adaptive and distributed coordination algorithms for mobile sensing networks, vol. 309, Lecture Notes in Control and Information Sciences (Springer, Berlin, 2005), pp. 43–62
38. P. Buser, *Geometry and Spectra of Compact Riemann Surfaces (Progress in Mathematics)* (Birkhauser, Switzerland, 1992)
39. Z. Butler, D. Rus, Controlling mobile sensors for monitoring events with coverage constraints, in *Proceedings of IEEE International Conference of Robotics and Automation* (2004), pp. 1563–1573

40. O.T. Carmichael, L.H. Kuller, O.L. Lopez, P.M. Thompson, A. Lu, S.E. Lee, J.Y. Lee, H.J. Aizenstein, C.C. Meltzer, Y. Liu, A.W. Toga, J.T. Becker, Ventricular volume and dementia progression in the cardiovascular health study. Neurobiol. Aging **28**(1), 389–397 (2007)
41. C. Carner, M. Jin, X. Gu, H. Qin, Topology-driven surface mappings with robust feature alignment, in *IEEE Visualization* (2005), pp. 543–550
42. J.M. Chambers, W.S. Cleveland, B. Kleiner, P.A. Tukey, *Graphical Methods for Data Analysis* (Wadsworth, Belmont, 1983)
43. K. Chen, N. Ayutyanont, J.B. Langbaum, A.S. Fleisher, C. Reschke, W. Lee, X. Liu, D. Bandy, G.E. Alexander, P.M. Thompson, L. Shaw, J.Q. Trojanowski, C.R. Jack, S.M. Landau, N.L. Foster, D.J. Harvey, M.W. Weiner, R.A. Koeppe, W.J. Jagust, E.M. Reiman, Characterizing Alzheimer's disease using a hypometabolic convergence index. Neuroimage **56**(1), 52–60 (2011)
44. T.M. Cheng, A.V. Savkin, Decentralized control of mobile sensor networks for asymptotically optimal blanket coverage between two boundaries. IEEE Trans. Ind. Inf. **9**, 365–376 (2012)
45. M.C. Chiang, A.D. Leow, A.D. Klunder, R.A. Dutton, M. Barysheva, S.E. Rose, K.L. McMahon, G.I. de Zubicaray, A.W. Toga, P.M. Thompson, Fluid registration of diffusion tensor images using information theory. IEEE Trans. Med. Imaging **27**(4), 442–456 (2008)
46. Y. Chou, N. Leporé, M. Chiang, C. Avedissian, M. Barysheva, K.L. McMahon, G.I. de Zubicaray, M. Meredith, M.J. Wright, A.W. Toga, P.M. Thompson, Mapping genetic influences on ventricular structure in twins. NeuroImage **44**(4), 1312–1323 (2009)
47. Y.Y. Chou, N. Lepore, P. Saharan, S.K. Madsen, X. Hua, C.R. Jack, L.M. Shaw, J.Q. Trojanowski, M.W. Weiner, A.W. Toga, P.M. Thompson, Ventricular maps in 804 ADNI subjects: correlations with CSF biomarkers and clinical decline. Neurobiol. Aging **31**(8), 1386–1400 (2010)
48. B. Chow, P. Lu, L. Ni, *Hamilton's Ricci Flow* (American Mathematical Society, Providence, 2006)
49. B. Chow, F. Luo, Combinatorial ricci flows on surfaces. J. Differ. Geom. 97–129 (2003)
50. G.E. Christensen, H.J. Johnson, Consistent image registration. IEEE Trans. Med. Imaging **20**(7), 568–582 (2001)
51. G.E. Christensen, R.D. Rabbitt, M.I. Miller, Deformable templates using large deformation kinematics. IEEE Trans. Image Process. **5**(10), 1435–1447 (1996)
52. M.K. Chung, K.M. Dalton, L. Shen, A.C. Evans, R.J. Davidson, Weighted fourier series representation and its application to quantifying the amount of gray matter. IEEE Trans. Med. Imaging **26**(4), 566–581 (2007)
53. M.K. Chung, S.M. Robbins, K.M. Dalton, R.J. Davidson, A.L. Alexander, A.C. Evans, Cortical thickness analysis in autism with heat kernel smoothing. NeuroImage **25**(4), 1256–1265 (2005)
54. E. Colin de Verdière, J. Erickson, Tightening non-simple paths and cycles on surfaces, in *Proceedings of the seventeenth annual ACM-SIAM symposium on discrete algorithm* (2006), pp. 192–201
55. E. Colin de Verdière, F. Lazarus, Optimal system of loops on an orientable surface, in *Proceedings of the 43rd symposium on foundations of computer science* (2002), pp. 627–636
56. E. Colin de Verdière, F. Lazarus, Optimal pants decompositions and shortest homotopic cycles on an orientable surface. J. ACM **54**(4), 1–1 (2007)
57. J. Cortes, S. Martinez, T. Karatas, F. Bullo, Coverage control for mobile sensing networks. IEEE Trans. Robot. Autom. **20**(2), 243–255 (2004)
58. J. Cotrina, N. Pla, Modeling surfaces from meshes of arbitrary topology. Comput. Aided Geom. Des. **17**(7), 643–671 (2000)
59. J. Cotrina, N. Pla, M. Vigo, A generic approach to free form surface generation, in *Proceedings of the seventh ACM symposium on solid modeling and applications* (2002), pp. 35–44
60. R. Cuingnet, E. Gerardin, J. Tessieras, G. Auzias, S. Lehericy, M.O. Habert, M. Chupin, H. Benali, O. Colliot, Automatic classification of patients with Alzheimer's disease from structural MRI: a comparison of ten methods using the ADNI database. Neuroimage **56**(2), 766–781 (2011)

61. A. Cvetkovski, M. Crovella, Hyperbolic embedding and routing for dynamic graphs, in *Proceedings of INFOCOM* (2009), pp. 1647–1655
62. E. D'Agostino, F. Maes, D. Vandermeulen, P. Suetens, A viscous fluid model for multimodal non-rigid image registration using mutual information. Med. Image Anal. **7**(4), 565–75 (2003)
63. A.M. Dale, B. Fischl, M.I. Sereno, Cortical surface-based analysis I: segmentation and surface reconstruction. Neuroimage **9**, 179–194 (1999)
64. Y.K. Demjanovich, Finite-element approximation on manifolds, in *Proceedings of the International Conference on the Optimization of the Finite Element Approximations (St. Petersburg, 1995)* **8**(9), 25–30 (1996)
65. R.S. Desikan, F. Segonne, B. Fischl, B.T. Quinn, B.C. Dickerson, D. Blacker, R.L. Buckner, A.M. Dale, R.P. Maguire, B.T. Hyman, M.S. Albert, R.J. Killiany, An automated labeling system for subdividing the human cerebral cortex on MRI scans into gyral based regions of interest. Neuroimage **31**(3), 968–980 (2006)
66. T.K. Dey, K. Li, J. Sun, D. Cohen-Steiner, Computing geometry-aware handle and tunnel loops in 3D models. ACM Trans. Gr. (TOG) **27**(3), 1–9 (2008)
67. P. Dierckx, On calculating normalized powell-sabin b-splines. Comput. Aided Geom. Des. **15**(1), 61–78 (1997)
68. I. Dryden, K. Mardia, *Statistical Shape Analysis* (Wiley, New Jersey, 1998)
69. Q. Du, V. Faber, M. Gunzburger, Centroidal Voronoi tessellations: applications and algorithms. SIAM Rev. **41**(4), 637–676 (1999)
70. Q. Du, M. Gunzburger, L. Ju, Advances in studies and applications of centroidal Voronoi tessellations. Numer. Math. Theory Methods Appl. **3**(2), 119–142 (2010)
71. Q. Du, M.D. Gunzburger, L. Ju, Constrained centroidal Voronoi tessellations for surfaces. SIAM J. Sci. Comput. **24**(5), 1488–1506 (2003)
72. Q. Du, D. Wang, Recent progress in robust and quality Delaunay mesh generation. J. Comput. Appl. Math. **195**(1), 8–23 (2006)
73. A. Ericsson, K. Astrom, An affine invariant deformable shape representation for general curves, in *Proceeding in IEEE International Conference on Computer Vision* (2003), pp. 1142–1149
74. Q. Fang, J. Gao, L.J. Guibas, Landmark-based information storage and retrieval in sensor networks, in *IEEE INFOCOM* (2006), pp. 1–12
75. G. Fejes, Tóth, A stability criterion to the moment theorem. Studia Scientiarum Mathematicarum Hungarica **38**(1–4), 209–224 (2001)
76. L. Ferrarini, W.M. Palm, H. Olofsen, M.A. van Buchem, J.H.C. Reiber, F. Admiraal-Behloul, Shape differences of the brain ventricles in Alzheimer's disease. NeuroImage **32**(3), 1060–1069 (2006)
77. B. Fischl, N. Rajendran, E. Busa, J. Augustinack, O. Hinds, B.T. Yeo, H. Mohlberg, K. Amunts, K. Zilles, Cortical folding patterns and predicting cytoarchitecture. Cereb. Cortex **18**(8), 1973–1980 (2008)
78. B. Fischl, M.I. Sereno, A.M. Dale, Cortical surface-based analysis. II: inflation, flattening, and a surface-based coordinate system. Neuroimage **9**(2), 195–207 (1999)
79. B. Fischl, M.I. Sereno, R.B. Tootell, A.M. Dale, High-resolution intersubject averaging and a coordinate system for the cortical surface. Hum. Brain Mapp. **8**(4), 272–284 (1999)
80. R. Flury, S. Pemmaraju, R. Wattenhofer, Greedy routing with bounded stretch, in *Proceedings of INFOCOM* (2009), pp. 1737–1745
81. R. Flury, R. Wattenhofer, Randomized 3D geographic routing, in *Proceedingsof INFOCOM* (2008), pp. 834–842
82. H. Frey, I. Stojmenovic, On delivery guarantees of face and combined greedy-face routing in ad hoc and sensor networks, in *Proceedings of MobiCom* (2006), pp. 390–401
83. G.B. Frisoni, N.C. Fox, C.R. Jack, P. Scheltens, P.M. Thompson, The clinical use of structural MRI in Alzheimer disease. Nat. Rev. Neurol. **6**(2), 67–77 (2010)
84. K.J. Friston, A.P. Holmes, K.J. Worsley, J.-P. Poline, C.D. Frith, R.S.J. Frackowiak, Statistical parametric maps in functional imaging: a general linear approach. Hum. Brain Mapp. **2**(4), 189–210 (1994)

85. S. Funke, N. Milosavljevic, Guaranteed-delivery geographic routing under uncertain node locations, in *Proceedings of INFOCOM* (2007), pp. 1244–1252
86. S. Funke, I. Rauf, Information brokerage via location-free double rulings, in *Proceedings of the 6th International Conference on Ad-Hoc, Mobile and Wireless Networks* (2007), pp. 87–100
87. M.E. Gage, Curve shortening on surfaces. Annales Scientifiques de L'Ecole Normale Supérieure **23**(2), 229–256 (1990)
88. G.A. Galperin, A concept of the mass center of a system of material points in the constan curvature spaces. Comm. Math. Phys. **154**(1), 63–84 (1993)
89. J. Gao, X.D Gu, F. Luo, Discrete ricci flow for geometric routing, in *Discrete Ricci Flow for Geometric Routing* (2016), pp. 556–563
90. F.P. Gardiner, N. Lakic, *Quasiconformal Teichmüller Theory*, vol. 76 (American Mathematical Society, 2000)
91. M. Garland, P.S. Heckbert, Surface simplification using quadric error metrics, in *Proceedings of the 24th Annual Conference on Computer Graphics and Interactive Techniques*, SIGGRAPH '97 (ACM Press/Addison-Wesley Publishing Company, New York, 1997), pp. 209–216
92. Allen Gersho, Asymptotically optimal block quantization. IEEE Trans. Inf. Theory **25**(4), 373–380 (1979)
93. G. Giorgetti, S. Gupta, G. Manes, Wireless localization using self-organizing maps, in *Proceedings of IPSN* (2007), pp. 293–302
94. R. Gormaz, B-spline knot-line elimination and Bézier continuity conditions, in *Curves and Surfaces in Geometric Design*, ed. by A.K. Peters (MA, Wellesley, 1994), pp. 209–216
95. M. Goswami, C.-C. Ni, X. Ban, J. Gao, X.D Gu, V. Pingali, Load balanced short path routing in large-scale wireless networks using area-preserving maps, in *Proceedings of the 15th ACM international symposium on mobile ad hoc networking and computing*, MobiHoc '14 (2004), pp. 63–72
96. M.A. Grayson, Shortening embedded curves. Ann. Math. **129**, 71–111 (1989)
97. C. Grimm, J.F. Hughes, Modeling surfaces of arbitrary topology using manifolds, in *SIGGRAPH* (1995), pp. 359–368
98. X. Gu, F. Luo, J. Sun, S.-T. Yau, Variational principles for Minkowski type problems, discrete optimal transport, and discrete Monge-Amperé equations (2013), arXiv:1302.5472
99. X. Gu, B.C. Vemuri, Matching 3D shapes using 2D conformal representations. MICCAI **1**, 771–780 (2004)
100. X. Gu, Y. Wang, T.F. Chan, P.M. Thompson, S.-T. Yau, Genus zero surface conformal mapping and its application to brain surface mapping. IEEE Trans. Med. Imaging **23**(8), 949–958 (2004)
101. X. Gu, Y. Wang, S.-T. Yau, Geometric compression using Riemann surface. Commun. Inf. Syst. **3**(3), 171–82 (2005)
102. X. Gu, S.-T. Yau, Global conformal surface parameterization, in *Proceedings Eurographics/SIGGRAPH symposium geometry processing* (Eurographics Association, 2003), pp. 127–137
103. X. Gu, Y. He, H. Qin, Manifold splines. Gr. Models **68**(3), 237–254 (2006)
104. X. Gu, S. Wang, J. Kim, Y. Zeng, Y. Wang, H. Qin, D. Samaras, Ricci flow for 3D shape analysis, in *ICCV* (2007)
105. X. Gu, S.-T. Yau, Global conformal surface parameterization, in *Proceedings of the Eurographics/ACM SIGGRAPH symposium on geometry processing* (2003), pp. 127–137
106. R. Guo, Local rigidity of inversive distance circle packing (2009), arXiv:0903.1401v2
107. B. Gutman, Y. Wang, J. Morra, A.W. Toga, P.M. Thompson, Disease classification with hippocampal shape invariants. Hippocampus **19**(6), 572–578 (2009)
108. S. Haker, S. Angenent, A. Tannenbaum, R. Kikinis, G. Sapiro, M. Halle, Conformal surface parameterization for texture mapping. IEEE Trans. Vis. Comput. Gr. **6**(2), 181–189 (2000)
109. R.S. Hamilton, Three manifolds with positive Ricci curvature. J. Differ. Geom. **17**, 255–306 (1982)
110. R.S. Hamilton, The Ricci flow on surfaces. Contemp. Math. **71**, 237–262 (1988)

111. X. Han, Xu Chenyang, J.L. Prince, A topology preserving level set method for geometric deformable models. IEEE Trans. Pattern Anal. Mach. Intell. **25**(6), 755–768 (2003)
112. A. Hatcher, P. Lochak, L. Schneps, On the teichmüller tower of mapping class groups. J. Reine Angew. Math **521**, 1–24 (2000)
113. H.C. Hazlett, H. Gu, B.C. Munsell, S.H. Kim, M. Styner, J.J. Wolff, J.T. Elison, M.R. Swanson, H. Zhu, K.N. Botteron, D.L. Collins, J.N. Constantino, S.R. Dager, A.M. Estes, A.C. Evans, V.S. Fonov, G. Gerig, P. Kostopoulos, R.C. McKinstry, J. Pandey, S. Paterson, J.R. Pruett, R.T. Schultz, D.W. Shaw, L. Zwaigenbaum, J. Piven, J. Piven, H.C. Hazlett, C. Chappell, S.R. Dager, A.M. Estes, D.W. Shaw, K.N. Botteron, R.C. McKinstry, J.N. Constantino, J.R. Pruett, R.T. Schultz, S. Paterson, L. Zwaigenbaum, J.T. Elison, J.J. Wolff, A.C. Evans, D.L. Collins, G.B. Pike, V.S. Fonov, P. Kostopoulos, S. Das, G. Gerig, M. Styner, C.H. Gu, C.H. Gu, Early brain development in infants at high risk for autism spectrum disorder. Nature **542**(7641), 348–351 (2017)
114. Y. He, X. Gu, H. Qin, Rational spherical splines for genus zero shape modeling, in *Proceedings of Shape Modeling International '05* (2005), pp. 82–91
115. Y. He, M. Jin, X. Gu, H. Qin, A C^1 globally interpolatory spline of arbitrary topology. In *Proceedings of the 3rd IEEE Workshop on Variational, Geometric and Level Set Methods in Computer Vision*. Lecture Notes in Computer Science, vol. 3752 (2005), pp. 295–306
116. Y. He, K. Wang, H. Wang, X. Gu, H. Qin, Manifold T-spline, in *Proceedings of Geometric Modeling and Processing* (2006), pp. 409–422
117. D. Healy, D. Rockmore, P. Kostelec, S. Moore, Ffts for the 2-sphere - improvements and variations. J. Fourier Anal. Appl. **9**(4), 341–385 (2003)
118. P. Henrici, *Applied and Computational Complex Analysis*, vol. 3 (Wiley, New Jersey, 1988)
119. G. Hermosillo, *Variational methods for multimodal image matching*. PhD thesis, Université de Nice (INRIA-ROBOTVIS), Sophia Antipolis, France (2002)
120. J. Hersberger, J. Snoeyink, Around and around: computing the shortest loop, in *The third Canadian conference on computational geometry* (1991), pp. 157–161
121. J. Hersberger, J. Snoeyink, Computing minimum length paths of a given homotopy class. Comput. Geom. Theory Appl. **4**(2), 63–97 (1994)
122. A.P. Holmes, R.C. Blair, J.D. Watson, I. Ford, Nonparametric analysis of statistic images from functional mapping experiments. J. Cereb. Blood Flow Metab. **16**(1), 7–22 (1996)
123. B.-W. Hong, S. Soatto, *Shape matching using multiscale integral invariants* (IEEE Trans. Pattern Anal. Mach, Intell, 2014)
124. A. Howard, M. Mataric, G.S. Sukhatme, Mobile sensor network deployment using potential fields: a distributed, scalable solution to the area coverage problem, in *Proceedings of the 6th international symposium on distributed autonomous robotics systems* (2002)
125. M.K. Hurdal, K. Stephenson, Cortical cartography using the discrete conformal approach of circle packings. NeuroImage **23**, S119–S128 (2004)
126. M.K. Hurdal, K. Stephenson, Discrete conformal methods for cortical brain flattening. NeuroImage **45**, S86–S98 (2009)
127. K. Im, J.M. Lee, U. Yoon, Y.W. Shin, S.B. Hong, I.Y. Kim, J.S. Kwon, S.I. Kim, Fractal dimension in human cortical surface: multiple regression analysis with cortical thickness, sulcal depth, and folding area. Hum. Brain Mapp. **27**, 994–1003 (2006)
128. C.R. Jack, M.A. Bernstein, B.J. Borowski, J.L. Gunter, N.C. Fox, P.M. Thompson, N. Schuff, G. Krueger, R.J. Killiany, C.S. Decarli, A.M. Dale, O.W. Carmichael, D. Tosun, M.W. Weiner, Update on the magnetic resonance imaging core of the Alzheimer's disease neuroimaging initiative. Alzheimers Dement **6**(3), 212–220 (2010)
129. C.R. Jack, R.C. Petersen, Y.C. Xu, P.C. O'Brien, G.E. Smith, R.J. Ivnik, B.F. Boeve, S.C. Waring, E.G. Tangalos, E. Kokmen, Prediction of AD with MRI-based hippocampal volume in mild cognitive impairment. Neurology **52**(7), 1397–1403 (1999)
130. M. Jin, N. Ding, W. Zeng, X. Gu, S.-T. Yau, Computing fenchel-nielsen coordinates in Teichmüller shape space. Comm. Inf. Syst. (CIS) **9**(2), 213–234 (2009)
131. M. Jin, J. Kim, F. Luo, X. Gu, Discrete surface Ricci flow. IEEE Trans. Vis. Comput. Gr. **14**(5), 1030–1043 (2008)

132. M. Jin, Y. Wang, S.-T. Yau, X. Gu, Optimal global conformal surface parameterization, in *Proceedings of the IEEE Visualization 2004* (IEEE Computer Society, 2004), pp. 267–274
133. M. Jin, W. Zeng, F. Luo, X. Gu, Computing Teichmüller shape space. IEEE Trans. Vis. Comput. Gr. **15**(3), 504–517 (2009)
134. M. Jin, N. Ding, Yang Yang, Computing shortest homotopic cycles on polyhedral surfaces with hyperbolic uniformization metric. Comput. Aided Des. **45**(2), 113–123 (2013)
135. M. Jin, F. Luo, X.D. Gu, Computing general geometric structures on surfaces using ricci flow. Comput. Aided Des. **39**(8), 663–675 (2007)
136. M. Jin, G. Rong, H. Wu, L. Shuai, X. Guo, Optimal surface deployment problem in wireless sensor networks, in *Proceedings of the 31st Annual IEEE Conference on Computer Communications (INFOCOM'12)* (2012), pp. 2345–2353
137. M. Jin, S. Xia, H. Wu, X. Gu, Scalable and fully distributed localization with mere connectivity, in *Proceedings of IEEE Conference on Computer Communications (INFOCOM)* (2011), pp. 3164–3172
138. A.A. Joshi, D.W. Shattuck, P.M. Thompson, R.M. Leahy, Surface-constrained volumetric brain registration using harmonic mappings. IEEE Trans. Med. Imaging **26**(12), 1657–1669 (2007)
139. P. Joshi, M. Meyer, T. DeRose, B. Green, T. Sanocki, Harmonic coordinates for character articulation. ACM Trans. Gr. **26**(3) (2007)
140. S.H. Joshi, R.P. Cabeen, A.A. Joshi, B. Sun, I. Dinov, K.L. Narr, A.W. Toga, R.P. Woods, Diffeomorphic sulcal shape analysis on the cortex. IEEE Trans. Med. Imaging **31**(6), 1195–1212 (2012)
141. S.H. Joshi, R.T. Espinoza, T. Pirnia, J. Shi, Y. Wang, B. Ayers, A. Leaver, R.P. Woods, K.L. Narr, Structural plasticity of the hippocampus and amygdala induced by electroconvulsive therapy in major depression. Biol. Psychiatry (2015)
142. J. Jost, R.R. Simha, *Compact Riemann Surfaces: An Introduction to Contemporary Mathematics* (Springer, Berlin, 1997)
143. P. Juang, H. Oki, Y. Wang, M. Martonosi, L.-S. Peh, D. Rubenstein, Energy-efficient computing for wildlife tracking: design tradeoffs and early experiences with zebranet. SIGARCH Comput. Archit. News **30**(5), 96–107 (2002)
144. F. Kälberer, M. Nieser, K. Polthier, Quadcover - surface parameterization using branched coverings. Comput. Gr. Forum **26**(10), 375–384 (2007)
145. L.V. Kantorovich, On a problem of Monge. Uspekhi Mat. Nauk. **3**, 225–226 (1948)
146. M. Kaplan, E. Cohen, Computer generated celtic design, in *Proceedings of the 14th Eurographics Workshop on Rendering Techniques* (2003), pp. 2–19
147. B. Karp, H.T. Kung, GPSR: greedy perimeter stateless routing for wireless networks, in *ACM Mobicom* (2000), pp. 243–254
148. D.G. Kendall, Shape manifolds, procrustean metrics, and complex projective spaces. Bull. Lond. Math. Soc. **16**(2), 81–121 (1984)
149. R. Kershner, *The number of circles covering a set* (Am. J, Math, 1939)
150. R. Kershner, The number of circles covering a set. Am. J. Math. **61**, 665–671 (1939)
151. B. Kim, J.L. Boes, K.A. Frey, C.R. Meyer, Mutual information for automated unwarping of rat brain autoradiographs. NeuroImage **5**(1), 31–40 (1997)
152. R. Kleinberg, Geographic routing using hyperbolic space, in *Proceedings of INFOCOM* (2007), pp. 1902–1909
153. E. Kranakis, H. Singh, J. Urrutia, Compass routing on geometric networks, in *Proceedings of Canadian Conference on Computational Geometry (CCCG)* (1999), pp. 51–54
154. W.S. Kremen, M.S. Panizzon, M.C. Neale, C. Fennema-Notestine, E. Prom-Wormley, L.T. Eyler, A. Stevens, C.E. Franz, M.J. Lyons, M.D. Grant, A.J. Jak, T.L. Jernigan, H. Xian, B. Fischl, H.W. Thermenos, L.J. Seidman, M.T. Tsuang, A.M. Dale, Heritability of brain ventricle volume: converging evidence from inconsistent results. Neurobiol. Aging **33**(1), 1–8 (2012)
155. F. Kuhn, R. Wattenhofer, Y. Zhang, A. Zollinger, Geometric ad-hoc routing: theory and practice, in *Proceedings of the 22nd ACM symposium on the principles of distributed computing* (2003), pp. 63–72

156. F. Kuhn, R. Wattenhofer, A. Zollinger, Worst-case optimal and average-case efficient geometric ad-hoc routing, in *Proceedings of MobiHOC* (2003), pp. 267–278
157. S. Kurtek, E. Klassen, J.C. Gore, Z. Ding, A. Srivastava, Elastic geodesic paths in shape space of parameterized surfaces. IEEE Trans. Pattern Anal. Mach. Intell. **34**(9), 1717–1730 (2012)
158. Y. Lai, M. Jin, X. Xie, Y. He, J. Palacios, E. Zhang, S. Hu, X. Gu, Metric-driven rosy fields design and remeshing. IEEE Trans. Vis. Comput. Gr. (TVCG) **15**(3), 95–108 (2010)
159. Y. Lao, L.A. Dion, G. Gilbert, M.F. Bouchard, G. Rocha, Y. Wang, N. Lepore, D. Saint-Amour, Mapping the basal ganglia alterations in children chronically exposed to manganese. Sci. Rep. **7**, 41804 (2017)
160. Y. Lao, B. Nguyen, S. Tsao, N. Gajawelli, M. Law, H. Chui, M. Weiner, Y. Wang, N. Lepore, A T1 and DTI fused 3D corpus callosum analysis in MCI subjects with high and low cardiovascular risk profile. Neuroimage Clin. **14**, 298–307 (2017)
161. Y. Lao, Y. Wang, J. Shi, R. Ceschin, M.D. Nelson, A. Panigrahy, N. Lepore, *Thalamic alterations in preterm neonates and their relation to ventral striatum disturbances revealed by a combined shape and pose analysis* (Brain Struct, Funct, 2014)
162. S.M. Lee, N.A. Clark, P.A. Araman, A shape representation for planar curves by shape signature harmonic embedding, in *Proceeding in IEEE Computer Society Conference on Computer Vision and Pattern Recognition (CVPR'06)* (2006), pp. 1940–1947
163. Y. Lee, S. Lee, Geometric snakes for triangular meshes. Comput. Gr. Forum Eurogr. **21**(3), 229–238 (2002)
164. R. Leech, D.J. Sharp, The role of the posterior cingulate cortex in cognition and disease. Brain **137**(Pt 1), 12–32 (2014)
165. T. Leighton, A. Moitra, Some results on greedy embeddings in metric spaces, in *Proceedings of the 49th IEEE annual symposium on foundations of computer science* (2008), pp. 337–346
166. A.D. Leow, S.C. Huang, A. Geng, J. Becker, S. Davis, A.W. Toga, P.M. Thompson, Inverse consistent mapping in 3D deformable image registration: its construction and statistical properties. Inf. Process. Med. Imaging **19**, 493–503 (2005)
167. B. Lévy, S. Petitjean, N. Ray, J. Maillot, Least squares conformal maps for automatic texture atlas generation. ACM Trans. Gr. **21**(3), 362–371 (2002)
168. B. Li, J. Shi, B.A. Gutman, L.C. Baxter, P.M. Thompson, R.J. Caselli, Y. Wang, Influence of APOE genotype on hippocampal atrophy over time - an N = 1925 surface-based ADNI study. PLoS ONE **11**(4), e0152901 (2016)
169. F. Li, J. Luo, W. Wang, Y. He, Autonomous deployment for load balancing k-surface coverage in sensor networks. IEEE Trans. Wirel. Commun. **14**(1), 279–293 (2015)
170. L. Li, T. Kunz, Localization applying an efficient neural network mapping, in *Proceedings of the 1st International Conference on Autonomic Computing and Communication Systems* (2007), pp. 1–9
171. X. Li, X. Gu, H. Qin, Surface mapping using consistent pants decomposition. IEEE Trans. Vis. Comput. Gr. **15**(4), 558–571 (2008)
172. H. Lim, J. Hou, Distributed localization for anisotropic sensor networks. ACM Trans. Sens. Netw. **5**(2), 11–37 (2009)
173. N. Litke, M. Droske, M. Rumpf, P. Schröder, An image processing approach to surface matching, in *Proceedings of the third eurographics symposium on geometry processing*, SGP '05 (Eurographics Association, Aire-la-Ville, Switzerland, 2005)
174. T. Liu, D. Geiger, Approximate tree matching and shape similarity, in *Proceeding of International Conference on Computer Vision* (1999), pp. 456–462
175. Xin Liu, Qingfeng Huang, Ying Zhang, Balancing push and pull for efficient information discovery in large-scale sensor networks. IEEE Trans. Mob. Comput. **6**, 241–251 (2007)
176. Y. Liu, W. Wang, B. Lévy, F. Sun, D.-M. Yan, L. Lin, C. Yang, On centroidal Voronoi tessellation - energy smoothness and fast computation. ACM Trans. Gr. **28**(4), 1–17 (2009)
177. S. Lloyd, Least squares quantization in pcm. IEEE Trans. Inf. Theory **28**(2), 129–137 (1982)
178. W.E. Lorensen, H.E. Cline, Marching cubes: a high resolution 3D surface construction algorithm. SIGGRAPH Comput. Gr. **21**(4), 163–169 (1987)

179. E. Luders, K.L. Narr, R.M. Bilder, P.R. Szeszko, M.N. Gurbani, L. Hamilton, A.W. Toga, C. Gaser, Mapping the relationship between cortical convolution and intelligence: effects of gender. Cereb. Cortex **18**, 2019–2026 (2008)
180. E. Luders, K.L. Narr, R.M. Bilder, P.M. Thompson, P.R. Szeszko, L. Hamilton, A.W. Toga, Positive correlations between corpus callosum thickness and intelligence. Neuroimage **37**, 1457–1464 (2007)
181. E. Luders, P.M. Thompson, F. Kurth, J.Y. Hong, O.R. Phillips, Y. Wang, B.A. Gutman, Y.Y. Chou, K.L. Narr, A.W. Toga, Global and regional alterations of hippocampal anatomy in long-term meditation practitioners. Hum. Brain Mapp. **34**(12), 3369–3375 (2013)
182. L.M. Lui, J. Kwan, Y. Wang, S.-T. Yau, Computation of curvatures using conformal parameterization. Commun. Inf. Syst. **8**(1), 1–16 (2008)
183. L.M. Lui, S. Thiruvenkadam, Y. Wang, T.F. Chan, P.M. Thompson, Optimzed conformal parameterization of cortical surfaces using shape based matching of landmark curves. SIAM J. Imaging Sci. **3**(1), 52–78 (2010)
184. L.M. Lui, W. Zeng, S.T. Yau, X. Gu, Shape analysis of planar multiply-connected objects using conformal welding. IEEE Trans. Pattern Anal. Mach. Intell. **36**(7), 1384–1401 (2014)
185. F. Luo, Geodesic length functions and teichmuller spaces. J. Differ. Geom. **48**(2), 275–317 (1998)
186. O. Lyttelton, M. Boucher, S. Robbins, A. Evans, An unbiased iterative group registration template for cortical surface analysis. Neuroimage **34**(4), 1535–1544 (2007)
187. L. Ma, D.Z. Chen, Curve shortening flow in a riemannian manifold (2003), arXiv:math.DG/0312463v1
188. M. Ma, Yuanyuan Yang, Adaptive triangular deployment algorithm for unattended mobile sensor networks. IEEE Trans. Comput. **56**, 847–946 (2007)
189. J. Maillot, H. Yahia, A. Verroust, Interactive texture mapping. *(Proceedings of SIGGRAPH 93 Computer Graphics)* (1993), pp. 27–34
190. E. Martinson, D. Payton, *Lattice formation in mobile autonomous sensor arrays*, vol. 3342, Lecture Notes in Computer Science (Springer, Berlin, 2005), pp. 98–111
191. H. Matsuda, Role of neuroimaging in Alzheimer's disease, with emphasis on brain perfusion SPECT. J. Nucl. Med. **48**(8), 1289–1300 (2007)
192. C.R. Meyer, J.L. Boes, B. Kim, P.H. Bland, K.R. Zasadny, P.V. Kison, K. Koral, K.A. Frey, R.L. Wahl, Demonstration of accuracy and clinical versality of mutual information for automatic multimodality image fusion using affine and thin plate spline warped geometric deformation. Med. Image Anal. **1**(3), 195–206 (1997)
193. K. Mikula, D. Sevcovic, Evolution of curves on surface driven by the geodesic curvature and external force. Appl. Anal. **85**, 345–362 (2006)
194. M.I. Miller, A. Trouve, L. Younes, On the metrics and Euler-Lagrange equations of computational anatomy. Annu. Rev. Biomed. Eng. **4**, 375–405 (2002)
195. J.W. Milnor, J.D. Stasheff, *Characteristic Classes* (Princeton University Press, Princeton, 1974)
196. J.W. Milnor, On the existence of a connection with curvature zero. Comment. Math. Helv. **32**, 215–223 (1958)
197. B. Leong, S. Mitra, B. Liskov, Path vector face routing: geographic routing with local face information, in *Proceedings of ICNP* (2005), pp. 147–158
198. F. Mokhtarian, A. Mackworth, A theory of multiscale, curvature-based shape representation for planar curves. IEEE Trans. Pattern Anal. Mach. Intell. **14**(8), 789–805 (1992)
199. M. Monje, M.E. Thomason, L. Rigolo, Y. Wang, D.P. Waber, S.E. Sallan, A.J. Golby, Functional and structural differences in the hippocampus associated with memory deficits in adult survivors of acute lymphoblastic leukemia. Pediatr. Blood Cancer **60**(2), 293–300 (2013)
200. J. Morra, Z. Tu, L.G. Apostolova, A.E. Green, C. Avedissian, S.K. Madsen, N. Parikshak, A.W. Toga, C.R. Jack, N. Schuff, M.W. Weiner, P.M. Thompson, Automated mapping of hippocampal atrophy in 1-year repeat MRI data from 490 subjects with Alzheimer's disease, mild cognitive impairment, and elderly controls. NeuroImage, **45**(1, Supplement 1), S3–S15 (2009)

201. D.J. Newman, The hexagon theorem. IEEE Trans. Inf. Theory **28**(2), 137–139 (1982)
202. Xinlai Ni, Michael Garland, John C. Hart, Fair morse functions for extracting the topological structure of a surface mesh. ACM Trans. Gr. **23**(3), 613–622 (2004)
203. F. Nielsen, R. Nock, Hyperbolic Voronoi diagrams made easy. ACM Comput. Res. Repos. (2009), abs/0903.3287,
204. A. Okabe, B. Boots, K. Sugihara, S.N. Chiu, *Spatial Tessellations: Concepts and Applications of Voronoi Diagrams*, 2nd edn. (Wiley, New Jersey, 1999)
205. D. Pantazis, A. Joshi, J. Jiang, D.W. Shattuck, L.E. Bernstein, H. Damasio, R.M. Leahy, Comparison of landmark-based and automatic methods for cortical surface registration. Neuroimage **49**(3), 2479–2493 (2010)
206. C. Papadimitriou, D. Ratajczak, On a conjecture related to geometric routing. Theor. Comput. Sci. **344**(1), 3–14 (2005)
207. H. Park, J.S. Park, J.K. Seong, D.L. Na, J.M. Lee, Cortical surface registration using spherical thin-plate spline with sulcal lines and mean curvature as features. J. Neurosci. Methods **206**(1), 46–53 (2012)
208. B. Patenaude, S.M. Smith, D.N. Kennedy, M. Jenkinson, A Bayesian model of shape and appearance for subcortical brain segmentation. Neuroimage **56**(3), 907–922 (2011)
209. G. Perelman, The entropy formula for the Ricci flow and its geometric applications. Technical Report, 11 November 2002, arXiv:math/02111590211159
210. G. Perelman, Finite extinction time for the solutions to the Ricci flow on certain three-manifolds. Technical Report, 17 July 2003, arXiv:math/0307245
211. G. Perelman, Ricci flow with surgery on three-manifolds. Technical Report, 10 March 2003, arXiv:math/0303109
212. G. Peyré, L. Cohen, Surface segmentation using geodesic centroidal tesselation, in *Proceedings of 2nd international symposium on 3D data processing, visualization, and transmission* (IEEE Computer Society, Washington, 2004), pp. 995–1002
213. G. Peyré, L. Cohen, Geodesic remeshing using front propagation. Int. J. Comput. Vis. **69**(1), 145–156 (2006)
214. M. Pievani, S. Galluzzi, P.M. Thompson, P.E. Rasser, M. Bonetti, G.B. Frisoni, APOE4 is associated with greater atrophy of the hippocampal formation in Alzheimer's disease. Neuroimage **55**(3), 909–919 (2011)
215. U. Pinkall, K. Polthier, Computing discrete minimal surfaces and their conjugate. Exp. Math. **2**(1), 15–36 (1993)
216. S.M. Pizer, D.S. Fritsch, P.A. Yushkevich, V.E. Johnson, E.L. Chaney, Segmentation, registration, and measurement of shape variation via image object shape. IEEE Trans. Med. Imaging **18**(10), 851–865 (1999)
217. S. Poduri, G.S. Sukhatme, Constrained coverage for mobile sensor networks, in *IEEE International Conference on Robotics and Automation* (2004), pp. 165–171
218. M.J.D. Powell, M.A. Sabin, Piecewise quadratic approximations on triangles. ACM Trans. Math. Softw. **3**(4), 316–325 (1977)
219. A.N. Pressley, *Elementary Differential Geometry* (Springer, Berlin, 2010)
220. S. Ratnasamy, B. Karp, L. Yin, F. Yu, D. Estrin, R. Govindan, S. Shenker, GHT: a geographic hash table for data-centric storage in sensornets, in *The 1st ACM Workshop on Wireless Sensor Networks ands Applications* (2002), pp. 78–87
221. J. Raven, J.C. Raven, J.H. Court, *Manual for Raven's progressive Matrices and Vocabulary Scales* (Oxford Psychologists Press, Oxford, 1998)
222. N. Ray, B. Vallet, W.C. Li, B. Lévy, N-symmetry direction field design. ACM Trans. Gr. **27**(2), 10:1–10:13 (2008)
223. M. Reuter, H.D. Rosas, B. Fischl, Highly accurate inverse consistent registration: a robust approach. Neuroimage **53**(4), 1181–1196 (2010)
224. D. Rey, G. Subsol, H. Delingette, N. Ayache, Automatic detection and segmentation of e-volving processes in 3D medical images: application to multiple sclerosis. Med. Image Anal. **6**(2), 163–179 (2002)

225. G. Rong, M. Jin, L. Shuai, X. Guo, Centroidal Voronoi tessellation in universal covering space of manifold surfaces. Comput. Aided Geom. Des. **28**(8), 475–496 (2011)
226. Guodong Rong, Yang Liu, Wenping Wang, Gu Xiaotian Yin, Xiaohu Guo David, Gpu-assisted computation of centroidal voronoi tessellation. IEEE Trans. Vis. Comput. Gr. **17**(3), 345–356 (2011)
227. D. Rueckert, L.I. Sonoda, C. Hayes, D.L. Hill, M.O. Leach, D.J. Hawkes, Nonrigid registration using free-form deformations: application to breast MR images. IEEE TMI **18**(8), 712–21 (1999)
228. R. Sarkar, X. Yin, J. Gao, F. Luo, X.D. Gu, Greedy routing with guaranteed delivery using Ricci flows, in *Proceedings of the 8th international symposium on information processing in sensor networks*, IPSN'09 (2009), pp. 121–132
229. R. Sarkar, W. Zeng, J. Gao, X.D. Gu, Covering space for in-network sensor data storage, in *Proceedings of the 9th ACM/IEEE International Conference on Information Processing in Sensor Networks* (2010), pp. 232–243
230. R. Sarkar, X. Zhu, J. Gao, Double rulings for information brokerage in sensor networks, in *ACM MobiCom* (2006), pp. 286–297
231. B. Schmitzer, C. Schnrr, Object segmentation by shape matching with wasserstein modes. J. Energy Minim. Methods Comput. Vis. Pattern Recognit. 123–136 (2013)
232. R. Schoen, S.-T. Yau, *Lectures on Harmonic Maps* (International Press, Austria, 1997)
233. B. Scholkopf, A.J. Smola, *Learning with Kernels: Support Vector Machines, Regularization, Optimization, and Beyond* (MIT Press, Cambridge, 2001)
234. M. Schwager, J. Mclurkin, D. Rus, Distributed coverage control with sensory feedback for networked robots, in *Proceedings of Robotics: Science and Systems* (2006)
235. T. Sebastian, P. Klein, B. Kimia, Shock based indexing into large shape databases, in *Proceeding in European Conference on Computer Vision* (2002), pp. 731–746
236. T.W. Sederberg, J. Zheng, A. Bakenov, A.H. Nasri, T-splines and T-NURCCs. ACM Trans. Gr. **22**(3), 477–484 (2003)
237. H.-P. Seidel, Polar forms and triangular B-spline surfaces, in *Euclidean Geometry and Computers*, 2nd edn., ed. by D.-Z. Du, F. Hwang (World Scientific Publishing Company, Singapore, 1994), pp. 235–286
238. H.-P. Seidel, An introduction to polar forms. IEEE Comput. Gr. Appl. **13**(1), 38–46 (1993)
239. M. Seppälä, T. Sorvali, *Geometry of Riemann surfaces and Teichmiller spaces, North-Holland Mathematics Studies* (North Holland, Amsterdam, 1991)
240. M. Seppala, T. Sorvali, *Geometry of Riemann Surfaces and Teichmüller Spaces, North-Holland Mathematics Studies* (North-Holland, Amsterdam, 1992)
241. Y. Shang, W. Ruml, Improved mds-based localization, in *Proceedings of INFOCOM* (2004), pp. 2640–2651
242. Y. Shang, W. Ruml, Y. Zhang, M.P.J. Fromherz, Localization from mere connectivity, in *Proceedings of MobiHoc* (2003), pp. 201–212
243. E. Sharon, D. Mumford, 2D-shape analysis using conformal mapping. Int. J. Comput. Vis. **70**, 55–75 (2006)
244. J. Shi, O. Collignon, L. Xu, G. Wang, Y. Kang, F. Lepore, Y. Lao, A.A. Joshi, N. Lepore, Y. Wang, Impact of early and late visual deprivation on the structure of the corpus callosum: a study combining thickness profile with surface tensor-based morphometry. Neuroinformatics **13**(3), 321–336 (2015)
245. J. Shi, N. Lepore, B.A. Gutman, P.M. Thompson, L.C. Baxter, R.J. Caselli, Y. Wang, Genetic influence of apolipoprotein E4 genotype on hippocampal morphometry: an N = 725 surface-based Alzheimer's disease neuroimaging initiative study. Hum. Brain Mapp. **35**(8), 3903–3918 (2014)
246. J. Shi, C.M. Stonnington, P.M. Thompson, K. Chen, B. Gutman, C. Reschke, L.C. Baxter, E.M. Reiman, R.J. Caselli, Y. Wang, Studying ventricular abnormalities in mild cognitive impairment with hyperbolic Ricci flow and tensor-based morphometry. Neuroimage **104**, 1–20 (2015)

247. J. Shi, P.M. Thompson, B. Gutman, Y. Wang, Surface fluid registration of conformal representation: application to detect disease burden and genetic influence on hippocampus. Neuroimage **78**, 111–134 (2013)
248. J. Shi, Y. Wang, R. Ceschin, X. An, Y. Lao, D. Vanderbilt, M.D. Nelson, P.M. Thompson, A. Panigrahy, N. Lepore, A multivariate surface-based analysis of the putamen in premature newborns: regional differences within the ventral striatum. PLoS ONE **8**(7), e66736 (2013)
249. J. Shi, W. Zhang, M. Tang, R.J. Caselli, Y. Wang, Conformal invariants for multiply connected surfaces: application to landmark curve-based brain morphometry analysis. Med. Image Anal. **35**, 517–529 (2017)
250. J. Shi, W. Zhang, Y. Wang, Shape analysis with hyperbolic Wasserstein distance. IEEE Conference on Computer Vision and Pattern Recognition (CVPR) **2016**, 5051–5061 (2016)
251. R. Shi, W. Zeng, Z. Su, J. Jiang, H. Damasio, Z. Lu, Y. Wang, S.T. Yau, X. Gu, Hyperbolic harmonic mapping for surface registration. IEEE Trans. Pattern Anal. Mach. Intell. **39**(5), 965–980 (2017)
252. Y. Shi, R. Lai, A.W. Toga, Cortical surface reconstruction via unified Reeb analysis of geometric and topological outliers in magnetic resonance images. IEEE Trans. Med. Imaging **32**(3), 511–530 (2013)
253. G. Simon, M. Molnr, L. Gnczy, B. Cousin, Dependable k-coverage algorithms for sensor networks, in *Proceedings of IMTC* (2007)
254. M. Siqueira, D. Xu, J. Gallier, L.G. Nonato, D.M. Morera, L. Velho, A new construction of smooth surfaces from triangle meshes using parametric pseudo-manifolds. Comput. Gr. **33**(3), 331–340 (2009)
255. A.M.-C. So, Y. Ye, Theory of semidefinite programming for sensor network localization, in *Proceedings of the sixteenth annual acm-siam symposium on discrete algorithms (SODA)* (2005), pp. 405–414
256. B. Springborn, P. Schröder, U. Pinkall, Conformal equivalence of triangle meshes, in *ACM SIGGRAPH 2008 papers*, SIGGRAPH '08 (ACM, New York, 2008), pp. 77:1–77:11
257. J. Stam, Flows on surfaces of arbitrary topology. ACM Trans. Gr. **22**(3), 724–731 (2003)
258. I. Stojmenovic, B. Vukojevic, A routing strategy and quorum based location update scheme for ad hoc wireless networks. Technical Report, Technical Report TR-99-09, University of Ottawa (1999)
259. M. Styner, I. Oguz, S. Xu, C. Brechbuhler, D. Pantazis, J.J. Levitt, M.E. Shenton, G. Gerig, Framework for the statistical shape analysis of brain structures using SPHARM-PDM. Insight J **1071**, 242–250 (2006)
260. M. Styner, J.A. Lieberman, R.K. McClure, D.R. Weinberger, D.W. Jones, Guido Gerig, Morphometric analysis of lateral ventricles in schizophrenia and healthy controls regarding genetic and disease-specific factors. Proc. Natl. Acad. Sci. USA **102**(13), 4872–4877 (2005)
261. Z. Su, Y. Wang, R. Shi, W. Zeng, J. Sun, F. Luo, X. Gu, *Optimal mass transport for shape matching and comparison* (IEEE Trans. Pattern Anal. Mach, Intell, 2015)
262. Z. Su, W. Zeng, Y. Wang, Z.-I. Lu, X. Gu, Shape classification using wasserstein distance for brain morphometry analysis, in *Information Processing in Medical Imaging, vol. 9123*, Lecture Notes in Computer Science, ed. by S. Ourselin, D.C. Alexander, C.F. Westin, M.J. Cardoso (Springer, Berlin, 2015), pp. 411–423
263. D. Sun, T.G. van Erp, P.M. Thompson, C.E. Bearden, M. Daley, L. Kushan, M.E. Hardt, K.H. Nuechterlein, A.W. Toga, T.D. Cannon, Elucidating a magnetic resonance imaging-based neuroanatomic biomarker for psychosis: classification analysis using probabilistic brain atlas and machine learning algorithms. Biol. Psychiatry **66**(11), 1055–1060 (2009)
264. G. Tan, M. Bertier, A.-M. Kermarrec, Visibility-graph-based shortest-path geographic routing in sensor networks, in *Proceedings of INFOCOM* (2009), pp. 119–1727
265. M. Tarini, K. Hormann, P. Cignoni, Claudio Montani. Polycube-maps. ACM Trans. Gr. **23**(3), 853–860 (2004)
266. P.M. Thompson, J.N. Giedd, R.P. Woods, D. MacDonald, A.C. Evans, A.W. Toga, Growth patterns in the developing human brain detected using continuum-mechanical tensor mapping. Nature **404**(6774), 190–193 (2000)

267. P.M. Thompson, K.M. Hayashi, G. de Zubicaray, A.L. Janke, S.E. Rose, J. Semple, D. Herman, M.S. Hong, S.S. Dittmer, D.M. Doddrell, A.W. Toga, Dynamics of gray matter loss in Alzheimer's disease. J. Neurosci. **23**(3), 994–1005 (2003)
268. P.M. Thompson, K.M. Hayashi, E.R. Sowell, N. Gogtay, J.N. Giedd, J.L. Rapoport, G.I. de Zubicaray, A.L. Janke, S.E. Rose, J. Semple, D.M. Doddrell, Y. Wang, T.G.M. van Erp, T.D. Cannon, A.W. Toga, Mapping cortical change in Alzheimer's disease, brain development, and schizophrenia. NeuroImage **23**(Supplement 1), S2–S18 (2004)
269. P.M. Thompson, A.D. Lee, R.A. Dutton, J.A. Geaga, K.M. Hayashi, M.A. Eckert, U. Bellugi, A.M. Galaburda, J.R. Korenberg, D.L. Mills, A.W. Toga, A.L. Reiss, Abnormal cortical complexity and thickness profiles mapped in Williams syndrome. J. Neurosci. **25**(16), 4146–4158 (2005)
270. P.M. Thompson, A.W. Toga, A framework for computational anatomy. Comput. Vis. Sci. **5**, 1–12 (2002)
271. P.M. Thompson, K.M. Hayashi, G.I. de Zubicaray, A.L. Janke, S.E. Rose, James Semple, Michael S. Hong, David H. Herman, David Gravano, David M. Doddrell, Arthur W. Toga, Mapping hippocampal and ventricular change in Alzheimer's disease. NeuroImage **22**(4), 1754–1766 (2004)
272. W. Thurston, Hyperbolic geometry and 3-manifolds, *Low-Dimensional Topology (Bangor, 1979)*, vol. 48, London Mathematical Society Lecture Note Series (Cambridge Univerity Press, Cambridge, 1982), pp. 9–25
273. W.P. Thurston, *Geometry and Topology of Three-Manifolds*. Princeton lecture notes (1976)
274. R. Tibshirani, Regression shrinkage and selection via the lasso. J. R. Stat. Soc. Ser. B (Methodol.) **58**(1), 267–288 (1996)
275. D. Tosun, J.L. Prince, A geometry-driven optical flow warping for spatial normalization of cortical surfaces. IEEE Trans. Med. Imag. **27**(12), 1739–1753 (2008)
276. M. Vaillant, J. Glaunés, Surface matching via currents. Inf. Process. Med. Imaging **19**, 381–392 (2005)
277. D.C. Van Essen, A population-average, landmark- and surface-based (PALS) atlas of human cerebral cortex. Neuroimage **28**(3), 635–662 (2005)
278. D.C. Van Essen, H.A. Drury, J. Dickson, J. Harwell, D. Hanlon, C.H. Anderson, An integrated software suite for surface-based analyses of cerebral cortex. J. Am. Med. Inf. Assoc. **8**(5), 443–459 (2001)
279. E.C.D. Verdière, F. Lazarus, Optimal pants decompositions and shortest homotopic cycles on an orientable surface. J. ACM **54**(4), 18 (2007)
280. N.J. Vilenkin, *Special Functions and the Theory of Group Representations* (American Mathematical Society, Providence, 1968)
281. C. Villani, *Topics in Optimal Transportation* (American Mathematical Society, Providence, 2003)
282. V. Vivekanandan, V.W.S. Wong, Ordinal mds-based localization for wireless sensor networks. Int. J. Sens. Netw. **1**(3/4), 169–178 (2006)
283. J. Wallner, H. Pottmann, Spline orbifolds. *Curves and Surfaces with Applications in CAGD* (2007), pp. 445–464
284. G. Wang, G. Cao, T.F. La Porta, Movement assisted sensor deployment. IEEE Trans. Mob. Comput. **5**(6), 640–652 (2006)
285. H. Wang, Y. He, X. Li, X. Gu, H. Qin, Polycube splines. Comput. Aided Des. **40**(6), 721–733 (2008)
286. W. Wang, D. Slepev, S. Basu, J.A. Ozolek, G.K. Rohde, A linear optimal transportation framework for quantifying and visualizing variations in sets of images. IJCV 254–269 (2013)
287. X. Wang, X. Ying, Y.-J. Liub, S.-Q. Xin, W. Wang, X. Gu, W. Mueller-Wittig, Y. He, Intrinsic computation of centroidal voronoi tessellation (cvt) on meshes. *ACM Symposium on solid and physical modeling* (2014)
288. X. Wang, G. Xing, Y. Zhang, C. Lu, R. Pless, C. Gill, Integrated coverage and connectivity configuration in wireless sensor networks, in *Proceedings of SenSys* (2003), pp. 28–39

289. Y. Wang, M.-C. Chiang, P.M. Thompson, Mutual information-based 3D surface matching with applications to face recognition and brain mapping, in *Proceedings of the Tenth IEEE International Conference on Computer Vision ICCV'05*, vol. 1 (2005), pp. 527–534
290. Y. Wang, W. Dai, T.F. Chan, S.-T. Yau, A.W. Toga, P.M. Thompson, Teichmüller shape space theory and its application to brain morphology, in *Proceedings of Medical Image Computing and Computer-Assisted Intervention* (2009), pp. 133–140
291. Y. Wang, X. Gu, T.F. Chan, P.M. Thompson, S.-T. Yau, Brain surface conformal parameterization with algebraic functions, in *Proceedings of Medical Image Computing and Computer-Assisted Intervention, Part II*, vol. 4191 (LNCS, 2006), pp. 946–954
292. Y. Wang, X. Gu, T.F. Chan, P.M. Thompson, S.-T. Yau, Brain surface conformal parameterization with the Ricci flow, in *IEEE international symposium on biomedical imaging: from nano to macro*, ISBI'07 (2007), pp. 1312–1315
293. Y. Wang, M. Gupta, S. Zhang, S. Wang, X. Gu, D. Samaras, P. Huang, High resolution tracking of non-rigid motion of densely sampled 3D data using harmonic maps. Int. J. Comput. Vis. **76**(3), 283–300 (2008)
294. Y. Wang, L.M. Lui, T.F. Chan, P.M. Thompson, Optimization of brain conformal mapping with landmarks. Proceedings of Medical Image Computing and Computer-Assisted Intervention, Part **II**, 675–683 (2005)
295. Y. Wang, L.M. Lui, X. Gu, K.M. Hayashi, T.F. Chan, A.W. Toga, P.M. Thompson, S.-T. Yau, Brain surface conformal parameterization using Riemann surface structure. IEEE Trans. Med. Imaging **26**(6), 853–865 (2007)
296. Y. Wang, J. Shi, X. Yin, X. Gu, T.F. Chan, S.-T. Yau, A.W. Toga, P.M. Thompson, Brain surface conformal parameterization with the Ricci flow. IEEE Trans. Med. Imaging **31**(2), 251–264 (2012)
297. Y. Wang, Y. Song, P. Rajagopalan, T. An, K. Liu, Y.Y. Chou, B. Gutman, A.W. Toga, P.M. Thompson, Surface-based TBM boosts power to detect disease effects on the brain: an N = 804 ADNI study. Neuroimage **56**(4), 1993–2010 (2011)
298. Y. Wang, L. Yuan, J. Shi, A. Greve, J. Ye, A.W. Toga, A.L. Reiss, P.M. Thompson, Applying tensor-based morphometry to parametric surfaces can improve MRI-based disease diagnosis. Neuroimage **74**, 209–230 (2013)
299. Y. Wang, J. Zhang, B. Gutman, T.F. Chan, J.T. Becker, H.J. Aizenstein, O.L. Lopez, R.J. Tamburo, A.W. Toga, P.M. Thompson, Multivariate tensor-based morphometry on surfaces: application to mapping ventricular abnormalities in HIV/AIDS. NeuroImage **49**(3), 2141–2157 (2010)
300. Y. Wang, S. Lederer, J. Gao, Connectivity-based sensor network localization with incremental delaunay refinement method, in *Proceedings of INFOCOM* (2009), pp. 2401–2409
301. M.W. Weiner, Expanding ventricles may detect preclinical Alzheimer disease. Neurology **70**(11), 824–825 (2008)
302. G. Werner-Allen, K. Lorincz, J. Johnson, J. Lees, M. Welsh, Fidelity and yield in a volcano monitoring sensor network, in *Proceedings of the 7th symposium on operating systems design and implementation*, OSDI '06 (2006), pp. 381–396
303. J. West, J.M. Fitzpatrick, M.Y. Wang, B.M. Dawant, C.R. Maurer, R.M. Kessler, R.J. Maciunas, C. Barillot, D. Lemoine, A. Collignon, F. Maes, P. Suetens, D. Vandermeulen, P.A. van den Elsen, S. Napel, T.S. Sumanaweera, B. Harkness, P.F. Hemler, D.L.G. Hill, D.J. Hawkes, C. Studholme, J.B.A. Maintz, M.A. Viergever, G. Malandain, X. Pennec, M.E. Noz, G.Q. Maguire, M. Pollack, C.A. Pelizzari, R.A. Robb, D. Hanson, R.P. Woods, Comparison and evaluation of retrospective intermodality brain image registration techniques. J. Comp. Assist. Tomogr. **21**(4), 554–68 (1997)
304. W. Chunlin, X. Tai, A level set formulation of geodesic curvature flow on simplicial surfaces. IEEE Trans. Vis. Comput. Gr. **16**(4), 647–662 (2010)
305. W. Hongyi, C. Wang, Nian-Feng Tzeng, Novel self-configurable positioning technique for multi-hop wireless networks. IEEE/ACM Trans. Netw. **13**(3), 609–621 (2005)
306. S. Xia, X. Yin, H. Wu, M. Jin, X.D Gu, Deterministic greedy routing with guaranteed delivery in 3D wireless sensor networks, in *Proceedings of the twelfth ACM international symposium on mobile ad hoc networking and computing*, MobiHoc '11 (2011), pp. 1–10

307. S. Xiang, L. Yuan, W. Fan, Y. Wang, P. M. Thompson, J. Ye, Bi-level multi-source learning for heterogeneous block-wise missing data. Neuroimage (2013)
308. S.-Q. Xin, G.-J. Wang, Applying the improved chen and han's algorithm to different versions of shortest path problems on a polyhedral surface. Comput. Aided Des. **42**, 942–951 (2010)
309. D.-M. Yan, B. Lévy, Y. Liu, F. Sun, W. Wang, Isotropic remeshing with fast and exact computation of restricted Voronoi diagram. *Computer Graphics Forum*, in *(Proceedings of symposium on geometry processing 2009)***28**(5), 1445–1454 (2009)
310. J. Yang, U. Yoon, H.J. Yun, K. Im, Y.Y. Choi, S.I. Kim, K.H Lee, J.-M Lee, Prediction of human intelligence using morphometric characteristics of cerebral cortex. WCECS, 1 (2011)
311. Q. Yang, S. Ma, Matching using schwarz integrals. Pattern Recognit. **32**(6), 1039–1047 (1999)
312. S. Yang, F. Dai, M. Cardei, J. Wu, F. Patterson, *On connected multiple point coverage in wireless sensor networks* (J. Wirel. Inf, Netw, 2006)
313. Y. Yang, M. Jin, H. Wu, 3D surface localization with terrain model, in *Proceedings of IEEE Conference on Computer Communications (INFOCOM)* (2014), pp. 46–54
314. Y. Yang, M. Jin, Y. Zhao, Wu Hongyi, Distributed information storage and retrieval in 3-D sensor networks with general topologies. IEEE/ACM Trans. Netw. **23**(4), 1149–1162 (2015)
315. F. Ye, H. Luo, J. Cheng, S. Lu, L. Zhang, A two-tier data dissemination model for large-scale wireless sensor networks, in *ACM MobiCom* (2002), pp. 148–159
316. B.T. Yeo, M.R. Sabuncu, T. Vercauteren, N. Ayache, B. Fischl, P. Golland, Spherical demons: fast diffeomorphic landmark-free surface registration. IEEE Trans. Med. Imaging **29**(3), 650–668 (2010)
317. X. Yin, M. Jin, X. Gu, Computing shortest cycles using universal covering space. Vis. Comput. **23**(12), 999–1004 (2007)
318. L. Ying, D. Zorin, A simple manifold-based construction of surfaces of arbitrary smoothness. ACM Trans. Gr. **23**(3), 271–275 (2004)
319. L. Yuan, Y. Wang, P.M. Thompson, V.A. Narayan, J. Ye, Multi-source feature learning for joint analysis of incomplete multiple heterogeneous neuroimaging data. Neuroimage **61**(3), 622–632 (2012)
320. W. Zeng, M. Jin, F. Luo, X. Gu, Canonical homotopy class representative using hyperbolic structure, in *IEEE International Conference on Shape Modeling and Applications* (2009), pp. 171–178
321. W. Zeng, L.M. Lui, X. Gu, S.-T. Yau, Shape analysis by conformal modules. Int. J. Methods Appl. Anal. (MAA) **15**(4), 539–556 (2008)
322. W. Zeng, L.M. Lui, L. Shi, D. Wang, W.C. Chu, J.C. Cheng, J. Hua, S.T. Yau, X. Gu, Shape analysis of vestibular systems in adolescent idiopathic scoliosis using geodesic spectra. Med. Image Comput. Comput. Assist. Interv. **13**(Pt 3), 538–546 (2010)
323. W. Zeng, J. Marino, K. Chaitanya Gurijala, X. Gu, A. Kaufman, Supine and prone colon registration using quasi-conformal mapping. IEEE Trans. Vis. Comput. Gr. **16**(6), 1348–1357 (2010)
324. W. Zeng, D. Samaras, X. Gu, Ricci flow for 3D shape analysis. IEEE Trans. Pattern Anal. Mach. Intell. **32**(4), 662–677 (2010)
325. W. Zeng, R. Shi, Y. Wang, S.-T. Yau, X. Gu, Teichmüller shape descriptor and its application to Alzheimer's disease study. Int. J. Comput. Vis. **105**(2), 155–170 (2013)
326. W. Zeng, R. Sarkar, F. Luo, X. Gu, J. Gao, Resilient routing for sensor networks using hyperbolic embedding of universal covering space, in *Proceedings of the 29th Conference on Information Communications*, INFOCOM'10 (2010), pp. 1694–1702
327. W. Zeng, X. Yin, Y. Zeng, Y. Lai, X. Gu, D. Samaras, 3D face matching and registration based on hyperbolic ricci flow, in *CVPR Workshop on 3D Face Processing* (2008), pp. 1–8
328. D. Zhang, M. Hebert, Harmonic maps and their applications in surface matching. IEEE Computer Society Conference on Computer Vision and Pattern Recognition **2**(1999), 524–530 (1999)
329. J. Zhang, Y. Fan, Q. Li, P.M. Thompson, J. Ye, Y. Wang, Empowering cortical thickness measures in clinical diagnosis of alzheimer's disease with spherical sparse coding. *13th IEEE international symposium on biomedical imaging: from nano to macro, 2017. ISBI 2017* (2017), pp. 446–450

330. J. Zhang, J. Shi, C.M. Stonnington, Q. Li, B.A. Gutman, K. Chen, E.M. Reiman, R.J. Caselli, P.M. Thompson, J. Ye, Y. Wang, Hyperbolic space sparse coding with its application on prediction of alzheimer's disease in mild cognitive impairment, in *19th International Conference on Medical Image Computing and Computer Assisted Intervention (MICCAI)* (2016)
331. J. Zhang, C.M. Stonnington, Q. Li, J. Shi, R.J. Bauer, B.A. Gutman, K. Chen, E.M. Reiman, P.M. Thompson, J. Ye, Y. Wang, Applying sparse coding to surface multivariate tensor-based morphometry to predict future cognitive decline, in 13th IEEE international symposium on biomedical imaging: from nano to macro, 2016. ISBI **2016**, 646–650 (2016)
332. M.-C. Zhao, J. Lei, M.- Y. Wu, Y. Liu, W. Shu, Surface coverage in wireless sensor networks, in *Proceedings of INFOCOM* (2009), pp. 109–117
333. Y. Zhao, H. Wu, M. Jin, Y. Yang, H. Zhou, S. Xia, Cut-and-sew: a distributed autonomous localization algorithm for 3d surface wireless sensor networks, in *Proceedings of the 14th ACM international symposium on mobile ad hoc networking and computing (MobiHoc'13)* (2013)
334. Y. Zhao, H. Wu, M. Jin, S. Xia, Localization in 3D surface sensor networks: challenges and solutions, in *Proceedings of the 31st Annual IEEE Conference on Computer Communications (INFOCOM'12)* (2012), pp. 55–63
335. H. Zhou, S. Xia, M. Jin, H. Wu, Localized algorithm for precise boundary detection in 3D wireless networks, in *IEEE ICDCS* (2010), pp. 744–753
336. H. Zhou, H. Wu, S. Xia, M. Jin, N. Ding, A distributed triangulation algorithm for wireless sensor networks on 2D and 3D surface, in *Proceedings of INFOCOM* (2011), pp. 1053–1061
337. Z. Zhou, S. Das, H. Gupta, Connected k-coverage problem in sensor networks, in *Proceedings of ICCCN* (2004), pp. 373–378
338. S.C. Zhu, A.L. Yuille, A flexible object recognition and modeling system. Int. J. Comput. Vis. **20**(3), 187–212 (1996)
339. G. Zou, J. Hua, Z. Lai, X. Gu, M. Dong, Intrinsic geometric scale space by shape diffusion. IEEE Trans. Vis. Comput. Gr. **15**(6), 1193–1200 (2009)
340. Y. Zou, K. Chakrabarty, Sensor deployment and target localization based on virtual forces, in *INFOCOM* (2003), pp. 1293–1303

Index

M. Jin et al., *Conformal Geometry*, https://doi.org/10.1007/978-3-319-75332-4

Printed in the United States
By Bookmasters